Springer

Tokyo
Berlin
Heidelberg
New York
Barcelona
Budapest
Hong Kong
London
Milan
Paris
Santa Clara
Singapore

K. Iwatsuki · P.H. Raven (Eds.)

Evolution and Diversification of Land Plants

With 55 Figures

 Springer

ISBN 4-431-70203-2 Springer-Verlag Tokyo Berlin Heidelberg New York

Library of Congress Cataloging-in-Publication Data

Evolution and diversification of land plants / K. Iwatsuki, P.H. Raven
(eds.).
p. cm.
Includes bibliographical references and index.
ISBN 4-431-70203-2
1. Plants—Evolution. 2. Plant diversity. 3. Plants—Phylogeny.
I. Iwatsuki, Kunio, 1934– . II. Raven, Peter H.
QK980.E86 1997
581.3′8—dc21   97-26061

Typesetting: Best-set Typesetter Ltd., Hong Kong
Printing and binding: Sokosha, Japan
SPIN: 10575489

Foreword

It is agreed by most scientists that there is a real possibility of mankind's being plunged into crisis around the middle of the twenty-first century because of insufficient crop production to support the explosively increasing world population and because of serious pollution of the global environment. Plants should play the major role in alleviating this crisis. Consequently, plant science will become of crucial importance in the next century.

The application of plant science to crop improvement and the production of plants resistant to environmental pollution will be achieved through complete understanding of the basic functions of plants. Fundamental understanding of individual species and the evolution of plants is also important.

The Botanical Society of Japan, which was established in 1882, decided to publish a series of special issues of the *Botanical Magazine, Tokyo* (now the *Journal of Plant Research*). The aim of this series was to present overviews of recent advances in various fields of plant sciences. As a result, The Botanical Society of Japan published *Controlling Factors in Plant Development* in 1978, *Regulation of Photosynthetic Processes* in 1991, and *Cellular and Molecular Biology in Plant Cell Cultures* in 1993. The contributions to these high-quality special issues were written by leading international scientists in each field, and these publications had a great impact worldwide.

In 1997 The Botanical Society of Japan is publishing two monographs following on from the previous publications. One is *Ginkgo Biloba—A Global Treasure*, edited by Drs. T. Hori, R.W. Ridge, W. Tulecke, P. Del Tredici, J. Trémouillaux-Guiller, and H. Tobe, commemorating the 100th anniversary of the discovery of *Ginkgo* sperm by Sakugoro Hirase, a Japanese botanist, who published his findings in the *Botanical Magazine, Tokyo* (now the *Journal of Plant Research*). The other is *Evolution and Diversification of Land Plants*; edited by Drs. K. Iwatsuki and P.H. Raven. Since elucidation of the mechanisms of evolution is a major goal of plant science and of biology in general, this topic is most fitting for a publication just prior to the beginning of the twenty first century.

I trust that these two books published by The Botanical Society of Japan will provide readers with information at the very forefront of plant science.

As the president of The Botanical Society of Japan, I would like to acknowledge the editors and distinguished contributors for devoting their valuable time to these publications.

April 1997

Atsushi Komamine
President
The Botanical Society of Japan

Preface

The 20th century has been a period of significant advances in our understanding of the evolution and diversification of plants. These advances have come about not only as a result of the deepening of the biosystematic investigation that had become one of the dominant themes in studies of plant systematics and evolution by mid-century, but especially as a result of the application of (1) a greatly deepened understanding of the fossil record; (2) the development of new and improved objective methods of analysis employed in the comparison of taxa; and (3) the addition of biochemical evidence based especially on sequencing nucleic acids. All of these approaches are well represented in the present volume, which affords at once a review and general outline of what has been accomplished and a set of signposts for the future.

In the interpretation of the fossil record, careful studies of the morphology and anatomy of living organisms, together with detailed examination especially of fossil flowers and fruits, has led to insights that were not imaginable just a few decades ago. This clear demonstration that the monocots are not in fact one of the two great evolutionary lines of angiosperms, but rather a derivative of a particular group of dicots, is only one of the surprises that have come about as a result of these investigations. The evolving picture of angiosperm evolution has been strengthened by the application of the sorts of macromolecule analyses mentioned above, and has in many cases, confirmed the validity of evolutionary relationships deduced from comparisons of newly discovered fossils and careful investigations of living taxa. In general, the application of cladistic methodology has provided a secure basis for making molecular comparisons and for viewing the overall characteristics of angiosperms in an integrated, verifiable context to which additional data and analyses can readily be added.

In consideration of all of these advances and of the widespread application of newly developed methodologies, it seems certain that we shall have a well-worked-out, truly evolutionary system for all plants (bryophytes and vascular plants) within a decade. This will be a truly Darwinian classification and will, in turn, form the basis for many additional insightful studies that, in principle, depend on a secure elucidation of relationships for their validity.

Thus, the kinds of results presented by Drs. Hasebe and Banks on the evolution of the MADS gene family and flowering plants will be applied even more widely as results concerning the overall genomic sequences of organisms are made available. The securely demonstrated and reliable relationships among different groups of organisms will provide a framework against which the evolution of individual genes or other features, such as pollen or chromosomes, can be compared. By such comparisons, and by placing those features within such a context, we will be able to understand to a degree that would never have been thought possible a decade or so ago, the evolution of genes, chromosomes, embryological features, and many other characteristics of plants in terms of their adaptive significance, rates of evolution, and other characteristics.

At the same time, the kinds of studies of plant populations that were pioneered by Bonner, Clements, Turesson, and, notably, by Clausen, Keck, and Heisey, will be understood better and more definitively when similar sorts of exact molecular comparisons are applied to our understanding of evolution at the population level. The nature of species of plants will be better understood as will the level of genetic diversification that characterizes individual populations, races, species, and other aspects of diversification at this level.

Onward then to a 21st century filled with deepened understanding and many exciting new discoveries! The rich assortment of papers presented in this volume serves as an outstanding guide to the major trends in these areas at present and reflects many of the kinds of investigation that will dominate this fundamental branch of science over the years to come.

Peter H. Raven
Missouri Botanical Garden
P.O. Box 299, St. Louis
MO 63166-0299, USA

Contents

Foreword V

Preface VII

Contributors XI

Origin and Diversification of Primitive Land Plants

1. Charting Diversity in Early Land Plants: Some Challenges for
   the Next Millennium
   D. Edwards 3

2. Morphological Diversity and Evolution of Vegetative
   Organs in Pteridophytes
   M. Kato and R. Imaichi 27

3. RNA Editing in Land Plants
   K. Yoshinaga 45

4. Phylogenetic Reconstruction of Some Conifer Families:
   Role and Significance of Permineralized Cone Records
   T.A. Ohsawa 61

Origin and Diversification of Angiosperms

5. Evolutionary Biology of Flowers: Prospects for the Next Century
   P.K. Endress 99

6. Fossil History of Magnoliid Angiosperms
   E.M. Friis, P.R. Crane, and K.R. Pedersen 121

7. Molecular Phylogenetic Relationships Among Angiosperms:
   An Overview Based on rbcL and 18S rDNA Sequences
   D.E. Soltis, C. Hibsch-Jetter, P.S. Soltis, M.W. Chase
   and J.S. Farris 157

 8. Evolution of MADS Gene Family in Plants
    M. HASEBE and J.A. BANKS 179

 9. Palynological Approaches to the Origin and
    Early Diversification of Angiosperms
    M. TAKAHASHI 199

10. Chromosomal Evolution of Angiosperms
    H. OKADA 209

Speciation and Mechanisms of Diversification

11. Mating Systems and Evolution in Flowering Plants
    K.E. HOLSINGER and J.E. STEINBACHS 223

12. Plant Speciation on Oceanic Islands
    D.J. CRAWFORD and T.F. STUESSY 249

13. Relations of Environmental Change to Angiosperm Evolution
    During the Late Cretaceous and Tertiary
    J.A. WOLFE 269

14. Modes and Mechanisms of Speciation in Pteridophytes
    C.H. HAUFLER 291

15. Speciation and Morphological Evolution in Rheophytes
    R. IMAICHI and M. KATO 309

Subject Index 319

Contributors

Jo Ann Banks
Department of Botany and Plant Pathology, Purdue University
Lilly Hall of Life Sciences, West Lafayette, IN 47907-1155, USA

Mark W. Chase
Laboratory of Molecular Systematics, Royal Botanic Gardens
Kew, Richmond, Surrey TW9 3AB, UK

Peter R. Crane
Department of Geology, The Field Museum
Roosevelt Road at Lake Shore Drive, Chicago, IL 60605, USA

Daniel J. Crawford
Department of Plant Biology, The Ohio State University
1735 Neil Avenue, Columbus, OH 43210, USA

Dianne Edwards
Earth Sciences, University of Wales/Cardiff
P.O. Box 914, Cardiff CF1 3YE, Wales, UK

Peter K. Endress
Institute of Systematic Botany, University of Zürich
Zollikerstrasse 107, 8008, Zürich, Switzerland

James S. Farris
Naturhistoriska Riksmusect, Molekylärsystematiska laboratoriet
P.O. Box 50087, S-104 05, Stockholm, Sweden

Else Marie Friis
Department of Palaeobotany, Swedish Museum of Natural History
P.O. Box 50007, S-104 05, Stockholm, Sweden

MITSUYASU HASEBE
National Institute for Basic Biology
38 Nishigonaka, Myoudaiji-cho, Okazaki 444, Japan

CHRISTOPHER H. HAUFLER
Department of Botany, Haworth Hall, University of Kansas
Lawrence, KS 66045, USA

KENT E. HOLSINGER
Department of Ecology and Evolutionary Biology, University of Connecticut
75 N. Eagleville Road, U-43, Storrs, CT 06269-3043, USA

CAROLA HIBSCH-JETTER
Department of Botany, Washington State University
Pullman, WA 99164-4238, USA

RYOKO IMAICHI
Department of Chemical and Biological Sciences
Faculty of Science, Japan Women's University
2-8-1 Mejirodai, Bunkyo-ku, Tokyo 112, Japan

MASAHIRO KATO
Department of Biological Sciences, Graduate School of Science
University of Tokyo
7-3-1 Hongo, Bunkyo-ku, Tokyo 113, Japan

TAKESHI ASAKAWA OHSAWA
Department of Biology, Faculty of Science, Chiba University
1-33 Yayoi-cho, Inage-ku, Chiba 263, Japan

HIROSHI OKADA
Botanical Gardens, Faculty of Science, Osaka City University
2000 Kisaichi, Katano, Osaka 576, Japan

KAJ RAUNSGAARD PEDERSEN
Department of Geology, University of Aarhus
DK-8000, Århus C, Denmark

DOUGLAS E. SOLTIS
Department of Botany, Washington State University
Pullman, WA 99164-4238, USA

PAMELA S. SOLTIS
Department of Botany, Washington State University
Pullman, WA 99164-4238, USA

JENNIFER E. STEINBACHS
Department of Ecology and Evolutionary Biology, University of Connecticut
75 N. Eagleville Road, U-43, Storrs, CT 06269-3043, USA

TOD F. STUESSY
Institut für Botanik, Universität Wien
Rennweg 14, A-1030, Vienna, Austria

MASAMICHI TAKAHASHI
Department of Biology, Faculty of Education, Kagawa University
1-1 Saiwai-cho, Takamatsu 760, Japan

JACK A. WOLFE
Department of Geosciences, University of Arizona
Tucson, AZ 85721, USA

KOICHI YOSHINAGA
Department of Chemistry, Faculty of Science, Shizuoka University
836 Oya, Shizuoka 422, Japan

Origin and Diversification of Primitive Land Plants

Charting Diversity in Early Land Plants: Some Challenges for the Next Millennium

DIANNE EDWARDS

1.1 Introduction

Space constraints prevent a detailed review of the current knowledge of diversity in early land plants. Instead I propose to present a brief and obviously idiosyncratic account of what I consider have been the major 20th-century benchmarks in furthering our understanding of the nature of pioneering land vegetation, and to outline some of the challenges I perceive in charting diversity in the next millennium. The bibliography is not exhaustive, but references chosen will lead the reader to the original sources.

1.2 Major Advances of the 20th Century

1.2.1 Rhynie Chert

In considering such benchmarks, the discovery of the Lower Devonian autochthonous Rhynie Chert assemblages, preserved in exquisite cellular detail by siliceous fluid from hot springs, must surely rank amongst the most noteworthy. Kidston and Lang's seminal descriptions [1–4] of vascular plants added credibility and respectability to the studies of 19th-century pioneers such as Penhallow and Dawson, who described fragmentary and morphologically simple coalified fossils, which for the most part could not be closely related to extant plants. In also describing algae, fungi, and cyanobacteria, Kidston and Lang [5] laid the foundations for the study of terrestrial ecosystems, a discipline which only recently has resurfaced as a major theme in middle Palaeozoic palaeontology [6,7]. Although discoveries of new vascular plants in the Chert have been few in the intervening years [8–10], the descriptions of unequivocal gametophytes by Prof. Remy and co-workers at Munster [11–16] provide evidence for ± isomorphic alteration of generations in early land plants, with those of *Aglaophyton*, *Horneophyton*, and *Nothia* possessing almost identical anatomy, particularly that associated with

Earth Sciences, University of Wales/Cardiff, P.O. Box 914, Cardiff CF1 3YE, Wales, UK

homoiohydry, to the sporophytes. As Kenrick [17] emphasised, the gameto-phytes of present-day homosporous vascular plants are considerably reduced when compared with such Lower Devonian examples although some similarities remain in the extant Lycopodiales. The rewards of routine sectioning and metic-ulous screening of large quantities of chert are evident in the recent detailed anatomical information on the vascular plants [18] and, most spectacularly, in major progress in charting diversity in early fungi by Taylor and co-workers. This includes not only descriptions of individuals and life cycles [19,20] but also mycoparisitism [21] and fungal interactions with other organisms, notably the discovery of the earliest lichen (with photobiont a green alga or cyanobacterium [22]) and arbuscular mycorrhiza in *Aglaophyton* [23]. Studies on Rhynie palaeo-environments centre on the extensive cores taken through the deposits and form the basis of a major project in Aberdeen University [10,24]. The locality itself now belongs to the Scottish nation and strenuous attempts are in progress to have it recognised as a World Heritage Site.

1.2.2 Silurian Land Plants

Lang's paper [25] focussed attention on far more fragmentary fossils in the "Downtonian" clastic, often marine, sediments from the Welsh Borderland and South Wales, a sequence now known to extend from the Late Silurian (Prídolí) into the basal Devonian (Lochkovian). His genus *Cooksonia* has become almost synonymous with the earliest vascular plants although it was not unequivocally proven to possess xylem until almost 50 years later [26,27]. The abundant naked branching axes that he described as *Hostimella* (*sic*) are now known to derive from a number of taxa with terminal sporangia [28], while further diversity in plants of such simple morphology has been revealed by the demonstration of in situ spores [29,30]. But perhaps even more significant was Lang's recognition of a new group of organisms in some extremely scrappy coalified fragments of presumed thalli composed of tubes. In erecting the Nematophytales, he empha-sised their distinction from "land plants of typical cellular and vascular construc-tion" and from the algae, although he believed that their nearest relatives were to be found in the latter group. He also included Dawson's *Prototaxites* in his new taxon. Some progress has been made in documenting anatomical information on both the *Prototaxites* and *Nematothallus* complexes, but their affinities today remain almost as conjectural as they were in Lang's time [31–33] and this pro-vides a challenge for the next century. Hueber's recent hypothesis that *Prototax-ites* was a fungus (Munster lecture, 1994) is an exciting development, raising both the possibility of the demonstration of differences in chemical composition of the fossils and the need to identify the sources of energy for what were sometimes very large organisms.

1.2.3 Classification and Phylogeny

Høeg [34] compiled an excellent catalogue of 50 years of progress in descriptions of early vascular plants, among the most significant being those of Kräusel and

Weyland (Germany), Leclercq and Stockmans (Belgium), Croft and Lang (Wales). He divided them into two orders, Rhyniales and Psilophytales, within the Psilophyta. However it was Banks [35] who set the taxonomic agenda for subsequent researchers in his erection of three major subdivisions, the Rhyniophytina, Zosterophyllophytina, and Trimerophytina. Like Høeg, he set apart the barinophytes as *incertae sedis*. This initiative prompted much more rigorous assessment of anatomical data (particularly in relation to water-conducting cells) an activity which was logically extended to character analysis as a prerequisite for cladistic phylogenetic approaches. Banks [36] himself, in reconsidering his classification 25 years on, found it "to have still some usefulness," a gross understatement in that the Zosterophyllophytina and Trimerophytina are today generally accepted as natural higher taxa. The Zosterophyllophytina in particular has been enlarged by numerous new taxa (Banks, 1968–6 genera; 1992–16) and strengthened by the demonstration of common tracheid construction (G-type [37,38]) which is also recorded in the earliest members of the Lycophytina. Indeed Niklas and Banks [39] have identified a level of organisation (radial symmetry and nonterminate spike) which they considered closest to the putative ancestral lycopods. Trimerophytes include further species of *Psilophyton* and also taxa which extend the original concept of the subdivision, particularly as regards xylem anatomy—a dilemma relating to problems of defining the limits of a group which is generally believed to be the stem group for the non-lycophyte, free-sporing tracheophytes.

The Rhyniophytina is the least satisfactory of the three subdivisions. Edwards and Edwards [40] reduced the original number of taxa, by recognising only those with in situ tracheids, and advocated the term "rhyniophytoid" for plants of rhyniophyte morphology, but in which tracheids have not been demonstrated. This procedure was followed by Banks [36] who retained *Cooksonia pertoni*, *Rhynia gwynne-vaughanii* and *Uskiella spargens*. However, the discovery of the S-type tracheids in *Rhynia* links it with *Stockmansella*, *Huyvena*, and *Sennicaulis* [41]. The two fertile genera also possess sporangia showing a putative abscission zone, but borne laterally in *Stockmansella* and terminally in *Rhynia*, thus combining characters of two of the subdivisions. The first phylogenetic analysis based on cladistics united lycopods and zosterophylls as sister groups in the Lycopodiopsida, while the second major clade of eutracheophytes, the Psilophytopsida, encompasses all the major groups of vascular plants, with the trimerophytes as its earliest members [41]. Kenrick and Crane coined the informal name "polysporangiophytes" to including the eutracheophytes and the protracheophytes, the latter including the Rhyniopsida with S-type tracheids and also plants with anomalous conducting tissue such as *Aglaophyton* [41]. A subsequent analysis of taxa produced a consensus tree with similar pattern for the zosterophylls and lycophytes, but the remaining early vascular plants formed part of a large unresolved basal polychotomy [42]. Banks [36] also listed a number of "aberrant" taxa that defied supra-generic classification based on his three subdivisions. Some, e.g. *Renalia*, might well conform given additional information, others, e.g. *Aglaophyton* and *Adoketophyton*, present new combinations [40,43]. The numerous Chinese representatives that fall into this category are discussed in greater detail later in this chapter. Their classification and phylogeny certainly present a significant challenge to future researchers.

Not strictly palaeobotanical, but of major relevance to the phylogeny of early land plants, has been the recognition that, of all the green algae, the charophytes are the sister group of the embryophytes [44–46]. Their synapomorphies mainly involving ultrastructure and biochemistry are unlikely to be seen in fossils, but coalified gyrogonites and vegetative thalli of the Charopsida recovered from Silurian marginal marine sediments, and vegetative cells of *Palaeonitella* in the fresh-water Lower Devonian Rhynie Chert demonstrate their existence at a time of rapid diversification of land vegetation [47]. An excellent all-embracing interdisciplinary survey of such matters is given in Graham [48], whose detailed researches on *Coleochaete* particularly *C. orbicularis* have contributed greatly to the hypothesis that the "common ancestor of all embryophytes might have been classified as a species of *Coleochaete*" [46].

1.2.4 Dispersed Spores and the Earliest Land Vegetation

Although spores had been isolated in the 1950s from rocks predating those with vascular plant megafossils, the significance of the non-trilete components of these assemblages became apparent only when Gray and Boucot [49] described and gave the first illustrations of tetrads in the earliest Silurian of New York State and, also for the first time, aired the possibility that they derived from bryophytes. In a subsequent series of papers Gray has argued very persuasively in support of this hypothesis [50–52]. The non-trilete component also include monads and dyads, and in 1984 Richardson et al. [53] erected the Anteturma Cryptosporites to include "non-marine sporomorphs (non-pollen grains) with no visible haptotypic features such as contact areas or tetrad marks." Richardson later extended the concept to include spores separating from dyads, which were alete but possessed distinct ± circular, usually smooth, contact areas [54]. The first dyads were actually described by Lang [25] in an elongate spore mass (<4.5 mm) from the Late Silurian of the Welsh Borderland although he did not name them as such. Instead he wrote that "Each one has the peculiarity of being bicellular, two hemispherical cells being separated by a transverse wall across the equator of the whole structure." This transverse structure has more recently been resolved by confocal microscopy as a single wall between adjacent spores [55]. Lang himself thought that the two cells represented the first stage of germination. The limited progress in tracing the parent plants of the cryptospores summarised later in this chapter demonstrates why such activity remains a major challenge.

1.2.5 Phytogeography

The accurate dating of assemblages is of paramount importance in palaeophytogeography and in the testing of phylogenetic hypotheses. Thus Banks [56] used the time of appearance of "plant biocharacters" to demonstrate the evolution of plants with more complex morphology and anatomy in Silurian and

Devonian times, an overview which fitted comfortably with his classification system. While such evidence derived from localities on the Old Red Sandstone Continent (i.e. Laurussia), Banks [57] stated that "a brief glance at the Siegenian-Emsian of Victoria, Australia indicates that it is closely related to floras of comparable age in the western hemisphere." Similarly what little was known of further extra-Laurussian occurrences in China also suggested global uniformity in early land vegetation [58]. However, complacency was shattered by the discovery that some of the *Baragwanathia* assemblages of Victoria were, on the evidence of graptolite dating, Late Silurian (Ludlow) and thus coeval with the *Cooksonia*-dominated assemblages of Laurentia [59]. Unequivocal *Baragwanathia* was first recorded in the present northern hemisphere in the Emsian of Canada [60]. More recently I have suggested the possibility of four phytochoria in the Early Devonian [61] with subsequent highest endemicity in China (see section 1.3.1). Charting the biodiversity in high-latitude Gondwana must also be a priority in future work [62].

1.3 Challenges for the Next Millennium

1.3.1 Palaeophytogeography: The Chinese Puzzle

Lower Devonian. In that much importance has been placed on the geographical bias of plant records towards the Laurussian palaeocontinent [61], the recent descriptions of extensive assemblages from southern China have given a major boost to both phytogeographic and evolutionary studies. The best-studied occurrences come from Yunnan, but Lower Devonian plants also occur in Guizhou, Sichuan, and Xinjiang [63]. Although a Lower Devonian plant was described from Yunnan by Halle [64,65] (*Drepanophycus spinaeformis* now *D. qujingensis* [66]), subsequent studies were not initiated until 1966 [67] with Hsü's description of a new species of the well-known cosmopolitan Lower Devonian genus *Zosterophyllum* (*Z. yunnanicum*), followed by the erection of a number of species of *Zosterophyllum* [68]. Subsequent studies, including some revision of Li and Cai's taxa [43], have shown that zosterophylls dominate species lists, which also contain a number of taxa that defy assignation to Banks' major divisions based on Lower Devonian plants [35,69] and that sometimes possess more derived characters than seen in coeval plants outside China. The recent discovery of Lower Devonian assemblages in Sichuan [70,71] reinforces these observations.

The species listed for Yunnan (Table 1.1) show that the least familiar and more diverse assemblages occur in the older Posongchong Formation of the Wenshan district, south-east Yunnan, recently confirmed as Siegenian (Pragian) on palynological evidence [72] (polygonalis-emsiensis [PE] zone: *Verrucosisporites polygonalis—Dictyotriletes emsiensis* spore biozone of Richardson and McGregor [73]). Members of *Zosterophyllum* itself (e.g. *Z. australianum*) remain inadequately described, but are the subject of ongoing investigations in Beijing (Hao and Li Cheng-Sen, personal communication). The three endemic genera

TABLE 1.1. Species lists from eastern and southern Yunnan Province, China

Age	Formation	Species
Emsian	Xujiachong	Zosterophyllum yunnanicum[a] Z. australianum Z Hsüa robusta R[b] Drepanophycus qujingensis L[a] new lycopod genus[b]
Pragian	Posongchong (Wenshan)	Z. contigua Z[a] Z. cf. australianum Z Discalis longistipa Z[b] Adoketophyton subverticillatum[b,c] Huia recurvata[b,c] Stachyophyton yunnanense[b,c] Gumuia zyzzata Z[b] Eophyllophyton bellum[b,c] Catenalis digitata[b,c] Yunia dichotoma T[b]

Z, zosterophyll; R, rhyniophyte; L, lycophyte; T, trimerophyte.
[a] Endemic species.
[b] Endemic genus.
[c] Of uncertain taxonomic position.

Demergotheca [74], *Gumuia*, and *Discalis* suggest that a southern China radiation of zosterophylls parallelled that noted for Laurussia in Pragian times [28]. From an evolutionary viewpoint, *Gumuia* is a representative of the terminate group of zosterophylls established by Niklas and Banks [39] while *Discalis longistipa* with its lax strobilus, spines covering the entire plant and circinate tips, shares characters with *Zosterophyllum* and *Sawdonia* and was cited as similar to a "potential ancestral lycophyte" in displaying nonterminate growth and radial symmetry.

Of the less familiar genera, *Stachyophyton* [75] and *Adoketophyton* [43] both possess strobili with sporangia axillary to bracts. In the latter these are fan shaped and arranged in four rows—the strobilus terminating smooth axes. In *Stachyophyton*, the spirally inserted bracts, with circular attachment points, bifurcate near their tips, and the subtending axes bear repeatedly branched (4–8 times), planated structures recently interpreted as leaves [76]. However, similarly organised branches bear strobili in the distal part of the plant [75], and not all of the branching axes in Geng's specimens are flattened in section. Wang and Cai [76] reported details of the xylem with allegedly exarch maturation in a short length of sterile specimen. The outline of the xylem is mainly elliptical with occasional indentation, and the tracheids are probably of G-type. Such compound strobili have no direct counterparts in extant plants, although sporangia similarly associated with bracts or leaves occur in barinophytes (axes smooth) and lycophytes (microphyllous), both of which possess G-type tracheids. The forked sporophyll resembles those of *Protolepidodendron scharyanum*, but microphylls are absent in the Chinese plant. The conclusion that both taxa should be designated *incertae sedis* (*Stachyophyton* [75,76]; *Adoketophyton* [68]) thus still holds. A taxon that appears to bridge zosterophylls and trimerophytes is *Huia recurvata* Geng [77] which has lateral adaxially recurved sporangia, that are borne spirally in loose spikes but that do not appear to split into two valves. A further non-zosterophyll feature is the centrarch axial xylem.

Catenalis digitata [78] is of interest in the disposition of its strobili ("fertile branchlets" of the authors), its sporangial arrangement and the interpretation by its originators that it was a plant from an aquatic or semi-aquatic habitat. The latter was inferred from the presumed flattened nature of the distal vegetative axes which also showed gross planar organisation (as has been seen in certain zosterophylls (e.g. *Gosslingia* [79]; *Sawdonia* [80])), and from the slightly sunken sporangia on one side of the strobilus, with all sporangia in all members of a digitate cluster showing similar orientation. Hao and Beck related such features to light harvesting and spore dispersal. They concluded that the plant was a tracheophyte anchored in shallow water or the inter-tidal zone and showed morphological convergence with certain members of the Phaeophyceae. Its presumed conducting tissues (tracheids were not isolated) and conventional trilete spores could also be considered evidence for full terrestrial status. The authors did not report a cuticle nor indicate the preservation state that yielded the sporangial wall cells.

Even more unusual is *Eophyllophyton bellum* [81,82]. The reconstruction shows part of a plant c. 47 cm tall with a spiny axial system comprising horizontal axes bearing much-branched, distally smooth, lateral systems that terminate in slightly recurved tips, and erect pseudomonopodial axes with occasional similar lateral structures, although more frequently vegetative and fertile leaves are borne in pairs. The vegetative leaves are laminate and fan-shaped, up to 5 mm long, comprising numerous segments each containing a vein. Fertile leaves are smaller, and bear numerous (8–18) sessile to short stalked sporangium-like spherical to reniform structures in abaxial rows. The segments themselves curve towards other member(s) of the pair or cluster producing a cupule-like appearance.

Permineralised axes associated with sterile leaves contain a terete, centrarch xylem strand composed of tracheids with G-type thickenings. Hao and Beck [82] interpreted the leaves as megaphylls—much-branched, vascularised lateral systems homologous to those described for *Psilophyton* and *Pertica* [83], but more advanced in being laminate. In considering the affinities of the Chinese plant, emphasis was placed on similarities (pseudomonopodial branching, centrarch xylem, laminar leaves, and arrangement of sporangia) with members of the trimerophyte–aneurophyte–seed-plant clade, and with some species of *Archaeopteris* (laminate fertile appendages with adaxial sporangia). However, in that *Eophyllophyton* predates the trimerophytes, the authors suggested that it represents a group of plants quite distinct from the zosterophyllophyte–lycophyte and trimerophyte–aneurophyte–seed-plant clades of Laurussia as a result of its evolution in isolation on the South China plate.

In contrast, the sparingly spiny axes of *Yunia dichotoma* are isotomously branched with little change in axis diameter in the daughter branches which are produced at wide angles (50°–70°) [84]. These sterile axes are consistently associated with vertically elongate sporangia, elliptical to ovoid in shape and dehiscing longitudinally into two equal valves. The surprising feature of this plant with such simple gross morphology lies in the anatomy of the xylem of the sterile axes. The predominantly terete xylem strand in unbranched axes is centrarch with abundant protoxylem comprising tracheids plus conspicuous parenchyma. The former may be aggregated into up to three peripheral strands or intermixed with the parenchyma. The annular, helical to scalariform and reticulate metaxylem tracheids have walls perforated by small pores connecting the secondary thickenings and thus may be of the G-type. The authors, despite its morphological simplicity, tentatively assigned the plant to the trimerophytes, while emphasising its close similarities with certain Carboniferous filicalean and zygopterid genera as support for the hypothesis that these two groups had their origins in the Trimerophytales.

Hao and Beck [78] sought explanation for the high percentage of endemics in the Posongchong Formation in the long isolation of the China plate during and prior to Early Devonian times [85], because palaeogeographic reconstructions place South China in an equatorial position with general climate and environmental features (e.g. day length) similar to those of Laurussia [86]. However, they also suggested that China might have been warmer because of a major ocean current flowing northwards along the eastern coast [87]. They further noted that the prevailing winds could not have transported spores westwards towards Laurussia.

While the presence of a high percentage of endemics clearly indicates that the South Chinese plants were isolated from those on other palaeocontinents in Pragian times, the taxa present which were common to those palaeocontinents can provide evidence for the timing of such reproductive isolation. It thus becomes vitally important that such taxa be unequivocally identified, e.g. *Z. myretonianum* [68] (one spike only), *Z. australianum*, and species of *Drepanophycus*. It is this activity which presents a major challenge to understanding the phytogeographic status of the southern China assemblages—and, more generally, the geographic radiations of early land plants.

Silurian. The Late Silurian assemblage of Xinjiang, northwest China, has similarities with that from coeval rocks in eastern Kazakhstan and hence derives from the Kazakhstan palaeocontinent rather than the South China plate [88]. The fossils show greater diversity and complexity than Prídolí examples from Laurussia, and have no confidently identified taxa in common with the *Baragwanathia* flora of the Ludlow of Australia [89]. Cai et al. [88] were reluctant to overemphasise any distinctions in the light of such limited data, but speculated on the existence of possibly three phytochoria in the Late Silurian. Such ambivalence highlights the need for more information from localities world-wide in the Silurian and Devonian, yet their discovery can introduce more problems than they solve. China also provides a classic example of this. *Pinnatiramosus* [90] from Guizhou Province is a quite remarkable fossil in its size, morphology and anatomy, and age. Originally thought to be Wenlock [90], recent palaeozoological evidence suggests a Llandovery age [91], although the ornamented (apiculate) spores recovered from the matrix are first recorded in Ludlow rocks of Laurussia [73]. Its organisation is axial, with profuse branching producing an extensive lateral pinnate structure, interpreted as upright branch systems borne on horizontal axes. Geng recovered some remarkably pitted tubular structures from the latter, which he compared with tracheids, but nevertheless interpreted the plant as non-vascular. Indeed if his three-dimensional reconstruction is correct, (there being no direct evidence for it), such a flattened pinnate, branching system would surely have required water to support it. Wang and Cai [92] described similar unevenly thickened fragments of tubes and tubes with annular thickenings from macerations at the locality, comparing the latter favourably with scalariform and bordered pitting of vascular plants. Detailed scanning electron microscopy (SEM) is required to substantiate this—the distribution of light and dark areas under light microscopy (LM) is not that usually associated with bordered pitting—especially as Wang and Cai concluded that on the basis of the presence of such tracheids, *Pinnatiramosus* was a primitive vascular plant.

1.3.2 Affinities of the Earliest Embryophytes

Ultrastructure. The value of dispersed spores in reconstructing land vegetation in the Upper Palaeozoic [93] has been amplified by more precise identification of the spore producers [94] and when in situ spores are unavailable, by the use of ultrastructure in the determination of broad affinities [95]. In older rocks, parallel studies are hampered by the dearth of in situ occurrences in early vascular plants and rhyniophytoids [30]. Ultrastructural studies via transmission electron microscopy (TEM), although in their infancy, demonstrate variation in exospore layering, although its relationship to extant pteridophytes and bryophytes remains equivocal [96,97]. In contrast, W. Taylor's pioneering work on dispersed cryptospores (tetrads and dyads) appears more promising, largely because he concentrated on specimens from rocks which have undergone relatively little diagenetic alteration (or thermal maturation). This was particularly the case for *Dyadospora* sp. [98] (Lower Silurian: Ohio, USA), which shows convincing microlayering in the inner (1.4 µm thick) region of the exospore, the outer region (<1.5 µm) being homogeneous. The lamination consists of 10–15 units, each 50–150 nm thick. While sporopollenin deposition on lamellae is considered a common character of all embryophytes, their presence is usually not apparent in the mature exospore [99]. Exceptions are recorded in the hepatics, where in the Sphaerocarpales in particular, mature organization resembles that in the fossil dyad. This led Taylor to suggest that the earliest land plants were related to this group of extant bryophytes. Variation in dyad ultrastucture is evidenced in his sections of the pseudodyad *Pseudodyadospora* in which the exospore is homogeneous and in *Segestrespora membranifera* where the wall is described as spongy

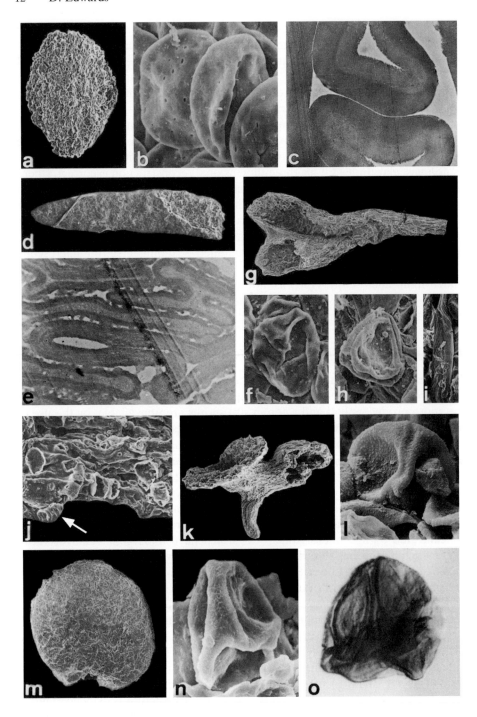

FIG. 1.1a–o. Silurian and Devonian cryptospores. Specimens are housed in the National Museum of Wales (NMW) Cardiff. **a**. Disc composed of hilate dyads, scanning electron micrograph (SEM). Ludford Lane, Shropshire. Upper Silurian (Prídolí), NMW 96 11G.1. ×47.5. **b**. Spores from **a**. Nature of shallow pits in exospore unknown. ×1125. **c**. Transmission electron micrograph (TEM) of equators of two adjacent spores from **a**, showing layering of exospore. ×6800. **d**. Elongate spore mass of hilate dyads (SEM). Ludford Lane, NMW 96 11G.2. ×38. **e**. TEM of several spores showing adpressed distal and proximal surfaces, bilayered exospore, and superficial granular material which sometimes extends between adjacent spores. ×8200. **f**. SEM of single dyad with highly folded wall. ×760. **g**. Axial fragment with bifurcating tip containing smooth tetrads (SEM). North Brown Clee Hill, Shropshire. Lower Devonian (Lochkovian), NMW 94 76G.1. (From [102], with permission) ×36. **h**. Single tetrad (cf *Tetrahedraletes*) from **g**. ×460. **i**. Strap-shaped structure with herringbone pattern from spore-bearing region in **g**. ×550. **j**. Part of surface of vegetative axis, showing broken blistered appearance. *Arrow* indicates banded tube. (From [102], with permission) ×475. **k**. Bifurcating axial fragment with bases of two terminal sporangia (SEM). North Brown Clee Hill, NMW 96 11G.3. ×62.5. **l**. Spore tetrad with ornamented ?envelope from **k**. ×1550. **m**. Discoidal structure with a cellular wrinkled layer (?cuticle) covering numerous tetrads. Ludford Lane, NMW 96 11G.4. ×37. **n**. SEM of single enveloped tetrad. ×875. **o**. LM of single enveloped tetrad. ×780

but not uniformly so. The latter dyad is enclosed in an envelope with muri on its surface [100]. His finding of the same ultrastructure in tetrads with similarly ornamented envelopes from the same localities led to his suggestion that the tetrads and dyads were produced by the same plant or closely related species. His published TEM sections [101] through an Upper Ordovician naked permanent tetrad, *Tetrahedraletes medinensis* Strother and Traverse emend. Wellman and Richardson, a taxon first recorded earlier in the Ordovician, showed far simpler exospore organisation. The four spores were described as being in an "extremely tight association" over 60%–80% of their proximal surfaces. Indeed the illustrations show little or no disruption in the organisation of the exospore between adjacent spores in these regions, while externally the junction is not marked by a separation suture (I use this phrase to describe the line [fissure] marking the junction between the four spores of a tetrad). The exospore itself was described as "essentially homogeneous" although "subtle" laminae are recorded as occasionally visible in the outer half. The exospore also appears more spongy in the immediate vicinity of the lumen, which is conspicuous only below the "prominent equatorial thickenings."

In ultrastructural studies in Cardiff we have adopted a comparative approach. In an attempt to eliminate some taphonomic influences which might prejudice the use of ultrastructure in the determination of affinity, we have concentrated on coeval palynomorphs (cryptospores and trilete spores) from the Upper Silurian and Lower Devonian in a restricted geographical area which is known to have undergone a similar diagenetic history. Our results show that variations in ultrastructure do exist, and while diagenetic alteration has obliterated much detail, thus limiting direct comparison with extant embryophytes, there can be a consistency of wall organisation in in situ spores of plants which are also united on morphological characters [96,97]. Further, wall ultrastructure can be used to detect relationships between the "pteridophytes" and cryptospore-producers.

Thus, for example, preliminary studies on a number of discrete discoidal spore masses containing dissociated hilate dyads attributable to *Laevolancis divello-media* indicate that the entire exospore is consistently bilayered, the outer always appearing darker (more electron dense) than the inner (Fig. 1.1a–c). The layering is visible but less distinct in both the unstained and unoxidised examples. Some variation in relative widths of the two layers exists between samples, with the outer often slightly narrower. The lumen is usually compressed except at the equator. In the discoidal masses, there is little or no extra sporal material, but in an elongate mass, with one intact rounded end, composed of spores which in light microscopy preparations would be assigned to *Laevolancis divellomedia*, this is abundant (Fig. 1.1d,f). High-resolution scanning electron microscopy (SEM) reveals that the distal surfaces are covered by sporadic, irregularly shaped gran-ular material which in TEM appears as highly staining outgrowths of the darker staining outer layer of a bilayered exospore, otherwise similar to that seen in spores in the discoidal masses. That "ornament" sometimes continues as bridges between adjacent spores (Fig. 1.1e).

Considering coeval presumed vascular plants, we have shown similarly bilay-ered exospores in in situ spores of *Cooksonia pertoni* ssp *pertoni* and *synorispora* [96,103] which are assigned to dispersed trilete taxa *Ambitisporites* sp and *Syn-orisporites verrucatus* respectively. Similarly shaped spore masses, from the same locality as the *Cooksonia pertoni* (Perton Lane), derive from the discoidal spo-rangia of *Pertonella langii* and contain *Retusotriletes* cf. *coronadus* where the exospore is homogeneous [97]. Thus it is a possibility that the discoidal laevigate spore masses derive from plants with affinity to *Cooksonia pertoni* on gross sporangial morphology and spore ultrastructure, the dyads possibly resulting from a meiotic abnormality where separation occurs after the first division of meiosis with sporopollenin deposition on the products of the second (see discus-sion in [29]). In this case, absence of an axial branching system might indicate an essentially parenchymatous succulent vegetative system showing greater similar-ity with the tissues of the Rhynie Chert plants than with *C. pertoni* with its sterome. On the other hand the parent plant could completely lack recalcitrant polymers and hence be closer to the bryophyte clade. Finally it should be remem-bered that *Hostinella* is abundant in many Prídolí localities and that the spore masses, perhaps originally surrounded by a sporangium wall of predominantly thin walled tissue, could once have been attached to them!

The discoidal laevigate dyad-containing spore masses also differ from *Cookson-ia pertoni* in that the extra-exosporal material described in the latter and tenta-tively compared with perispore, or less likely fossilised intralocular fluid [96], is absent. It seems unlikely that it was "taphonomically" leached away in the absence of a sporangial wall as extra-sporal material persists in the elongate, only partially surrounded spore-mass described above. However, the latter is more discretely organised and shows greater similarity with that attached to retusoid spores recovered from Lower Devonian bivalved sporangia (*Resilitheca* [104] Fig. 1.2g,h), where the body of the exospore was not bilayered. Such variation augers well for estimations of diversity based on spore characters. Indeed our preliminary studies on several species of the hilate dyad *Cymbohilates*, distinguished on the basis of variation in the nature of the distal apiculate ornament, show at least three types of exospore ultrastructure, two resembling presumed vascular plant spores and the third reminiscent of the layering seen in Taylor's *Dyadospora* [98] and hence of possible bryophyte affinity.

In situ spores. Such preliminary studies show that spore ultrastructure has some potential in determining relationships particularly between coeval plants, but clearly the best progress will be made in the discovery of in situ cryptospores. In this we have made a start, although representatives of such spores have been recorded mainly in Lower Devonian fossils. Exceptions are the loosely associated laevigate hilate dyads recorded in coalified *Salopella*-like sporangia from the late Silurian of the Welsh Borderland [106]. The sporangia (5.0–5.4 mm long and 0.47–0.87 mm wide) are parallel-sided with rounded tips, and are distinguished from the axes (apart from their spores) because they readily separate from the matrix but leave no smooth imprint on the rock: the axes adhere more strongly. Anatomical evidence for the sporangium wall is absent. The dyads themselves may be large (39–68 μm; x = 48), isomorphic or slightly anisomorphic and with a narrow equatorial crassitude and circular amb. They were described as ?*Archaeozonotriletes* (now *Laevolancis*) cf. *divellomedium*. Separated hilate monads with smooth circular contact areas and a distal ornament of irregular grana have been recorded on the inner surface of half of a longitudinally split, fusiform sporangium [106]. Similar ornament was recorded on the trilete *Aneurospora* sp. recovered from *Salopella marcensis* at the same locality. Such similarities again suggest that the separated dyads may result from a deviation from normal meiosis, but in this case ultrastructural evidence for such a relationship has not yet been obtained.

Undissociated laevigate dyads have been recorded in a single cup-shaped sporangium found among the mesofossils in the fluvial Lower Devonian rocks of the Welsh Borderland (Fig. 1.2e,f). The minute sporangium (0.8 mm long and 0.46 mm wide) terminates an unbranched axis, 1.3 mm long, which is essentially a corrugated sleeve of cuticle. The sporangium has a truncated tip with central depression and parallel sides. Cells are not visible, but the material covering the spores, comprises more than one layer. The dyads (c. 50 μm diameter) are deeply invaginated on the distal surfaces, possess a separation suture faintly visible as a ± wavy line, and show no tendency to split into two monads (Fig. 1.2f).

Obligate tetrads, closest morphologically on SEM examination to *Tetrahedraletes medinensis* isolated from Ordovician rocks, have recently been recorded in a bifurcating sporangium terminating a short length of unbranched axis (Fig. 1.1g) in which conventional cellular organisation (e.g. parenchyma or sclerenchyma) has not been observed [102]. Instead, the axis, U-shaped in cross section, and

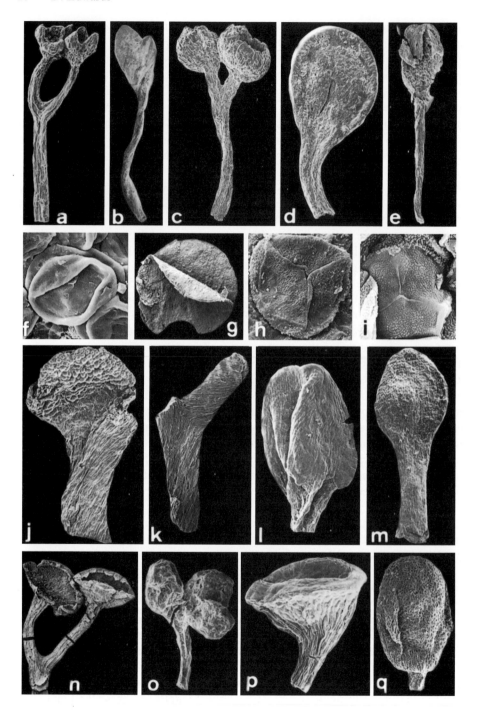

FIG. 1.2a–q. Silurian and Devonian mesofossils. Figures **i–l** are from Ludford Lane (Prídolí), the remainder from North Brown Clee Hill (Lochkovian) **a**. Unnamed rhyniophytoid with terminal sporangia splitting distally into three valves. NMW 96 5G.6. (From [105], with permission) ×45. **b**. *Tortilicaulis offaeus* with bifurcating terminal sporangium. NMW 93 98G.13. (From [106], with permission) ×28. **c**. Rhyniophytoid with globose terminal sporangia (*?Cooksonia cambrensis*). NMW 94 60G.16. (From [107], with permission) ×43. **d**. Unnamed terminal sporangium with thickened convex margins and presumed dehiscence into two valves. NMW 96 11G.5. ×22. **e**. Unnamed sporangium terminating smooth unbranched axis with in situ dyads. NMW 96 11G.6. ×26. **f**. Laevigate dyad from **e**, showing distal surface of one member. ×500. **g**. *Resilitheca salopensis*: isolated dehisced sporangium. NMW 94 60G.12. (From [108], with permission) ×15. **h**. Retusoid spore of *Resilitheca* in **g**. (From [108], with permission) ×1150. **i**. Proximal surface of spore from new plant illustrated in **j**. ×1350. **j**. Unnamed bivalved sporangium, with prominent cells in wall, terminating twisted bifurcating axis. NMW 96 11G.7. ×25. **k**. Sterile axis of taxon illustrated in **j**. NMW 94 60G.2. (From [107], with permission) ×19. **l**. Unnamed terminal sporangium. NMW 94 60G.7. (From [107], with permission) ×60. **m–q**. Diversity in shape of L. Devonian terminal sporangia. **m**. Unnamed sporangium NMW 94 60G.15. (From [107], with permission) ×17. **n**. Cf *Cooksonia pertoni* NMW 94 60G.14. ×15. **o**. Unnamed taxon NMW 96 11G.8. ×82. **p**. *C. pertoni* NMW 96 5G.8. (From [105], with permission) ×60. **q**. Unnamed taxon NMW 96 11G.9. ×38

hence possibly incomplete, is limited by a homogeneous region bearing superficial, irregularly sized, cup-shaped outgrowths (Fig. 1.1j), which probably represent the basis of hollow blisters (? derived from cuticle). Diamond-shaped outlines of cells with transverse superficial striations are associated with the depression between the two sporangial cavities, but it is impossible to distinguish individual cells in fortuitous transverse fractures of the wall. Internal tissues show few discrete cells except for fragments of tubes variously orientated, and usually with annular thickenings. Such tubes may be superficial. The lumens of some ?cells/tubes show irregularly distributed peg-shaped ingrowths. Within the sporangial cavities, a single strap-shaped structure with striations showing a herring-bone arrangement (Fig. 1.1i) has been observed crossing some spores close to the sporangium wall. The spores themselves, in the form of tetrads, are fragmentary and entirely smooth-walled (Fig. 1.1h). Conspicuous indentations occasionally occur between the constituent spores, but there is no separation suture—the contacts being superficially marked by a depression in a continuous layer (in SEM), as was seen in the Ordovician tetrads which Taylor [98,101] sectioned. Such similarities persuade us that a non-enveloped tetrad characterizes the Lower Devonian plant, although the possibility that a tightly adhering laevigate envelope [52] was present cannot be completely eliminated. Unfortunately insufficient material was available to allow examination by LM or TEM.

In toto, such anatomy is not characteristic of either axial vascular plants or rhyniophytoids preserved as compression fossils, but Professor Duckett related some of their features (e.g. tetrads and banded tubes) to extant liverworts.

However, it cannot be overemphasised that the fossil has no exact extant counterpart, and since 1995, our discovery of similarly thickened banded tubes fused to the surface of sporangia and axes of *Tortilicaulis* [106] has led to the suggestion that some of the tubes belong to pathogens or saprobes, while Graham et al. [109] compared them with bryophyte epidermal cells.

Possibly enveloped tetrads, with smooth or ornamented surfaces have been recorded at the base of incomplete but probably originally ellipsoidal sporangia borne terminally on bifurcating, typically rhyniophytoid axes. That in Figure 1.1k shows the bases of two such sporangia, containing tetrads apparently enclosed within an envelope with granular ornament (Fig. 1.1l).

Unequivocally, but less regularly, ornamented envelopes (Fig. 1.1n), their presence confirmed by light microscopy (Fig. 1.1o), surround tetrads in a discoidal spore mass enclosed within a wrinkled amorphous sheet of material (Fig. 1.1m). This mass definitely derives from a sporangium of a different shape from those just described, and we are currently sectioning the tetrads in an attempt to detect relationships via their ultrastructure.

We are thus slowly accumulating data on the nature of the earliest land vegetation and the phylogeny of the ancestral land plants, but more detailed information on spores and envelopes is needed to allow the testing of hypotheses, e.g. Gray [52].

1.3.3 *Siluro-Devonian Mesofossils* (Fig. 1.2)

To end on a self-indulgent note, the past 10 years has seen the revelation of a new facet of early land vegetation—that of very small plants—ground hugging and comparable in stature to the "moss forests" of today, but with plants possessing branching sporophytes. Some are vascular plants, but in the majority failure to demonstrate tracheids leaves their status equivocal. In some cases the fragmentary fossils are probably parts of much larger plants, e.g. sporangia of zosterophylls, but it seems unlikely that the minute branching axes with terminal sporangia have a similar origin. To date only two major productive localities have been found in the Welsh Borderland and it is a challenge for the next millennium to discover more. Both localities are characterised by a dark grey, fine-grained rock, presumably indicative of a reducing environment, but the Silurian Ludford Lane sediments were deposited in a marginal marine environment and the Lochkovian Brown Clee Hill facies is fluvial. At Ludford Lane, in particular, the best fossils were obtained from fresh, recently excavated rock (Fig. 1.2i–l). At both localities, the small fossils, usually a few millimetres long, show varying degrees of compression although cells with thicker walls, e.g. sterome (peripheral support tissues) and sporangial wall, retain their three-dimensional shape. Such records and their significance have been recently reviewed [107]. Anatomically, the most useful data include surface features, particularly stomata, sterome, and sporangial characters including dehiscence and in situ spores. Those mesofossils illustrated here demonstrate their diverse morphology (Fig. 1.2a–d,m–q). Many are unnamed because they are known only from one or two specimens. Their isolation requires dissolution of large volumes of rock, and apart from *Cooksonia*

(Fig. 1.2n,p), *Tortilicaulis* (Fig. 1.2b), and *Resilitheca* (Fig. 1.2g,h), examples of remaining taxa are few and we await accumulation of sufficient examples to allow demonstration of variability and reliable circumscription in diagnoses of new taxa.

A particular challenge relates to elucidation of the nature of conducting tissue, if indeed present. The distribution of the major types of conducting cells (G- and S-type tracheids) was based on the detailed investigations of pyritised xylem by Paul Kenrick in Pragian fossils [37,41], with the earliest direct evidence for the G-type in an indeterminate pyritised fragment from Lochkovian rocks [110]. Current investigations of the coalified mesofossils provide little evidence for water-conducting cells but continue to be a major thrust of our studies. Those in *Cooksonia pertoni* have led to the suggestion that the earliest conducting cells had heavily thickened walls and that subsequent evolution involved the development of lateral pits allowing lateral movement of water to peripheral metabolising tissues as plants increased in height [111], and in the first formed xylem, provision for extension growth.

The advantages of an endohydric conducting system even to a plant of such small stature as *Cooksonia* have been emphasised by Raven [112,113], and coupled with cuticle and stomata resulted in the acquisition of homoiohydry. Such a physiological strategy consequently "allowed" the evolution of increased size and the exploitation of much drier environments as responses to environmental pressures relating to competition for light, space, water, and mineral nutrients. Thus noting a parallel in the animal kingdom [114] it might be argued that 1) the critical events in the evolution of physiology, reproductive biology, and "bauplan" of the earliest vascular plants, particularly as they relate to the capacity for "open" sporophytic growth and the biochemical and anatomical innovations connected with homoiohydry, occurred in plants of small size; 2) such pioneering plants were preadapted to respond to increasing environmental pressures following initial success; and 3) such plants would be under-represented in the fossil record, partly due to their high surface to volume ratio which would facilitate rapid decay [114]. The continuing documentation of spores in such plants [107] might result in detection of their existence in earlier times by analysis of dispersed spore assemblages, although experience with *Cooksonia pertoni* [115] warns that spore morphology may change while gross morphology does not.

As regards estimates of diversity and change through time, the existence of such productive localities severely distorts the record. As an example, the number of species in the middle Lochkovian is increased at least twofold by the addition of the North Brown Clee Hill discoveries.

1.4 The Way Forward

It is unfortunate that as we approach the end of the century in which we have learnt so much about the history of plant life on our planet, it is extremely difficult to justify such "blue-sky" activity in that it does little to change man's quality of life by improving his environment or by creating wealth. Yet in our current preoccupation with major global changes in climate, and concerns relating to man's catastrophic interference in a variety of ecosystems, it is perhaps appropriate that we should revisit the middle Palaeozic as another time of profound global change during the establishment of terrestrial biotas and concommitant modification of the physical environment. As successive waves of vegetation, composed of plants with ever-increasing morphological and anatomical complexity and with more versatile physiology and reproductive biology, radiated into new habitats [116,117], there resulted major changes in surface processes, such as rock weathering, fluvial, and even marine sedimentation [118,119] as well as in the composition of the atmosphere [120]. Their proper understanding demands integrated activity between palaeontologists, physiologists, sedimentologists, climatologists, geochemists, and geophysicists, always underpinned by the continued rigorous descriptions of the green plants, whose diversification on land truly changed the face of planet Earth.

Acknowledgments. New research presented here has been financed by grants from the NERC (spore ultrastructure: GR9/1441) and the Leverhulme Trust, which are gratefully acknowledged. I also thank Lindsey Axe and Charles Wellman for cheerful assistance.

References

1. Kidston R, Lang WH (1917) On Old Red Sandstone plants showing structure, from the Rhynie Chert Bed, Aberdeenshire. Part I. *Rhynia gwynne-vaughani*, Kidston and Lang. Trans R Soc Edinburgh 51:761–784
2. Kidston R, Lang WH (1920) On Old Red Sandstone plants showing structure, from the Rhynie Chert Bed, Aberdeenshire. Part II. Additional notes on *Rhynia gwynne-vaughani*, Kidston and Lang; with descriptions of *Rhynia major*, n.sp. and *Hornea lignieri*, n.g., n.sp. Trans R Soc Edinburgh 52:603–627
3. Kidston R, Lang WH (1920) On Old Red Sandstone plants showing structure, from the Rhynie Chert Bed, Aberdeenshire. Part III. *Asteroxylon mackiei*, Kidston and Lang. Trans R Soc Edinburgh 52:643–680
4. Kidston R, Lang WH (1921) On Old Red Sandstone plants showing structure, from the Rhynie Chert Bed, Aberdeenshire. Part IV. Restorations of the vascular cryptogams, and discussion of their bearing on the general morphology of the Pteridophyta and the origin of the organisation of land plants. Trans R Soc Edinburgh 52:831–854
5. Kidston R, Lang WH (1921) On Old Red Sandstone plants showing structure, from the Rhynie Chert Bed, Aberdeenshire. Part V. The Thallophyta occurring in the peat-bed; the succession of the plants throughout a vertical section of the bed, and the conditions of accumulation and preservation of the deposit. Trans R Soc Edinburgh 52:855–902
6. Gray J, Boucot AJ (1994) Early Silurian nonmarine animal remains and the nature of the early continental ecosystem. Acta Palaeont Polonica 38:303–328

7. Edwards D, Selden P, Richardson JB, Axe L (1995) Coprolites as evidence for plant-animal interaction in Siluro-Devonian terrestrial ecosystems. Nature 377:329–331

8. Lyon AG (1964) Probable fertile region of *Asteroxylon mackiei* Kidston and Lang. Nature 203:1082–1083

9. Lyon AG, Edwards D (1991) The first zosterophyll from the Lower Devonian Rhynie Chert, Aberdeenshire. Trans R Soc Edinburgh: Earth Sci 82:323–332

10. Powell CL (1994) The palaeoenvironments of the Rhynie Cherts. PhD thesis, University of Aberdeen

11. Remy W, Remy R (1980) *Lyonophyton rhyniensis* nov. gen. et nov. spec., ein Gametophyt aus dem Chert von Rhynie (Unterdevon, Schottland). Argumenta Palaeobot 6:37–72

12. Remy W (1982) Lower Devonian gametophytes: relation to the phylogeny of land plants. Science 215:1625–1627

13. Remy W, Hass H (1991) Ergänzende Beobachtungen an *Lyonophyton rhyniensis*. Argumenta Palaeobot 8:1–27

14. Remy W, Hass H (1991) *Kidstonophyton discoides* nov. gen., nov. spec., ein Gametophyt aus dem Chert von Rhynie (Unterdevon, Schottland). Argumenta Palaeobot 8:29–45

15. Remy W, Hass H (1991) *Langiophyton mackiei* nov. gen., nov. spec., ein Gametophyt mit Archegoniophoren aus dem Chert von Rhynie (Unterdevon, Schottland). Argumenta Palaeobot 8:69–117

16. Remy W, Hass H (1991) Gametophyten und sporophyten im Unterdevon—fakten und spekulationen. Argumenta Palaeobot 8:193–223

17. Kenrick P (1994) Alternation of generations in land plants: new phylogenetic and palaeobotanical evidence. Biol Rev 69:293–330

18. Remy W, Hass H (1996) New information on gametophytes and sporophytes of *Aglaophyton major* and inferences about possible environmental adaptations. Rev Palaeobot Palynol 90:175–193

19. Taylor TN, Remy W, Hass H (1992) Fungi from the Lower Devonian Rhynie Chert: Chytridiomycetes. Am J Bot 79:1233–1241

20. Remy W, Taylor TN, Hass H (1994) Early Devonian fungi: a blastocladalean fungus with sexual reproduction. Am J Bot 81:690–702

21. Hass H, Taylor TN, Remy W (1994) Fungi from the Lower Devonian Rhynie Chert: mycoparasitism. Am J Bot 81:29–37

22. Taylor TN, Hass H, Remy W, Kerp H (1995) The oldest fossil lichen. Nature 378:244

23. Taylor TN, Remy W, Hass H, Kerp H (1995) Fossil arbuscular mycorrhizae from the Early Devonian. Mycologia 87:560–573

24. Trewin NH (1994) Depositional environment and preservation of biota in the lower Devonian hot-springs of Rhynie, Aberdeenshire, Scotland. Trans R Soc Edinburgh: Earth Sci 84:433–442

25. Lang WH (1937) On the plant-remains from the Downtonian of England and Wales. Philos Trans R Soc London B 227:245–291

26. Edwards D, Bassett MG, Rogerson ECW (1979) The earliest vascular land plants: continuing the search for proof. Lethaia 12:313–324

27. Edwards D, Davies KL, Axe L (1992) A vascular conducting strand in the early land plant Cooksonia. Nature 357:683–685

28. Edwards D, Davies MS (1990) Interpretations of early land plant radiations: "facile adaptationist guesswork" or reasoned speculation? In: Taylor PD, Larwood GP

(eds) Major Evolutionary Radiations. Syst Assoc Spec Vol 42, Clarendon, Oxford, pp 351–376

29. Fanning U, Richardson JB, Edwards D (1991) A review of in situ spores in Silurian land plants. In: Blackmore S, Barnes SH (eds) Pollen and spores, patterns of diversification. Syst Assoc Spec Vol 44. Clarendon, Oxford, pp 25–47

30. Edwards D, Richardson JB (1996) Review of in situ spores in early land plants. In: Jansonius J, Mcgregor DC (eds) Palynology: principles and applications, Vol. 1, principles. American Association of Stratigraphic Palynologists Foundation, Publishers Press, Salt Lake City, pp 391–407

31. Burgess ND, Edwards D (1988) A new Palaeozoic plant closely allied to *Prototaxites* Dawson. Bot J Linn Soc 97:189–203

32. Strother PK (1988) New species of *Nematothallus* from the Silurian Bloomsburg Formation of Pennsylvania. J Paleontol 62:967–982

33. Gensel PG, Johnson NG, Strother PK (1990) Early land plant debris (Hooker's "waifs and strays"?). Palaios 5:520–547

34. Høeg OA (1967) Psilophyta. In Boureau E (ed) Traité de Paléobotanique. Vol. II. Masson et Cie, Paris, pp 191–352

35. Banks HP (1968) The early history of land plants. In: Drake ET (ed) Evolution and environment: a symposium presented on the occasion of the one hundredth anniversary of the foundation of the Peabody Museum of Natural History at Yale University. Yale University Press, New Haven, pp 73–107

36. Banks HP (1992) The classification of early land plants—revisited. Palaeobotanist 41:36–50

37. Kenrick P, Edwards D (1988) The anatomy of Lower Devonian *Gosslingia breconensis* Heard based on pyritized axes, with some comments on the permineralization process. Bot J Linn Soc 97:95–123

38. Kenrick P, Edwards D, Dales RC (1991) Novel ultrastructure in water-conducting cells of the Lower Devonian plant *Sennicaulis hippocrepiformis*. Palaeontology 34:751–766

39. Niklas KJ, Banks HP (1990) A reevaluation of the Zosterophyllophytina with comments on the origin of lycopods. Am J Bot 77:274–283

40. Edwards D, Edwards DS (1986) A reconsideration of the Rhyniophytina, Banks. In: Spicer RA, Thomas BA (eds) Systematic and taxonomic approaches in palaeobotany. Syst Assoc Spec Vol 31, Clarendon, Oxford, pp 199–220

41. Kenrick P, Crane PR (1991) Water-conducting cells in early land plants: implications for the early evolution of tracheophytes. Bot Gaz 152:335–356

42. Gensel PG (1992) Phylogenetic relationships of the zosterophylls and lycopsids: evidence from morphology, paleoecology and cladistic methods of inference. Ann Missouri Bot Gard 79:450–473

43. Li C-S, Edwards D (1992) A new genus of early land plants with novel strobilar construction from the lower Devonian Posongchong formation, Yunnan Province, China. Palaeontology 35:257–272

44. Mishler BD, Churchill SP (1984) A cladistic approach to the phylogeny of the "bryophytes". Brittonia 36:406–424

45. Mishler BD, Churchill SP (1985) Transition to a land flora: phylogenetic relationships of the green algae and bryophytes. Cladistics 1:305–328

46. Graham LE, Delwiche CF, Mishler B (1991) Phylogenetic connections between the "green algae" and the "bryophytes". Adv Bryol 4:213–244

47. Ishchenko TA, Ishchenko AA (1982) New records of charophytes from the Upper Silurian of Podolia (in Russian). Nauk Dumka, Kiev, pp 21–32
48. Graham LE (1993) Origin of Land Plants. John Wiley, New York
49. Gray J, Boucot AJ (1971) Early Silurian spore tetrads from New York: earliest New World evidence for vascular plants? Science 173:918–921
50. Gray J (1984) Ordovician-Silurian land plants: the interdependence of ecology and evolution. In: Bassett MG, Lawson JD (eds) Autecology of Silurian organisms. Spec Pap Palaeontol 32:281–295
51. Gray J (1985) The microfossil record of early land plants: advances in understanding of early terrestrialization, 1970–1984. Philos Trans R Soc London B 309:167–195
52. Gray J (1991) *Tetrahedraletes, Nodospora*, and the "cross" tetrad: an accretion of myth. In: Blackmore S, Barnes SH (eds) Pollen and spores, patterns of diversification. Syst Assoc Spec Vol 44. Clarendon, Oxford, pp 49–87
53. Richardson JB, Ford JH, Parker F (1984) Miospores, correlation and age of some Scottish Lower Old Red Sandstone sediments from the Strathmore region (Fife and Angus). J Micropalaeontol 3:109–124
54. Richardson JB (1988) Late Ordovician and Early Silurian cryptospores and miospores from northeast Libya. In: El-Arnauti A, Owens B, Thusu B (eds) Subsurface palynostratigraphy of northeast Libya. Garyounis Univ Publ, Benghazi, pp 89–109
55. Shute CH, Hemsley AR (1995) Lang's mysterious dyads. Microscop Anal, pp 39–40
56. Banks HP (1981) Time of appearance of some plant biocharacters during Siluro-Devonian time. Can J Bot 59:1292–1296
57. Banks HP (1975) Palaeogeographic implications of some Silurian-Early Devonian floras. In: Campbell KSW (ed) Gondwana Geology. Australian National University Press, Canberra, pp 75–97
58. Lee H-H, Tsai C-Y (1978) Devonian floras of China. III. Papers for the International Symposium on the Devonian System 1978. Nanking Inst Geol Palaeont, Academica Sinica, Nanking, pp 1–14
59. Garratt MJ, Rickards RB (1984) Graptolite biostratigraphy of early land plants from Victoria, Australia. Proc Yorkshire Geol Soc 44: 377–384
60. Hueber FM (1983) A new species of *Baragwanathia* from the Sextant Formation (Emsian), Northern Ontario, Canada. Bot J Linn Soc 86:57–79
61. Edwards D (1990) Constraints on Silurian and Early Devonian phytogeographic analysis based on megafossils. In: McKerrow WS, Scotese CR (eds) Palaeozoic palaeogeography and biogeography. Geol. Soc., London, Mem. 12, pp 233–242
62. Morel E, Edwards D, Iñigez Rodriguez M (1995) The first record of *Cooksonia* from South America in Silurian rocks of Bolivia. Geol Mag 132:449–452
63. Li X-X (1995) (ed) Fossil floras of China through the geological ages. Guangdong Science and Technology, Guangzhou
64. Halle TG (1927) Fossil plants from south-western China. Palaeontologica Sinica, Ser A 1:1–26
65. Halle TG (1936) On *Drepanophycus, Protolepidodendron* and *Protopteridium*, with notes on the Palaeozoic flora of Yunnan. Palaeontologica Sinica, Ser A 1:1–38
66. Li C-S, Edwards D (1995) A re-investigation of Halle's *Drepanophycus spinaeformis* Göpp. from the Lower Devonian of Yunnan Province, southern China. Bot J Linn Soc 118:163–192

67. Hsü J (1966) On plant remains from the Devonian of Yunnan and their significance in the identification of the stratigraphical sequence of this region (in Chinese with English abstract). Acta Bot Sinica 14:50–69
68. Li X-X, Cai C-Y (1977) Early Devonian *Zosterophyllum* remains from southwest China (in Chinese with English abstract). Acta Palaeontologica Sinica 16:12–34
69. Banks HP (1975) Reclassification of psilophyta. Taxon 24:401–413
70. Geng B-Y (1992) Studies on Early Devonian flora of Sichuan (in Chinese with English abstract). Acta Phytotax Sinica 30:197–211
71. Geng B-Y (1992) *Amplectosporangium*—a new genus of plant from the Lower Devonian of Sichuan, China (in Chinese with English abstract). Acta Bot Sinica 34:450–455
72. Wang Y (1994) Lower Devonian miospores from Gumu in the Wenshan district, southeastern Yunnan (in Chinese with English abstract). Acta Micropalaeontol Sinica 11:319–332
73. Richardson JB, McGregor DC (1986) Silurian and Devonian spore zones of the Old Red Sandstone Continent and adjacent regions. Geol Surv Can, Bull 364:1–79
74. Li C-S, Edwards D (1996) *Demersatheca* Li et Edwards, gen. nov., a new genus of early land plants from the Lower Devonian, Yunnan Province, China. Rev Palaeobot Palynol 93:77–88
75. Geng B-Y (1983) *Stachyophyton* gen. nov. from the Lower Devonian of Yunnan and its significance (in Chinese with English abstract). Acta Bot Sinica 25:574–579
76. Wang Y, Cai C-Y (1996) Further observations on *Stachyophyton yunnanensis* Geng from Posonchong Formation (Siegenian) of SE Yunnan, China (in Chinese with English abstract). Acta Palaeontol Sinica 35:99–108
77. Geng B-Y (1985) *Huia recurvata*—A new plant from Lower Devonian of southeastern Yunnan, China (in Chinese with English abstract). Acta Botanica Sinica 27:419–426
78. Hao S-G, Beck CB (1991) *Catenalis digitata* gen. et sp. nov., a plant from the Lower Devonian (Siegenian) of Yunnan, China. Can J Bot 69:873–882
79. Edwards D (1970) Further observations on the Lower Devonian plant, *Gosslingia breconensis* Heard. Philos Trans R Soc London B 258:225–243
80. Rayner RJ (1983) New observations on *Sawdonia ornata* from Scotland. Trans R Soc Edinburgh: Earth Sci 74:79–93
81. Hao S-G (1988) A new Lower Devonian genus from Yunnan, with notes on the origin of leaf (in Chinese with English abstract). Acta Bot Sinica 30:441–448
82. Hao S-G, Beck CB (1993) Further observations on *Eophyllophyton bellum* from the Lower Devonian (Siegenian) of Yunnan, China. Palaeontographica 230B:27–41
83. Gensel PG (1984) A new Lower Devonian plant and the early evolution of leaves. Nature 309:785–787
84. Hao S-G, Beck CB (1991) *Yunia dichotoma*, a Lower Devonian plant from Yunnan, China. Rev Palaeobot Palynol 68:181–195
85. Bambach RK, Scotese CR, Ziegler AM (1980) Before Pangaea: the geographics of the Palaeozoic world. Am Scientist 68:26–38
86. Fang W, Van Der Voo R, Liang Q (1989) Devonian paleomagnetism of Yunnan Province across the Shan Thai—South China suture. Tectonics 8:939–952
87. Ziegler AM, Bambach RK, Parrish JT, Barrett SF, Gierlowski EH, Parker WC, Raymond A, Sepkoski JJ (1981) Paleozoic biogeography and climatology. In: Niklas KJ (ed) Paleobotany, paleoecology and evolution. 2. Praeger, New York, pp 231–266

88. Cai C-Y, Dou Y-W, Edwards D (1993) New observations on a Prídolí plant assemblage from north Xinjiang, northwest China, with comments on its evolutionary and palaeogeographical significance. Geol Mag 130:155–170

89. Tims JD, Chambers TC (1984) Rhyniophytina and Trimerophytina from the early land flora of Victoria, Australia. Palaeontology 27:265–279

90. Geng B-Y (1985) Anatomy and morphology of *Pinnatiramosus*, a new plant from the Middle Silurian (Wenlockian) of China (in Chinese with English abstract). Acta Botanica Sinica 28:664–670

91. Cai C, Ouyang S, Wang Y, Fang Z, Rong J, Geng L, Li X (1996) An early Silurian vascular plant. Nature 379:592

92. Wang Y, Cai C-Y (1995) Late Llandovery tubulary macerals from the Hanjiadian Formation of Fenggang, Guizhou, and their palaeobotanical significance (in Chinese with English abstract). Acta Micropalaeontol Sinica 12:1–11

93. Phillips TL, Peppers RA, Dimichele WA (1985) Stratigraphic and interregional changes in Pennsylvanian coal-swamp vegetation: environmental inferences. Int J Coal Geol 5:43–109

94. Scott AC, Hemsley AR (1993) The spores of the Dinantian lycopsid cone *Flemingites scotii* from Pettycur, Fife, Scotland. Spec Pap Palaeontology 49:31–41

95. Hemsley AR (1993) A review of Palaeozoic seed-megaspores. Palaeontographica B 229:135–166

96. Edwards D, Davies KL, Richardson JB, Axe L (1995) The ultrastructure of spores of *Cooksonia pertoni*. Palaeontology 38:153–168

97. Edwards D, Davies KL, Richardson JB, Wellman CH, Axe L (1996) Ultrastructure of *Synorisporites downtonensis* and *Retusotriletes* cf. *coronadus* in spore masses from the Prídolí of the Welsh Borderland. Palaeontol 39:783–800

98. Taylor WA (1995) Spores in earliest land plants. Nature 373:391–392

99. Scott RJ (1994) Pollen exine—the sporopollenin enigma and the physics of pattern. In: Scott RJ, Stead AD (eds) Molecular and cellular aspects of plant reproduction. Society for Experimental Biology Seminar Series 53:49–81

100. Taylor WA (1996) Ultrastructure of lower Paleozoic dyads from southern Ohio. Rev Palaeobot Palynol 92:269–279

101. Taylor WA (1995) Ultrastructure of *Tetrahedraletes medinensis* (Strother and Traverse) Wellman and Richardson, from the Upper Ordovician of southern Ohio. Rev Palaeobot Palynol 85:183–187

102. Edwards D, Duckett JG, Richardson JB (1995) Hepatic characters in the earliest land plants. Nature 374:635–636

103. Rogerson ECW, Edwards D, Davies KL, Richardson JB (1993) Identification of in situ spores in a Silurian *Cooksonia* from the Welsh Borderland. In: Collinson ME, Scott AC (eds) Studies in palaeobotany and palynology in honour of Professor WG Chaloner, FRS. Spec Pap Palaeontol 49:17–30

104. Edwards D, Fanning U, Davies KL, Axe L (1995) Exceptional preservation in Lower Devonian coalified fossils from the Welsh Borderland: a new genus based on reniform sporangia lacking thickened borders. Bot J Linn Soc 117:233–254

105. Edwards D, Abbot GD, Raven JA (1996) Cuticles of early land plants: a palaeoecophysiological evaluation. In: Kerstiens G (ed) Plant cuticles. Bios Scientific, Oxford, pp 1–31

106. Edwards D, Fanning U, Richardson JB (1994) Lower Devonian coalified sporangia from Shropshire: *Salopella* Edwards et Richardson and *Tortilicaulis* Edwards. Bot J Linn Soc 116:89–110

107. Edwards D (1996) New insights into early land ecosystems: a glimpse of a Lilliputian world. Rev Palaeobot Palynol 90:159–174
108. Edwards D, Fanning U, Davies KL, Axe L, Richardson JB (1995) Exceptional preservation in Lower Devonian coalified fossils from the Welsh Borderland: a new genus based on reniform sporangia lacking thickened borders. Bot J Linn Soc 117:233–254
109. Graham LE, Cook ME, Kroken SB (1996) Resistant moss and liverwort sporangial epidermis resembles enigmatic microfossils attributed to earliest land plants. Am J Bot 83:110–111
110. Fanning U, Edwards D, Richardson JB (1992) A diverse assemblage of early land plants from the Lower Devonian of the Welsh Borderland. Bot J Linn Soc 109:161–188
111. Edwards D, Abbott GD, Raven JA (1996) Cuticles of early land plants: a palaeoecophysiological evaluation. In: Kerstiens G (ed) Plant cuticles—an integrated functional approach. BIOS Scientific, Oxford, pp 1–31
112. Raven JA (1984) Physiological correlates of the morphology of early vascular plants. Bot J Linn Soc 88:105–126
113. Raven JA (1993) The evolution of vascular plants in relation to quantitative functioning of dead water-conducting cells and stomata. Biological Reviews 68:337–363
114. Fortey RA, Briggs DEG, Wills MA (1996) The Cambrian evolutionary "explosion": decoupling cladogenesis from morphological disparity. Biol J Linn Soc 57:13–33
115. Fanning U, Richardson JB, Edwards D (1988) Cryptic evolution in an early land plant. Evol Trends Plants 2:13–24
116. Edwards D, Selden PA (1993) The development of early terrestrial ecosystems. Bot J Scotland 46:337–366
117. Gray J (1993) Major Paleozoic land plant evolutionary bio-events. Palaeogeogr Palaeoclimatol Palaeoecol 104:153–169
118. Beerbower R (1985) Early development of continental ecosystems. In: Tiffney BH (ed) Geological factors and the evolution of plants. Yale University Press, New Haven, pp 47–91
119. Algeo TJ, Berner RA, Maynard JB, Scheckler SE (1995) Late Devonian oceanic anoxic events and biotic crises: "rooted" in the evolution of vascular land plants? GSA Today vol 5(3) 45:64–66
120. Berner RA (1993) Paleozoic atmospheric CO_2: importance of solar radiation and plant evolution. Science 261:68–70

Chapter 2

Morphological Diversity and Evolution of Vegetative Organs in Pteridophytes

Masahiro Kato[1] and Ryoko Imaichi[2]

2.1 Introduction

The primitive vascular plants, or pteridophytes, are nonseed plants that reproduce by free spores. Pteridophytes were the earliest and are most primitive among the three major evolutionary grades of vascular plants. Early vascular plants appeared more than 400 million years ago, and subsequently diversified into a variety of groups. Seed plants evolved from progymnosperms, a group of pteridophytes most closely related to Pteropsida [1]. At present, there are four groups of pteridophytes, i.e., Psilopsida, Lycopsida, Equisetopsida, and Pteropsida or Filicopsida [2].

During the long evolutionary history from the origin of vascular plants to the present, morphological diversification and elaboration took place. The early plants had very simple bodies consisting of only branched axes and, in some groups, lateral appendages as well [3]. Derived from these structures were various organs including stems, roots, and leaves, with which the vascular plants became increasingly adapted to terrestrial habitats. Pteridophytes, with about 12000 extant species, are a minor constituent of the present land flora when compared with flowering plants, with some 235000 species. Irrespective of their numbers, the morphological diversity of pteridophytes is as great as, or even greater than, that of flowering plants, with special reference to basic organs. Recently it has become clearer that certain pteridophytes possess unique organs in addition to the three general organs (stems, roots, leaves) that are shared with seed plants. Furthermore, it is likely that the general organs had multiple origins in pteridophytes. This chapter reviews recent information on the morphological diversity and evolution of pteridophytes, with emphasis on the organs specific to this group of plants.

[1] Department of Biological Sciences, Graduate School of Science, University of Tokyo 7-3-1 Hongo, Bunkyo-ku, Tokyo 113, Japan
[2] Department of Chemical and Biological Sciences, Faculty of Science, Japan Women's University, 2-8-1 Mejirodai, Bunkyo-ku, Tokyo 112, Japan

2.2 Root-producing Organ (Rhizophore)

The rhizophore is a leafless, root-producing axial organ that is unique to *Selaginella* of the microphyllous class Lycopsida (Fig. 2.1). The rhizophore may be converted to a leafy shoot at an early developmental stage under experimental conditions, and sometimes even in nature [4–9].

This axis is a historically controversial structure [10]. Long ago Nägeli and Leitgeb [11] and Bruchmann [12] designated it a rhizophore (Wurzelträger), i.e., a root-producing organ. This concept was adopted by Bower [13] and others, while some other authors interpreted the axis as a transformed stem (see Schoute [10]). Later, Webster and Steeves [14–16] and Webster and Jagels [17] argued that the rhizophore is the proximal portion of a root (aerial root) that branches to subterranean roots, and that it is rather a stemlike root in such primitive plants as *Selaginella* in which the root and the shoot are morphologically not as distinctly differentiated as in advanced plants.

Nägeli and Leitgeb's rhizophore concept has been revived in recent work [18,19]. In *Selaginella uncinata* (Desv.) Spring, rhizophore primordia are initiated exogenously at the branching point of the second youngest lateral shoot (Fig. 2.2a). The rhizophore apex has a tetrahedral apical cell with three lateral cutting faces and is not capped. When the rhizophore is about 1 mm long, the apical cell becomes indistinguishable and then a pair of root primordia is initiated endogenously from the inner cells of the rhizophore apex (Fig. 2.2b). Subsequently a root cap is formed from the distal face of the root apical cell. In later developmen-

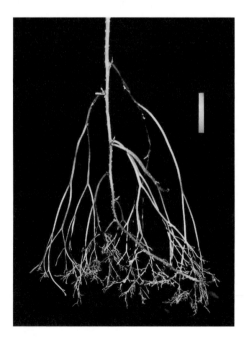

FIG. 2.1. Branched rhizophores of *Selaginella delicatula* arising from stem and producing roots at apices of ultimate branches. *Scale bar* = 10 cm

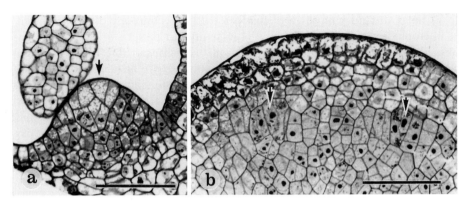

FIG. 2.2a,b. Longitudinal sections of rhizophores of *Selaginella uncinata*. **a.** Young rhizophore with tetrahedral apical cell (*arrow*). **b.** Apical portion of rhizophore where two root apical cells (*arrows*) are initiated inside. *Scale bars* = 50 μm

tal stages, while the roots elongate, the distal cells of the rhizophore (calyptra) covering the roots stretch, disorganize, and finally disappear.

The rhizophore is unbranched in the temperate species, *S. uncinata*, of species of moderate size. Imaichi and Kato [19] demonstrated that the rhizophore is branched in three tropical species of large size. In *S. delicatula* (Desv.) Alston, *S. caudata* (Desv.) Spring, and *S. plana* (Desv.) Alston, the rhizophore is initiated exogenously, has a tetrahedral apical cell, and is capless, as in *S. uncinata*. The rhizophore is usually three or four times dichotomously branched in *S. delicatula* and *S. plana* and four or five times in *S. caudata*. In the rhizophores of *S. delicatula* dichotomous branching involves loss of an original apical cell, followed by formation of two new apical cells on the flank of the apical meristem. A pair of roots is formed endogenously from inner cells below the dermal layer at the apex of each ultimate rhizophore branch.

Thus the concept of the rhizophore is revised as a root-producing, leafless, capless axial organ which is autonomously dichotomously branched in large-sized species and depauperately unbranched in small-sized species such as *S. uncinata* [18,19]. The rhizophore seems not to be a transformation of the stem or root, nor an intermediate structure between the stem and root, but can be categorized as a fundamental organ that coordinates with the stem and root.

The revised definition of the *Selaginella* rhizophore may be comparable with the rhizomorph, a rootlet-producing, supporting organ which is ubiquitous in the Carboniferous lepidodendrids [20]. The rhizomorph was also dichotomously branched in most lepidodendrids, although it differs from the *Selaginella* rhizophore in that it developed from the lower branch of the first dichotomy of an embryonic axis, produced many roots successively along an apparently indeterminate length, and was covered by a special protective structure called an apical plug [3,21].

A similar and much more problematical structure is the corm of *Isoetes*, another microphyllous genus. It shows a bipolar growth pattern in that leaves are formed on the upper portion and roots on the basal portion. From the basal meristem located internally at the basal portion, are produced roots, which are embedded in the parenchymatous (cortical) tissue simultaneously produced from the meristem [22,23]. Karrfart [24] argued that the basal meristem and the lateral meristem continuing to it can be regarded as a primary thickening meristem, although they have been considered as cambium [23,25]. Rothwell and Pryor [26] opined that most of the xylem of the stigmarian axis (rhizomorph) was similarly produced by the primary thickening meristem. The basal parenchymatous (cortical) tissue of *Paurodendron* that, like a root cap, covers young roots and perhaps peels off from the outermost layer was called a protective plug by Karrfart [27]. Rothwell and Erwin [21] speculated that *Selaginella*, which grows unidirectionally and has rhizophores (aerial roots in their paper) along the length of stems, is most primitive in the lycophytes, and the lepidodendrid rhizomorph is comparable with the root-bearing basal portion of the *Isoetes* corm. Further developmental investigation of the *Isoetes* corm will be useful for understanding the phylogenesis of the organ.

The homology and phylogenetic relationships among the *Selaginella* rhizophore, the lepidodendrid rhizomorph, and the basal portion of the *Isoetes* corm still remain uncertain in details. Nevertheless, it seems very likely that at least the former two, like the stem, leaf, and root, are basic organs composing plant bodies. It is noteworthy that the plants having such organs belong to the heterosporous microphyllous group of pteridophytes.

2.3 Psilotalean "Rhizome"

The subterranean "rhizome" (axis) of *Psilotum* and *Tmesipteris* (Psilotales) is one of the most enigmatic organs in vascular plant morphology. It is the only organ initiated in embryos, and produces upright, leafy, green aerial stems (Fig. 2.3) [28]. The "rhizome" is not homologous with the root, because the apical meristem is naked and not capped as the root, nor with the stem (rhizome) because of its nonleafy nature. The subterranean "rhizome" bears rhizoids and characteristically shows irregular branching.

A developmental investigation on the "rhizome" of *Psilotum nudum* (L.) P. Beauv. was made by Takiguchi et al. [29]. Her research focused on the cellular pattern of the apical meristem, which she found to differ greatly from that of the other organs of vascular plants. The following is a brief account of her discovery; the results of her investigation will be fully described elsewhere [29].

The apical meristem has a single, reversely tetrahedral apical cell, which produces derivative cells from three lateral cutting faces. In 44% of derivative cells, the first to the third division walls are oblique to preexisting walls to produce reversely tetrahedral new apical cells (Fig. 2.4a). In 37% of derivative cells, the first and second division walls are oblique, but the third division wall

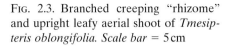

Fig. 2.3. Branched creeping "rhizome" and upright leafy aerial shoot of *Tmesipteris oblongifolia. Scale bar* = 5 cm

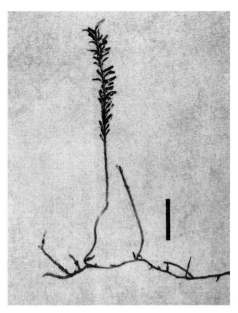

is parallel to the second. Several percent of the latter derivative cells later produce apical cells by oblique divisions. Thus, roughly 50% of the derivative cells produce apical cells. These derivative cells appear to have no spatial relation to the original apical cell nor to neighboring derivative cells producing new apical cells.

Almost all new apical cells produce their own derivative cells, and some secondary new apical cells are sometimes initiated from these derivative cells in a similar manner. The apices of the single subterranean "rhizome" usually have several primary and also secondary new apical cells around the original apical cell.

The total number of new apical cells per epidermal cell population at a post cell-division stage is estimated to be roughly twofold or more the number of rhizome branches per equivalent cell population. Thus, roughly half of the new apical cells give rise to rhizome branches or dormant buds, while the other half become inactive and unidentifiable as apical cells early in development.

The original and new apical cells show various relative meristematic activities, which produce a variety of "rhizome" branching, i.e., isotomous (Fig. 2.4b) or anisotomous dichotomy or monopodial (pseudomonopodial) branching, or even apparent unbranching with dormant lateral buds. Thus, the irregular branching that is characteristic of the subterranean "rhizome" of *Psilotum* is due to the disordered production of new apical cells in the apical meristem.

The ill-organized apical meristem, as well as the nonstem (nonleafy) and nonroot (uncapped) nature, may indicate that the subterranean "rhizome" cannot be categorized as a stem (rhizome) or a root, but is an independent axis.

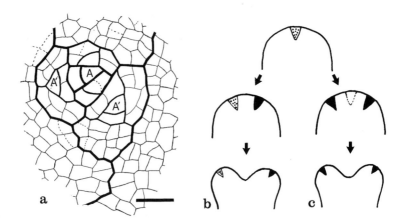

Fig. 2.4a–c. *Psilotum nudum*. **a.** Camera-lucida drawing of front-viewed apical meristem of "rhizome." The original apical cell (*A*) produces derivative cells or segments (outlined by *heavy lines*) in spiral. New apical cells (*A'*) with three faces are initiated in segments by oblique divisions (*light lines*). *Scale bar* = 50 μm. **b, c.** Diagrammatic illustrations showing two ways of dichotomous branching. **b.** "Rhizome". **c.** Aerial stem. *Dotted triangles*, original apical cells; *solid triangles*, daughter apical cells

Comparable organs might be axes of morphologically primitive early plants such as Rhyniopsida and Trimerophytopsida.

2.4 Root

The root exists in almost all extant vascular plants except the primitive Psilopsida and extremely specialized plants such as some aquatic plants. Irrespective of the fact that it is a general organ in vascular plants, its phylogenetic origin is still uncertain, mostly because there are very scanty fossil records of root precursors and initial roots. Banks et al. [30] suggested that some appendages branched from the stem of Lower Devonian *Psilophyton dawsonii* (Trimerophytopsida) might represent adventitious root precursors. Their suggestion is based on the evidence that changes in the size of the stele of the species during the branching of certain appendages are comparable with changes in the size of the stele of *Botryopteris antiqua* (Pteropsida) where it produced adventitious roots. Morphologically similar appendages of *Psilophyton crenulatum* were also suggested to represent root precursors [31]. There are also some inferences that there were roots or root-like branches in lycophytes [32].

Psilotum and *Tmesipteris* (Psilopsida) are regarded as the most primitive and simplest of extant vascular plants [28]. They lack roots throughout their life cycle. The only embryonic organ is the subterranean "rhizome" with rhizoids [33–36]. No adventitious roots are produced along the "rhizome" or aerial shoot. It is believed that roots have not originated in *Psilotum* and *Tmesipteris* [2].

Two branching patterns of roots are recognized in vascular plants [37–39]. In one pattern, lateral roots are formed endogenously from more or less differentiated cells below the root apex (Fig. 2.5a). They are usually initiated at the pericycle in pteridophytes and at the endodermis in seed plants [37]. Cell divisions take place in certain cells of the tissue, from which is produced a root with the apical meristem protected by a newly produced root cap. Adventitious roots are produced from the stem in a similar manner. This pattern of root branching exists in megaphyllous groups, i.e., ferns, *Equisetum* (horsetails), gymnosperms, and angiosperms [25,28,37–39].

In the other pattern, daughter roots branch dichotomously, involving root apices (Fig. 2.5b). In *Selaginella uncinata*, prior to the initiation of root branching, the root apical cells become indistinguishable in the apical meristem [18]. The root apical meristem divides into two daughter meristems, in each of which a new apical cell appears. The new apical cell, like the original apical cell, is a tetrahedral cell with three lateral cutting faces and a distal cutting face. As the apical cells produce derivative cells, dichotomous branching of roots takes place.

This pattern of root branching occurs exclusively in the microphyllous group of Selaginellales, Lycopodiales and Isoetales [37,39]. Little evidence is available on root branching in fossil plants. *Paurodendron* from the Upper Carboniferous, which seems to be related to Isoetales, had dichotomously branched rootlets [40]. For the homology of the isoetalean rootlets, see below.

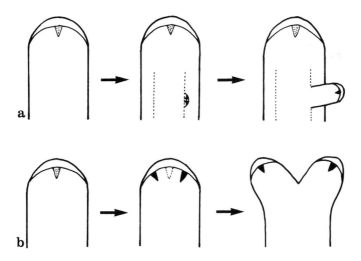

FIG. 2.5a,b. Diagrammatic illustrations showing two patterns of root initiation and branching. **a.** Endogenous origin and lateral branching in megaphyllous plants (Pteropsida, Equisetopsida, and seed plants). **b.** Exogenous origin and terminal dichotomous branching in microphyllous plants (Lycopsida). *Dotted triangles*, old apical cells; *solid triangles*, new apical cells; *broken lines*, pericycle or endodermis

The cellular pattern involved in dichotomous root branching is comparable with that in the dichotomous branching of some other organs, e.g., the stems of some ferns and of *Psilotum nudum* (aerial stem) (Fig. 2.4c) [41–43] and the *Selaginella* rhizophores [19]. The branching of this pattern differs from the dichotomous branching of aerial stems of *Psilotum* described by Roth [44] in which an original apical cell persists to become a daughter apical cell, while another daughter apical cell is newly formed. It also differs from a dichotomy involving establishment of two new apical cells by division of one cell, which is the definition of Bierhorst [45] and others.

The branching patterns of roots were used to classify the major groups of vascular plants by Kato [39]. Kato pointed out that lateral branching of endogenous origin occurs in Pteropsida, Equisetopsida, Gymnospermopsida, and Angiospermopsida, while dichotomous branching of exogenous origin exists in Lycopsida. Based on the distribution of these character states as well as some other characters, Kato [39] proposed the biphyletic classification of vascular plants in which the subdivisions Lycophytina and Pterophytina are recognized in the division Tracheophyta. It is hypothesized that the roots are at least biphyletic in origin in vascular plants [18,25,39].

Rothwell and Erwin [21] and Bateman et al. [20] considered that the rootlets (or lateral appendages) produced along the shoot-like rhizomorph in the lepidodendrids are homologous with leaves on shoots and are not true roots, as in other vascular plants. However, the roots, or rootlets in their sense, of extant Isoetaceae are very similar with true roots of Lycopodiaceae and Selaginellaceae in having root caps and branching dichotomously. The proposed non-homology of the rootlets and the true roots merits further investigation.

2.5 Apical Meristem

Pteridophytes, including early plants, are at the first evolutionary grade of vascular plants and have the longest evolutionary history. There are more than two axial organs in pteridophytes (stem, root, psilotalean rhizome, rhizophore). In order to understand the evolution of the axial organs of vascular plants, Philipson [46] compared the apical meristem organization of stems. Here, we attempt to compare various organs of pteridophytes and discuss the evolution of their apical meristems.

2.5.1 Stem

The stem of the vascular plants is an axis that produces leaves as lateral appendages and can branch autonomously. Its phylogenesis is much better understood than that of the root, which was described above. Comparative morphology suggests that the stems evolved from axes of primitive vascular plants such as Trimerophytopsida and Zosterophyllopsida [3,47,48]. The ancient axes are called telomes (and also mesomes) [48]. It is likely that the stems appeared independently along at least two phyletic clades. In the microphyllous clade, the leaves were derived from enations borne on the aerial axes of Zosterophyllopsida, and the stems were derived from the axes [e.g. 48]. In the megaphyllous clade, the leaves were derived from lateral distal axes, and main axes were involved in the origin of the stem.

Lycopsida. *Selaginella* of the heterosporous microphyllous group has single apical cells at the shoot apex [7,18,19,37,49,50]. However, Popham [51] stated that the genus may have a group of apical cells or unstratified apical meristem without identifiable initial cells as well. It merits detailed investigation. The single apical cells differ from each other in the number of cutting faces from which derivative cells are cut off. The apical cells are described as two-faced in *S. kraussiana* [37], *S. speciosa* [49], *S. uncinata* [18]; three-faced in *S. delicatula* [19]; and four-faced in *S. martensii* and some other species [7,50,52].

Another heterosporous genus *Isoetes* has a different shoot apex organization [53,54]. Although it had long been interpreted to have a single apical cell or a group of initial cells, it has been demonstrated that there is a cap-like surface layer of somewhat columnar meristematic cells [53,54]. Philipson [46] states that the meristem conforms to that of Newman's simplex type of apex that is commonly found in gymnosperms.

Lycopodium, a homosporous genus of Lycopsida, has a shoot apex organization similar to that of *Isoetes* [37,46,55]. There is a surface layer of columnar initial cells in the zoned apical meristem. Even if a large central cell is found in the apex, it has parallel sides and a broad base and is not tetrahedral. Philipson [46] also placed this meristem within Newman's simplex type. A similar apical meristem is described for *Asteroxylon*, an early lycophyte [56].

Pteropsida. In megaphyllous ferns, the shoot apices have single apical cells. Recognition of this meristem organization was formulated as the Apical Cell Theory by the 19th-century botanists, e.g. Nägeli and Hofmeister. This concept is applicable to a great diversity of leptosporangiate ferns [45,57], while some workers [e.g. 58,59] emphasized a zoned organization for understanding morphogenetic activity in the shoot apex. The apical meristem structure of eusporangiate ferns, i.e., whether single apical cells or plural initials, has been more controversial than that of leptosporangiate ferns [e.g. 60,61]. Although it is recorded by some authors that plural initials exist at the shoot apex of those ferns, critical developmental studies have shown that *Botrychium* (Ophioglossaceae) and *Angiopteris* and other genera (Marattiaceae) have single apical cells (Fig. 2.6) [61,62, references cited therein].

Psilopsida. In *Psilotum nudum* of the Psilopsida, both aerial shoots and subterranean "rhizomes" have single apical cells that are tetrahedral and have three cutting faces [29,44,63]. However, the initiation of the apical cells differs between the two stems (Fig. 2.4b,c), and the subterranean "rhizome" is an independent organ that is not categorized as a stem (for details, see "Psilotalean rhizome").

Fɪɢ. 2.6. Front-viewed apical meristem of *Angiopteris lygodii-folia* with apical cell (*asterisk*). *Scale bar* = 100 μm

Equisetopsida. In *Equisetum*, the only extant genus of the class Equisetopsida, the shoot apex has a single large apical cell [26,37]. This meristem structure seems to have long persisted, because the Carboniferous fossils, *Calamites* and *Spheno-phyllum*, had similar single apical cells [64,65].

2.5.2 Rhizophore

The rhizophores of *Selaginella uncinata* have single apical cells, which are tetra-hedral with three equal cutting faces (Fig. 2.2a) [18]. The branched rhizophores of *S. delicatula*, *S. plana*, and *S. caudata* have also single tetrahedral apical cells [19]. This cell configuration persists through branching of the rhizophores in these species. The tetrahedral apical cells produce the terete rhizophores. In comparison, the shoot apex is dorsoventrally compressed and the apical cell is sagitally elongate, even if it has three cutting faces [19].

2.5.3 Root

The roots differ in branching pattern among vascular plant groups, suggesting that the roots are polyphyletic [18,25,39, see "root" in this chapter]. Nevertheless, the roots have single tetrahedral apical cells in the majority of pteridophytes including Pteropsida, Equisetopsida, and Lycopsida (*Selaginella*) [18,19,25, 37,38,66]. The roots of *Lycopodium* and those (rootlets) of *Isoetes* have plural initial cells, which cannot be distinguished from the surrounding cells, as do the stems [37,38].

2.5.4 Phylogenetic Implications

Seed plants, which are more advanced in their reproductive characters than pteridophytes, also have complex apical meristems in the stems and roots [46,66–68]. Their meristem contains plural initial cells. The shoot apical meristem of angiosperms is the most elaborate and consists of plural meristematic cell groups - or zones.

As described above, pteridophytes excluding some lycopods (*Isoetes, Lycopodium*) have simpler apical meristem structures than seed plants. It is stressed here that all axial organs (stem, root, rhizophore, psilotalean "rhizome") of most pteridophytes have single apical cells in the apical meristems. These organs probably have been derived independently of each other from the phylogenetically non- or ill-differentiated axes of early plants. The extant pteridophytes are classified into four classes [e.g. 28,47,69] or four classes in two subdivisions [39], and the early vascular plants (pteridophytes) may be biphyletic [70]. Recent cladistic analyses using morphological data [71–73] and molecular phylogenetic analysis [74] supported the hypothesis that the pteridophytes are a paraphyletic group and consist of several groups, of which at least lycophytes are basal in the clade of vascular plants. The wide distribution of single apical cells not only in various organs but also in systematically remotely related plant groups seems to support the hypothesis that the meristem with a single apical cell is a primitive or plesiomorphic one, from which evolved the meristem with plural initial cells or the zoned meristem of seed plants. Philipson [46] hypothesized that the apical meristems of Pteropsida and seed plants arose independently. But this hypothesis is not supported by recent molecular phylogenetic [74,75] as well as morphological [2,3,39] and cladistic analyses [71–73,76].

Furthermore, it is inferred that the axes of early vascular plants had likewise single apical cells. Because of the ill-organized apical meristem and simple morphology, as described above, the psilotalean "rhizomes" might be most comparable with those ancient axes. However, there is no evidence for an apical cell in *Rhynia gwynne-vaughanii* (Rhyniopsida) [32].

In this context it is interesting to survey the apical meristem organization of the gametophytic generation and nonvascular plants. There are single apical cells in the axial gametophyte of *Psilotum nudum* [77,78] and bryophytes, especially leafy ones [60,79–83]. The sporophytes of certain mosses also have an apical cell [60]. From the cellular pattern it is likely that the apical cells of bryophyte gametophytes, like those of pteridophyte sporophytes, function as central cells producing derivative cells and leaves. In certain bryophytes, the apical cell structure is strictly related to phyllotaxis. Leaves are arranged on a longitudinal line corresponding with one of the cutting faces of the apical cell, because leaves are immediately produced by young derivative cells from the apical cell [60,84,85]. Species having apical cells with two cutting faces show distichous phyllotaxis (two rows of leaves), while those with three cutting faces show tristichous phyllotaxis. Such a strict correlation does not exist in pteridophytes. Some authors [86–88] described that in certain moss gametophytes and fern sporophytes (stems, roots), dividing apical cells are quite commonly encountered, moderating the idea that the apical cell is a quiescent one. Thus the apical meristem with a single apical cell is widely distributed in axial and leafy gametophytes of bryophytes and sporophytes of certain bryophytes, even though the structure differs in details. Phylogenetically, the bryophytes and pteridophytes are suggested to be monophyletic (e.g., Mishler et al. [76] and references cited therein), but it is uncertain whether their apical cell structures are of common origin or multiple origin. The wide occurrence of single apical cells in land plants seems to suggest that the single apical cell organization arose in various organs of early land plants.

2.6 Conclusions

Generally, the body of vascular plants is composed of three organs: stem, leaf, and root. This review shows that in the pteridophytes or primitive vascular plants, there are two more organs that coordinate with them, i.e., the psilotalean "rhizome" and the rhizophore. Furthermore, microphyllous leaves are unique to Lycopsida, because the leaves of all other pteridophytes and seed plants are megaphyllous. It is also likely that the roots of pteridophytes are biphyletic. Thus the morphological diversity of pteridophytes is much greater than that of seed plants with special reference to basic organs. The remarkable diversity of seed plants, particularly angiosperms, is due to morphological flexibility producing a variety of transformations of the three basic organs. For instance, the flower of angiosperms is interpreted to be a reproductive shoot consisting of the stem and floral organs comparable with reproductive and vegetative leaves. The carpel is interpreted to be a conduplicate megasporophyll, a heterochronic modification [89,90]. Intercalary growth contributes to the great morphological diversity of angiosperms [67,91].

The phylogenesis of organs has been investigated by comparative morphology and paleobotany, as described above. As regards leaf evolution, it is very likely that microphyllous leaves were derived from enations on axes, and megaphyllous leaves from lateral axes branched from main axes [3]. However, understanding of the morphology and origin of microphylls is much weaker than that for megaphylls, with special reference to development and morphogenesis, partly because leaf development has been investigated mainly for megaphyllous plants [see 68,92]. The phylogenesis of the rhizophore and psilotalean "rhizome" also remains to be made clear. Recently morphological evolution has aroused increasing interest particularly from molecular biologists [e.g. 93,94].

Evidence of morphological diversification contributes to understanding the phylogenetic systematics of pteridophytes. There are distinct differences between lycopods and ferns and horsetails in sporangia (attached on the adaxial side of leaves versus the abaxial side or the margin, transverse versus longitudinal dehiscence), leaves (microphylls versus megaphylls), roots (dichotomous versus lateral branching), and others. These differences exist in both fossil and living members. Based on these differences, a biphyletic classification system of pteridophytes was proposed [39]. Psilopsida are still a systematically problematic group. Recent molecular phylogenetic research suggests that Psilopsida are in the megaphyllous clade [e.g. 74,75], although they have some characters of microphyllous plants [39].

The evolution of organisms can be described in terms of clade and grade [95]. The clades of primitive and advanced plants have been and will continue to be clarified by molecular and other phylogenetic analyses. On the other hand, the traditional classifications of land plants into bryophytes, pteridophytes, gymnosperms, and angiosperms at least partly reflect the major grades of land plants. The morphological evolution from nonvascular to vascular grades and from free-sporing to ovular grades, and from naked-ovular to carpellary grades is a most salient aspect of land plant evolution.

Acknowledgments. We are grateful to Yukiko Takiguchi for her help with Fig. 2.4a, and Hiroyuki Akiyama for his help with literature on bryophyte morphology. We are also indebted to David E. Boufford for correcting the English of the manuscript. This study was supported in part by Grants-in-Aid for Scientific Research from the Ministry of Education, Science and Culture.

References

1. Beck CB, Wight DC (1988) Progymnosperms. In: Beck CB (ed) Origin and evolution of gymnosperms. Columbia University Press, New York, pp 1–84
2. Eames AJ (1936) Morphology of vascular plans. Lower groups. McGraw-Hill, New York
3. Stewart WN, Rothwell GW (1993) Paleobotany and the evolution of plants, 2nd edn. Cambridge University Press, Cambridge
4. Williams S (1931) An analysis of the vegetative organs of *Selaginella grandii* Moore, together with some observations on abnormalities and experimental results. Trans R Soc Edinburgh 57:1–21
5. Cusick F (1954) Experimental and analytical studies of pteridophytes. XXV. Morphogenesis in *Selaginella willldenowii* Baker-II. Angle-meristem and angle-shoots. Ann Bot (NS) 18:171–181
6. Webster TR (1969) An investigation of angle-meristem development in excised stem segments of *Selaginella martensii*. Can J Bot 47:717–722
7. Siegert A (1974) Die Verzweigung der Selaginellen unter Berücksichtigung der Keimungsgeschichte. Beitr Biol Pflanzen 50:21–112
8. Wochok ZA, Sussex IM (1975) Morphogenesis in *Selaginella*. III. Meristem determination and cell differentiation. Develop Biol 47:376–383
9. Jernstedt JA, Cutter EG, Lu P (1994) Independence of organogenesis and cell pattern in developing angle shoots of *Selaginella martensii*. Ann Bot 74:343–355
10. Schoute JC (1938) Morphology. In: Verdoorn F (ed) Manual of pteridology. Martinus Nijhoff, The Hague, pp 1–64
11. Nägeli C, Leitgeb H (1868) Entstehung und Wachstum der Wurzeln. Beitr Wissenschaftl Bot Leipzig 4:124–158
12. Bruchmann H (1905) Von den Wurzelträgern der *Selaginella kraussiana* A. Br. Flora 95:150–166
13. Bower FO (1935) Primitive land plants. Macmillan, London
14. Webster TR, Steeves TA (1963) Morphology and development of the root of *Selaginella densa* Rydb. Phytomorphology 13:367–376

15. Webster TR, Steeves TA (1964) Developmental morphology of the root of *Selaginella kraussiana* A. Br. and *Selaginella wallacei* Hieronym. Can J Bot 42:1665–1676

16. Webster TR, Steeves TA (1967) Developmental morphology of the root of *Selaginella martensii* Spring. Can J Bot 45:395–404

17. Webster TR, Jagels R (1977) Morphology and development of aerial roots of *Selaginella martensii* grown in moist containers. Can J Bot 55:2149–2158

18. Imaichi R, Kato M (1989) Developmental anatomy of the shoot apical cell, rhizophore and root of *Selaginella uncinata*. Bot Mag Tokyo 102:369–380

19. Imaichi R, Kato M (1991) Developmental study of branched rhizophores in three *Selaginella* species. Am J Bot 78:1694–1703

20. Bateman RM, DiMichele WA, Willard DA (1992) Experimental cladistic analysis of anatomically preserved arborescent lycopsids from the Carboniferous of Euramerica: an essay on paleobotanical phylogenetics. Ann Missouri Bot Gard 79:500–559

21. Rothwell GW, Erwin DM (1985) The rhizomorph apex of *Paurodendron*: implications for homologies among the rooting organs of Lycopsida. Am J Bot 72:86–98

22. Karrfart EE, Eggert DA (1977) The comparative morphology and development of *Isoetes* L. II. Branching of the base of the corm in *I. tuckermanii* A. Br. and *I. nuttallii* A. Br. Bot Gaz 138:357–368

23. Paollilo JD Jr (1982) Meristems and evolution: developmental correspondence among the rhizomorphs of the lycopsids. Am J Bot 69:1032–1042

24. Karrfart EE (1984) The origin and early development of the root-producing meristem of *Isoetes andicola* L. D. Gomez. Bot Gaz 145:372–377

25. Bierhorst DW (1971) Morphology of vascular plants. Macmillan, New York

26. Rothwell GW, Pryor JS (1991) Developmental dynamics of arborescent lycophytes—apical and lateral growth in *Stigmaria ficoides*. Am J Bot 78:1740–1745

27. Karrfart EE (1984) Further observations on *Nathorstiana* (Isoetales). Am J Bot 71:1023–1030

28. Gifford EM, Foster AS (1989) Morphology and evolution of vascular plants, 3rd edn. WH Freeman, New York

29. Takiguchi Y, Imaichi R, Kato M (1997) Cell division patterns in the apices of subterranean axis and aerial shoot of *Psilotum nudum* (Psilotaceae): morphological and phylogenetic implications on the subterranean axis. Am J Bot 84:588–596

30. Banks HP, Leclercq S, Hueber FM (1975) Anatomy and morphology of *Psilophyton dawsonii*, sp. n. from the late Lower Devonian of Quebec (Gaspé) and Ontario, Canada. Paleontogr Am 8(48):7–127

31. Doran JB (1980) A new species of *Psilophyton* from the Lower Devonian of northern New Brunswick, Canada. Can J Bot 58:2241–2262

32. Edwards D (1994) Towards an understanding of pattern and process in the growth of early vascular plants. In: Ingram DS, Hudson A (eds) Shape and form in plants and fungi. Academic, London, pp 39–59

33. Holloway JE (1918) The prothallus and young plant of *Tmesipteris*. Trans Proc New Zealand Inst 50:1–44

34. Holloway JE (1921) Further notes on the prothallus, embryo, and young sporophyte of *Tmesipteris*. Trans Proc New Zealand Inst 53:386–422

35. Holloway JE (1939) The gametophyte, embryo and developing sporophyte of *Psilotum triquetrum* Sw. Ann Bot (NS) 3:313–336

36. Bierhorst DW (1954) The gametangia and embryo of *Psilotum nudum*. Am J Bot 41:274–281

37. Guttenberg H von (1966) Histogenese der Pteridophyten. Gebrüder Borntraeger, Berlin-Nikolassee (Handbuch der Pflanzenanatomie, Band 7, Teil 2)

38. Ogura Y (1972) Comparative anatomy of vegetative organs of the pteridophytes. Gebrüder Borntraeger, Berlin Stuttgart (Handbuch der Pflanzenanatomie, Band 7, Teil 3)
39. Kato M (1983) The classification of major groups of pteridophytes. J Fac Sci Univ Tokyo Sect III 13:263–283
40. Phillips TL, Leisman GA (1966) *Paurodendron*, a rhizomorphic lycopod. Am J Bot 53:1086–1100
41. Gottlieb JE, Steeves TA (1961) Development of the bracken fern, *Pteridium aquilinum* (L.) Kuhn. III. Ontogenetic changes in the shoot apex and in the pattern of differentiation. Phytomorphology 11:230–242
42. Hagemann W, Schulz U (1978) Wedelanlegung und Rhizomverzweigung bei einigen Gleicheniaceae. Bot Jahrb Syst Pflanzengesch Pflanzengeogr 99:380–399
43. Mueller RJ (1982) Shoot morphology of the climbing fern *Lygodium* (Schizaeaceae): general organography, leaf initiation, and branching. Bot Gaz 143:319–330
44. Roth I (1963) Histogenese der Luftsprosse und Bildung der "dichotomen" Verzweigungen von *Psilotum nudum*. Adv Frontiers Plant Sci New Delhi 7:157–180
45. Bierhorst DW (1977) On the stem apex, leaf initiation, and early leaf ontogeny in filicalean ferns. Am J Bot 64:125–152
46. Philipson WR (1990) The significance of apical meristem in the phylogeny of land plants. Plant Syst Evol 173:17–38
47. Smith GM (1955) Cryptogamic botany, vol 2. Bryophytes and pteridophytes, 2nd edn. McGraw-Hill, New York
48. Zimmermann W (1959) Die Phylogenie der Pflanzen, 2nd edn. Gustav Fischer, Stuttgart
49. Hagemann W (1980) Über den Verzweigungsvorgang bei *Psilotum* und *Selaginella* mit Anmerkungen zum Begriff der Dichotomie. Plant Syst Evol 133:181–197
50. Dengler NG (1983) The developmental basis of anisophylly in *Selaginella martensii*. I. Initiation and morphology of growth. Am J Bot 70:181–192
51. Popham RA (1951) Principal types of vegetative shoot apex organization in vascular plants. Ohio J Sci 51:249–270
52. Jacobs WP (1988) Development of procambium, xylem, and phloem in the shoot apex of *Selaginella*. Bot Gaz 149:64–70
53. Bhambie S (1957) Studies in pteridophytes, I. The shoot apex of *Isoetes coromandeliana* L. J Ind Bot Soc 36:491–502
54. Sam SJ (1984) The structure of the apical meristem of *Isoetes engelmannii, I. riparia* and *I. macrospora* (Isoetales). Bot J Linn Soc 89:77–84
55. Freeberg JA, Wetmore RH (1967) The Lycopsida—a study in development. Phytomorphology 17:78–91
56. Hueber FM (1992) Thoughts on the early lycopsids and zosterophylls. Ann Missouri Bot Gard 79:474–499
57. Hébant-Mauri R (1994) Cauline meristems in leptosporangiate ferns: structure, lateral appendages, and branching. Can J Bot 71:1612–1624
58. McAlpin BW, White RA (1974) Shoot organization in the Filicales: the promeristem. Am J Bot 61:562–579
59. White RA (1979) Experimental investigations of fern sporophyte development. In: Dyer AF (ed) The experimental biology of ferns. Academic, London, pp 505–549
60. Gifford EM (1983) Concept of apical cells in bryophytes and pteridophytes. Annu Rev Plant Physiol 34:419–440
61. Imaichi R (1986) Surface-viewed shoot apex of *Angiopteris lygodiifolia* Ros. (Marattiaceae). Bot Mag Tokyo 99:309–317

62. Imaichi R, Nishida M (1986) Developmental anatomy of the three-dimensional leaf of *Botrychium ternatum* (Thunb.) Sw. Bot Mag Tokyo 99:85–106
63. Bierhorst DW (1954) The subterranean sporophytic axes of *Psilotum nudum*. Am J Bot 41:732–739
64. Good CW, Taylor TN (1971) The ontogeny of Carboniferous articulates: calamite leaves and twigs. Palaeontogr B 113:137–158
65. Good CW, Taylor TN (1972) The ontogeny of Carboniferous articulates: the apex of *Sphenophyllum*. Am J Bot 59:617–626
66. Cutter EG (1971) Plant anatomy. Experiment and interpretation. 2. Organ. Arnold, London
67. Esau K (1965) Plant anatomy, 2nd edn. Wiley, New York
68. Steeves TA, Sussex IM (1989) Patterns in plant development, 2nd edn. Cambridge University Press, Cambridge
69. Kramer KU, Green PS (eds) (1990) The families and genera of vascular plants, vol 1. Pteridophytes and gymnosperms. Springer-Verlag, Berlin Heidelberg New York
70. Banks HP (1975) Evolution and plants of the past. Wadsworth, Bermont
71. Bremer K, Humphries CJ, Mishler BD, Churchill SP (1987) On cladistic relationships in green plants. Taxon 36:339–349
72. Crane PR (1990) The phylogenetic context of microsporogenesis. In: Blackmore S, Know RB (eds) Microspores: Evolution and ontogeny. Academic, London, pp 11–41
73. Garbary DJ, Renzaglia KS, Duckett JG (1993) The phylogeny of land plants: a cladistic analysis based on male gametogenesis. Plant Syst Evol 188:237–269
74. Manhart JR (1995) Chloroplast 16S rDNA sequences and phylogenetic relationships of fern allies and ferns. Am Fern J 85:182–192
75. Raubeson LA, Jansen RK (1992) Chloroplast DNA evidence on the ancient evolutionary split in vascular land plants. Science 255:1697–1699
76. Mishler BD, Lewis LA, Buchheim MA, Renzaglia KS, Garbary DJ, Delwiche CF, Zechman FW, Kantz TS, Chapman RL (1994) Phylogenetic relationships of the "green algae" and "bryophytes." Ann Missouri Bot Gard 81:451–483
77. Bierhorst DW (1953) Structure and development of the gametophyte of *Psilotum nudum*. Am J Bot 40:649–658
78. Whittier DP (1975) The origin of the apical cell in *Psilotum* gametophytes. Am Fern J 65:83–85
79. Berthier J (1972) Recherches sur la structure et la déloppement de l'apex du gametophytes feullé des mosses. Rev Bryol Lichénol 38:421–551
80. Crandall-Stotler B (1984) Musci, hepatics and anthocerotes—an essay on analogues. In: Schuster RM (ed) New manual of bryology, vol 2. Hattori Botanical Laboratory, Nichinan, pp 1093–1129
81. Schuster RM (1984) Comparative anatomy and morphology of the Hepaticae. In: Schuster RM (ed) New manual of bryology, vol 2. Hattori Botanical Laboratory, Nichinan, pp 760–891
82. Schuster RM (1984) Morphology, phylogeny and classification of the Anthocerotae. In: Schuster RM (ed) New manual of bryology, vol 2. Hattori Botanical Laboratory, Nichinan, pp 1071–1091
83. Schofield WB, Hébant C (1984) The morphology and anatomy of the moss gametophyte. In: Schuster RM (ed) New manual of bryology, vol 2. Hattori Botanical Laboratory, Nichinan, pp 627–657
84. Crandall-Stotler B (1980) Morphogenetic designs and a theory of bryophyte origins and divergence. BioScience 30:580–585

85. Basile DV (1990) Morphoregulatory role of hydroxyproline-containing proteins in liverworts. In: Chopra RN, Bhalta SC (eds) Bryophytes: physiology and biochemistry. CRC Press, Boca Raton, pp 225–243
86. Hébant C, Hébant-Mauri R, Barthonnet J (1978) Evidence for division and polarity in apical cells of bryophytes and pteridophytes. Planta 138:49–51
87. Gifford EM, Kurth E (1983) Quantitative studies of the vegetative shoot apex of *Equisetum scirpoides*. Am J Bot 70:74–79
88. Gifford EM, Polito VS (1981) Mitotic activity at the shoot apex of *Azolla filiculoides*. Am J Bot 68:1050–1055
89. Takhtajan A (1976) Neoteny and the origin of flowering plants. In: Beck CB (ed) Origin and early evolution of angiosperms. Columbia University Press, New York, pp 207–219
90. Cronquist A (1988) The evolution and classification of flowering plants, 2nd edn. New York Botanical Garden, New York
91. Doyle JA, Hickey LJ (1976) Pollen and leaves from the Mid-Cretaceous Potomac Group and their bearing on early angiosperm evolution. In: Beck CB (ed) Origin and early evolution of angiosperms. Columbia University Press, New York, pp 139–206
92. Wardlaw CW (1965) Organization and evolution in plants. Longmans, London
93. Coen ES (1991) The role of homeotic genes in flower development and evolution. Annu Rev Plant Physiol Plant Mol Biol 42:241–279
94. Purugganan MD, Rounsley SD, Schmidt RJ, Yanofsky MF (1995) Molecular evolution of flower development: diversification of the plant MADS-box regulatory gene family. Genetics 140:345–356
95. Futuyma DJ (1986) Evolutionary biology, 2nd edn. Sinauer Associates, Sunderland

CHAPTER 3

RNA Editing in Land Plants

KOICHI YOSHINAGA

3.1 Introduction

The discovery of messenger RNA (mRNA) more than 30 years ago led to the proposition of the central dogma of molecular biology: that DNA functions as a template for RNA and subsequently RNA determines the sequence of amino acids within protein. However, the nucleotide sequence of every mRNA molecule is not a simple copy of the sequence of its DNA template. Intervening sequences that interrupt genes are removed at the RNA level by RNA splicing, and individual nucleotides within some mRNA are posttranscriptionally inserted, deleted, or altered in the sequence by RNA editing. By the process of editing, genetic information that is not found in the genomic template can be created in the mRNA after transcription (for review, see [1–7]). Thus, certain transcripts in plant mitochondria and in chloroplasts [8], which cannot be translated because of the lack of AUG initiation codons in the coding sequence, are rendered translatable by converting the cytidine (C) residues into uridine (U), probably by modification of specific bases.

RNA editing was originally detected in the kinetoplast genetic system of trypanosomes [9], and was subsequently observed in nucleus-encoded mRNA [10] and mRNA encoded by mitochondrial genes from higher plants [11–13]. More recently, several chloroplast transcripts of higher plants have been shown to be subjected to RNA editing [8,14–16].

Several types of RNA editing have been described. (1) In mitochondrial transcripts of trypanosomes, a few or hundreds of U are inserted or deleted [1,9]. (2) In mitochondrial transcripts of a slime mold, single C are added at multiple positions [17]. (3) UAA and UGA stop codons are reconstituted by polyadenylation in mitochondrial transcripts of vertebrates and in nuclear transcripts [18]. As mentioned, (4) C is converted to U in nuclear transcripts of mammals [6], in mitochondrial transcripts of virtually every protein-coding gene of higher plants

Department of Chemistry, Faculty of Science, Shizuoka University, 836 Oya, Shizuoka 422, Japan

[7,19,20], and in transcripts of several chloroplast genes. (5) U to C conversion occurs in mitochondrial transcripts of seed plants [19] and ferns [21] and in chloroplast transcripts of the hornwort [22].

3.2 RNA Editing in Plant Mitochondria

When the first protein-coding gene from plant mitochondria was analyzed [23], three CGG codons, which normally code Arg, were observed within the coding region of the cytochrome oxidase subunit II (*cox2*) of maize, where Trp is conserved in yeast, beef, and *Neurospora* polypeptides. Therefore, the codon CGG in mitochondria was expected to code for Trp instead of Arg using a nonstandard codon assignment. However, it has been found that the CGG codon can be converted to UGG, which is the actual codon for Trp by RNA editing of the conversion of C to U [11–13]. Thus, plant mitochondria use the universal genetic code. This is the first evidence of RNA editing in plant mitochondria that has been achieved by comparing nucleotide sequences of complementary DNA (cDNA) and genomic DNA. The cDNA has been synthesized with reverse transcriptase from specific oligonucleotides, using mitochondrial RNA as a template. Direct sequencing of the peptide later confirmed that mitochondrial proteins arise from edited mRNA [24–26]. Almost all the editing is conversion from C to U, and the amino acids are different from the expected ones. These events have been found not only in the CGG codon, but also in many codons within each of the genes (Fig. 3.1). These events also occur in many mitochondrial genes such as *cob*, *cox2*, *cox3*, *nad3*, *nad4*, and *rps12*, and in many higher plants both in monocots (maize, rice, wheat) and in dicots (pea, soybean, *Oenothera*). RNA editing has also been observed in mitochondrial transcripts of gymnosperms, pteridophytes, and bryophytes [27], but not in those of chlorophytes [21]. Almost all RNA editing has been confirmed in protein-coding regions, but a few editing events have been described in rRNA [28], tRNA [29], intron sequences [30,31], ribosomal binding sites [32], and pseudogenes [33,34].

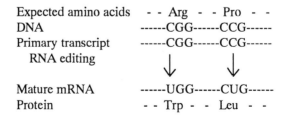

FIG. 3.1. Schematic representation of a typical RNA editing. The amino acids Arg and Pro are changed to Trp and Leu, respectively, by RNA editing of the conversion of C to U

3.3 RNA Editing in Chloroplasts

The process of RNA editing in chloroplasts is strikingly similar to that occurring in plant mitochondria, although the editing in chloroplasts of higher plants is more restricted in a few genes and in a few sites. It converts C into U, except in a hornwort where extensive conversion from U to C has been observed [22]. The examples of chloroplast RNA editing are listed in Table 3.1. Typical RNA editing in chloroplasts is the creation of an initiation codon by converting ACG to AUG [8,14,35–37], suggesting that the translation will be delayed until the completion of the editing. For example, the tobacco and spinach *psbL* gene has an ACG codon instead of the initiator AUG codon that is observed in the homologous position of the other eight plant species. Other events of RNA editing are found in internal codons.

Several peptides are shorter than predicted from their gene structure. An AUG start codon is created by the conversion of C to U in *ndhD* transcripts of tobacco, spinach, and snapdragon. The alignment of nucleotide and deduced

TABLE 3.1. RNA editing in chloroplasts

Genes	Products	Editing sites	Species	Reference
atpA	ATPase	2[a]	Tobacco	74
atpF		1	Tobacco	74
rpl2	50S ribosomal protein	1[b]	Maize	8
rpoB	RNA polymerase	4	Maize, Barley	52
		3	Rice	64
		2	Tobacco	65
psbL	Photosystem II	1[b]	Tobacco	14
		1[b]	Bell pepper	35
		1	Spinach	36
petB	Cytochrome	1	Maize	45
		1	Tobacco	66
ndhA	NADH dehydrogenase	4	Maize	15
ndhB		6	Maize	16
		2	Tobacco	65
		9	Barley	68
ndhD		1[b]	Tobacco	37
		1[b]	Spinach	37
		1[b]	Snapdragon	37
rbcL	Rubisco	20[c]	Hornwort	22
IRF170	Unknown	2	Maize	46

[a] Contains silent conversion.
[b] RNA editing has been observed at the initiation codon.
[c] Contains seven editing sites of U to C conversion.

amino acid sequences from different plant species suggests that the functional start codon of the transcripts of maize, barley, and rice is not the originally proposed start codon, but instead is the second AUG codon, which is located at the 15th position downstream from the originally proposed start codon [37].

3.4 The Role of RNA Editing

RNA editing could be involved in some control steps of genetic expression. For example, the conversion of U to C in the genomic RNA of hepatitis virus (HDV) controls the HDV life cycle. By editing, further HDV replication is inhibited, which makes it possible for viral RNA to be packed. This editing system is provided by the host cell [38]. In the cytoplasmic system of mammals, a 55-kDa form of apolipoprotein B (apo B) is synthesized in the liver, while a smaller form, apo B48, is produced in the small intestine by RNA editing, which converts C to U thus forming a stop codon in the apo B mRNA. The apo B48 is required for the absorption of dietary lipid [39]. Tissue- and stage-specific modulation of RNA editing has been observed in *psbF* and *psbL* transcripts from spinach plastids. A significant portion of the editing sites remain unedited in seeds and in roots; complete editing is observed in the same transcripts from all other plastid types including etioplasts and chloroplasts [36]. Fully edited *psbL* transcripts are also found in leaves and in ripe fruits such as bell peppers [35].

3.4.1 Editing That Synthesizes Functional Protein

The amino acids specified after RNA editing are generally evolutionarily better conserved than the amino acids deduced from the genomic sequences. Therefore, RNA editing in many open reading frames seems to be essential to ensure the synthesis of functionally competent proteins. A requirement of the editing process has been shown by using transgenic plants [40]. Edited and unedited coding sequences of wheat *atp9* were fused to the coding sequences of yeast *cox4* transit peptide and introduced into tobacco. When the chimeric peptides encoded from the unedited sequence were imported into the mitochondria, an organelle function of male fertility was impaired, but the peptides encoded from the edited sequence were able to function. It has also been shown in transcripts of mitochondrial *atp6* that RNA editing is essential for the production of functional protein [41].

Stop codons, required to form the appropriate C terminus of functional protein, are created by the editing of mitochondrial transcripts of *atp6* and *atp9* [4]. In the case of wheat, this modification has been confirmed by protein sequencing [24].

3.4.2 Editing That Switches on the Splicing

The relationship between RNA editing and gene expression has been studied in the mitochondrial system of rice [41] and wheat [42]. The precursor transcripts before being subjected to splicing have contained significantly more unedited nucleotides than mature transcripts. Iwabuchi et al. [41] proposed a model of gene regulation whereby primary transcripts subjected to processing to become the mature size could then be edited into functionally mature mRNA. However, an alternative model is also possible whereby the processing steps are delayed until the completion of the RNA editing of primary transcripts, because partially edited sequences have been identified only in the sequences that have not been processed as opposed to mature mRNA (Fig. 3.2).

The following observations in mitochondria and chloroplasts support the latter model. In *cox2* and *nad1* mRNA in plant mitochondria, RNA editing at any of their editing sites has been shown to precede intron removal [43,44]. Similar results have been reported in the transcripts of the *psbB* operon in maize chloroplasts [45]. In the same chloroplasts, editing events are already completed in partially spliced and even in unspliced transcripts of IRF170, which is located within the *rps4-rps14* cluster [46]. Ruf et al. [46] concluded that editing of IRF170-encoded chloroplast transcripts is an early processing step that precedes both splicing and cleavage to monocistronic mRNA. Furthermore, several events of RNA editing have been shown in intron sequences of *nad1* and *nad2* mRNA in *Oenothera* mitochondria. The editing allows additional base pairings in the

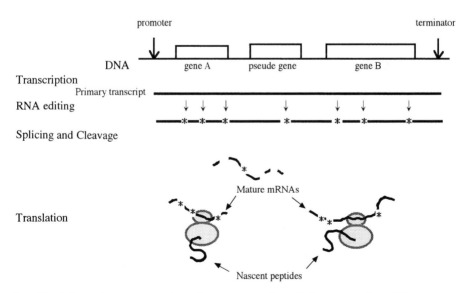

FIG. 3.2. Schematic representation of gene expression of plant organelles. Primary transcripts are subjected to RNA editing in a sequence-specific manner. Only edited transcripts are then spliced and used for translation, except in a few cases

secondary structure, which are required for group II splicing [31,47]. In fact, it has been shown in an in vitro system that the *Oenothera* group II intron of *nad1* cannot be spliced from the primary transcripts without previous editing [48]. However, an exception is known: the incomplete editing of *rps12* transcripts results in the synthesis of polymorphic polypeptides [49].

3.4.3 Editing That Switches on the Translation

Initiation codons are created by RNA editing in mitochondrial transcripts [50,51] and in chloroplast transcripts [8,14,35–37]. This type of editing must be an important control system of organelle gene expression because the translation cannot start until the start site is edited. The situation is similar in maize chloroplast *rpoB* transcripts, where four C to U editing sites are clustered within 150 nucleotides of the 5'-terminal region of the *rpoB* mRNA [52], and is also similar in hornwort chloroplast *rbcL* transcripts, where two stop codons are turned into sense by editing within the 134 nucleotides of the 5'-terminal region of the transcripts [22].

RNA editing occurs not only in the coding region but also in the intervening sequence, which seems to switch on the translation. An example is the editing upstream of the *rps14* open reading frame, which may improve the ribosome binding site [32]. Mispairing in the acceptor stem of bean and potato mitochondrial Phe-tRNA is corrected by the editing of the conversion of C to U [29]. It is also essential for the production of functional tRNA.

In contrast to switching on the translation, a switching-off of the synthesis of useless peptides is shown in pseudogenes. The introduction of a stop codon by RNA editing in an open reading frame has been demonstrated in the mitochondrial transcripts of the *rps19* pseudogene of *Oenothera* [33], although the *rps14* pseudogene of *Arabidopsis* already contains a stop codon in the putative coding region and the codon is not edited [34]. RNA editing creates a stop codon close to the initiation codon of the *rpl16* open reading frame of *Petunia* mitochondria [53], suggesting the creation of a pseudogene from what appears to be a normal gene in the DNA sequence. The functional gene will already be established in the nuclear genome because many ribosomal protein genes of mitochondria have been transferred into the nucleus.

3.5 The Mechanisms of RNA Editing

A model has been proposed for RNA editing in kinetoplastid mitochondria involving "guide" RNA molecules transcribed from maxicircle mitochondrial DNA [54]. Plant organelle RNA editing seems to be somewhat simpler than the phenomenon observed in trypanosome mitochondria and is remarkably similar to the editing of a single residue in the apoB mRNA observed in animal tissues [10,55]. The C to U editing of apoB involves a site-specific reaction by a cytidine deaminase [56]. Using wheat mitochondrial extracts, an in vitro system for the editing of *atp9* mRNA has been reported to be able to edit correctly the expected C residues. The reaction is sensitive to protease and nuclease, suggesting that both protein(s) and nucleic acid(s) are necessary for the reaction [57]. It is not known whether the editing reaction in mitochondria is a simple base modification by deamination, a base substitution, or a nucleotide substitution. However, evidence for a site-specific cytidine deamination reaction involving C to U RNA editing has been shown using mitochondrial lysate [58]. The editing sites are limited where the editing has been observed in the *cox2* transcripts of the pea. This result suggests that editing will be by deamination in both mitochondria and chloroplasts during C to U conversion. However, the mechanism of reverse editing from U to C is not known.

How can RNA editing in organella be controlled by nuclear-encoded factors? Both the extent of transcript abundance and the extent of editing of *nad3* mRNA in *Petunia* mitochondria are under the control of a single nuclear gene. However, the control by this nuclear gene is limited to the *nad3* mRNA, and the other transcripts examined are not under its control [59]. The extent of RNA editing of *atp6* transcripts in rice mitochondria is also affected by the processing, which is controlled by its nuclear gene [41]. Possible factors involving the process of RNA editing are RNA-binding proteins in chloroplasts. Five nuclear-encoded proteins have been identified in tobacco chloroplasts, which bind specifically to premature RNA before the formation of polysomes [60], but their precise functions remain unknown.

How can editing sites be recognized? If there is a guide RNA that is complementary to the segments of the edited mRNA, it can guide the editing sites as has been shown in the trypanosome mitochondrial system [54]. However, no such antisense transcript can be detected in the plant mitochondrial system [61] or in tobacco plastids [62]. The screenings of common nucleotide sequences that can determine editing sites have been carried out. However, the sequences that are surrounding an editing site are not sufficient determinants for the editing process in chloroplasts by themselves [16,52].

The RNA editing that occurs in plant mitochondria and in chloroplasts takes place at similarly selected editing sites. Editing occurs in the transcripts of chloroplast *ndhA* and *ndhB* and in mitochondrial *nad1* and *nad2* genes at sites that are homologous at the protein level [15,16]. The type of RNA editing in plant mitochondria and chloroplasts is also similar whether it is the conversion of C to U or U to C, suggesting that the editing processes of both the plant's organelles share common components and mechanistic steps. However, the mechanisms are not completely identical. When a chimeric gene containing a *Petunia cox2* exon was introduced into the tobacco chloroplast genome, no editing was observed in the transcript even though the exon contained seven sites that are edited in mitochondria [63]. In a reciprocal approach, a rice chloroplast sequence containing editing sites has been shown not to be edited in the mitochondria [64]. Therefore, the editing mechanisms of the organelles must require some organ-specific factor(s). The mechanism will also require site-specific factor(s) [65]. Light-dependent processes, however, do not influence the editing machinery, because complete editing is observed in the etioplasts of spinach *psbF* and *psbL* transcripts [36], in non-photosynthetic chromoplasts and chloroplasts of bell pepper *psbL* [35], and in non-photosynthetic proplastids of tobacco *petB* transcripts [66].

3.6 Evolution of RNA Editing in Green Plants

RNA editing in plant organelles has recently been described, and data on its phylogenetic distribution, frequency, and general characteristics are still quite limited in higher plants. However, a few cases have been described in the mitochondrial transcripts of gymnosperm [67], in all the major groups of land plants [21,27], and in the chloroplast transcripts of a hornwort [22].

The number of editing sites in transcripts of the mitochondrial *cox3* from various plants are shown in Table 3.2. The highest frequency of RNA editing ever observed is shown in the pteridophytes *Isoetes*, where 48 C residues corre-

TABLE 3.2. Number of editing sites in transcripts of the 381 nucleotides upstream of the mitochondrial *cox3* from various green plants

Plant	Genus	CU	UC	Reference
Angiosperm	*Triticum*	9	0	21
	Oenothera	7	0	21
Gymnosperm	*Ephedra*	14	0	27
	Gnetum	0	0	27
	Picea	16	0	21
	Ginkgo	21	0	21
	Cycas	19	0	21
Pteridophyta	*Osmunda*	10	2	21
	Asplenium	6	2	21
	Angiopteris	4	0	27
	Psilotum	10	0	21
	Ophioglossum	16	0	27
	Equisetum	7	0	21
	Isoetes	48	0	27
	Lycopodium	1	0	21
Bryophyta	*Marchantia*	0	0	21
	Pellia	12	1	27
	Tetraphis	1	0	27
	Physcomitrella	0	0	21
	Ceratodon	1	0	27
	Sphagnum	0	0	21
	Anthoceros	5	5	27
Chlorophyta	*Chara*	0	0	27
	Coleochaete	0	0	21
	Stichococcus	0	0	21

sponding to 39% of C are converted to U. In general, higher frequency of RNA editing is seen in gymnosperms and pteridophytes than in angiosperms and bryophytes, but no editing is found in chlorophytes. The conversion of U to C (reverse type) is found in pteridophytes and bryophytes. The differences between the gnetopsids *Gnetum* and *Ephedra*, the two lycopsids *Isoetes* and *Lycopodium*, and also the two liverworts *Marchantia* and *Pellia*, show that RNA editing frequency does not correlate with the phylogenetic position of the species.

In chloroplast transcripts, RNA editing events in higher plants are rather rare when compared to mitochondrial transcripts. The event can only be found in one or a few sites within the transcripts from each gene. Nine sites for *ndhB* transcripts (0.4% conversion) are the maximum number of sites known to date [68]. However, 20 sites (1.4% conversion) are found within the 1428 nucleotides of *rbcL* transcripts from the hornwort *Anthocheros formosae*, which contain 13 sites for the conversion of C to U and 7 sites for the conversion of U to C [22], although no editing has been observed in the liverwort *Marchantia polymorpha* [69]. Also, 8 sites (1.7% conversion) are found in the 480 nucleotides upstream of the fern *Pteridum chlL* (Yamada, in preparation), and 25 editing sites in the 12 genes and open reading frames (ORFs) of the gymnosperm *Pinus* have disappeared during evolution to the angiosperms *Nicotiana* [70].

In addition, the alignment of nucleotide sequences and amino acid sequences deduced from genomic DNA sequences shows the possible occurrence of RNA editing in the chloroplasts of primitive land plants (Fig. 3.3). For example, the hornwort *Megaceros enigmaticus* contains two nonsense codons within the putative coding region, and this alignment suggests that there are six editing sites in the 5'-part of the *rbcL* transcripts in this chloroplast. One of the nonsense codons is at the 41st position, which is exactly the same site as that of *A. formosae*, which has been subjected to the conversion of U to C. In the *M. enigmaticus* transcripts, the conversion of C to U at the 13th, 20th, 36th, and 40th codons, and the conversion of U to C at the 27th and 41st codons, can be expected. Although the 10th codon AGT coding Ser in *Isoetes* and *Selaginella*, the 23th codon AAT coding Asn in *Isoetes*, the 30th codon AAG coding Lys in *Isoetes*, and 50th codon GCC and GCT coding Ala in *Selaginella* and *Megaceros*, respectively, are not identical with those of *Anthoceros*, they will not be subjected to RNA editing; this is because they must change A to G at the 10th codon, A to C at the 23th, A to G at the 30th, and G to C at the 50th codon to adjust the amino acids to be the same as those of *Anthoceros*. These types of RNA editing have not been observed in any organelles to date. Also, the chemical properties of the corresponding amino acids are similar enough not to change the functional property of the protein.

The unusual initiation codon ACG of the *rbcL* sequence has been found in the lycopodophytes *Selaginella* and *Isoetes*, suggesting that the initiation codon AUG is created [71]. This will be the same type of typical RNA editing that has been found in the chloroplast transcripts of *rps2* in maize [8] and *psbL* in tobacco [14]. In both cases the genomic sequences begin with ACG, which is edited into AUG in the mRNA; this editing is limited to its initiation codon.

```
                          10                        20                        30
Iso    AcGTCACCACCAAACGGAGACTAAAGCGAGTGTTGGATTCAAAGCTGGCGTTAAAGATTACAGATTAAATTATTATACTCCTGATTATAAG
       Thr.........................Ser...............................Asn.................Lys
Sel    AcGTCACCGCAAACGGAAACCAAAGCAAGTGTTGGATTCAAGGCGTTAAAGATTACAGATTACAGATTACACCCCCGACTACGAA
       Thr.........................Ser
Meg                          AGGTGTTGGATcTAAAGCTGGTGTTAAGGATGATcATAGATTAAACCTACTATACTCtTGATTATGAA
                                    Ser...............His..................Leu
Ant-g  ATGTCACCACAAACGGAGACTAAAGCAGGTGTTGGATTTAAAGCTGGTGTTAAAGATTATAGATTAACCcATTATACCCCTGATTACGAG
                                                                            His
Ant-c  ATGTCACCACAAACGGAGACTAAAGCAGGTGTTGGATTTAAAGCTGGTGTTAAAGATTATAGATTAACCTATTATACCCTGATTACGAG
       MetSerProGlnThrGluThrLysAlaGlyValGlyPheLysLeuGlyValLysAspTyrArgLeuThrTyrTyrProAspTyrGlu
Mar    ATGTCACCACAAACGGAGACTAAAGCAGGTGTTGGATTCAAAGCTGGTGTTAAAGATTACAGATTATCGGATTATGAG

                          40                        50                        60
Iso    ACCAAAGACACCGATATTCTGGCAGCATTCCGAATGACTCCCAACCCGGAGTACCACCTGAGGAAGCAGGAGCCGCAGTAGCTGCTGAA
Sel    ACCAAGGATACCGATATATTGGCAGCATTCCGAATGACCCCGCAACCCGGCGTTCCCGCCGCAACCCGGAAGCAGGGGCCGCGGTAGCCGCGGAG
                                                                            Ala
Meg    ACCAAGGATACTGATAcTTTGGCAGCGTcTtGAATGACTCCTCAACCAGGGGTGCCAGCTGCCGAAGAAGCAGGAGCCGCAGTAGCCGCTGAA
                  Thr...............Ser***
Ant-g  ACCAAGGATACTGATATTTTGGCAGCGTcTtGAATGACTCCTtAACCAGGGGTGCCACCTGCCGAAGAAGCAGGAGCCGCAGTAGCTGCTGAA
                  Thr...............Ser***
Ant-c  ACCAAGGATACTGATATTTTGGCAGCGTTCGAATGACTCCTCAACCAGGGGTGCCACCTGCCGAAGAAGCAGGAGCCGCAGTAGCTGCTGAA
       ThrLysAspThrAspIleLeuAlaAlaPheThrProGlnProGlyValProProGluGluAlaGlyAlaAlaValAlaAlaGlu
Mar    ACCAAGGATACGGATATTTTAGCAGCATTTAGAATGACTCCTCCTCAGCCTGGAGTTCCAGCGGAGAAGCAGCAGGCAACGCAGTTGCTGCTGAA
                                                                            Asn
```

The observations of RNA editing in the mitochondria of plants, in the chloroplasts of primitive land plants, and the possible occurrence suggest that the event arose in the mitochondria and chloroplasts of the first land plants. The RNA editing process in plant organelles suggests that the mechanism of RNA editing was acquired to facilitate the adaptation of plants to land. RNA editing, in addition to the acquisition of introns [72], can increase the mutation rate of peptides, which provides the ability for plants to adapt to various environmental circumstances.

3.7 Conclusion

The amount of chloroplast transcripts is rather abundant compared to nuclear transcripts, suggesting that gene expression in chloroplasts is mainly controlled by the processing steps and by its translation. The order of the processing steps is RNA editing, splicing, and then cleavage in monocistronic mRNA. RNA editing can provide competent substrates for the splicing reactions and the translation reactions in both mitochondria and chloroplasts, except in a few cases. Therefore some control factors, which are not presently known but are probably nuclear-encoded factors, are key substances in the gene expression of the organelles.

The protein products of plant mitochondrial genes cannot be accurately predicted from their genomic sequences, because RNA editing modifies almost all mRNA sequences post-transcriptionally. Therefore, phylogenetic analysis, based on the mitochondrial nucleic acid sequences, requires the comparison of the protein-specifying information deduced from the edited mRNA sequences. The situation is similar in the chloroplast genes of primitive land plants because 19 amino acids deduced from the mRNA sequence differ from those predicted by the DNA sequence of *rbcL* [73]. This difference, corresponding to 4.0% of the large subunit of *Rubisco* of the hornworts, can be compared to the sequence of the fern *Angiopteris lygodiifolia* (DNA database accession number X58429).

FIG. 3.3. Alignment of the 5'-end of *rbcL* nucleotide sequences and its deduced amino acid sequences. *Numbers* indicate amino acid position from the N-terminal; *dots* indicate the same amino acid as that deduced from the cDNA of the hornwort *Anthoceros formosae*; *asterisks* indicate nonsense codons. Edited nucleotides are shown with lowercase letters, and the codons are underlined in *A. formosae*. The expected editing sites are also shown with lowercase letters and are underlined. Nucleotide sequences were obtained from the DNA data bank: lycopodophytes *Isoetes* (Iso) L11054 and *Selaginella* (Sel) L11280; anthocerophytes *Megaceros* (Meg) L13481 and *Anthoceros* genomic DNA (Ant-g) D43695 and cDNA (Ant-c) D43696; liverworts *Marchantia* (Mar) X04465

Acknowledgments. The author thanks Dr. Koh-ichi Kadowaki of the National Institute of Agrobiological Resources, Tsukuba, and Dr. Kyouji Yamada and Dr. Tatsuya Wakasugi of Toyama University for helpful suggestions. The author is supported by a grant-in-aid for scientific research (no. 08640886) from the Ministry of Education of Japan.

References

1. Stuart K (1991) RNA editing in mitochondrial mRNA of tripanosomatids. Trends Biochem Sci 16:68–72
2. Cattaneo R (1992) RNA editing in C/Uhloroplast and brA/I?in. Trends Biochem Sci 17:4–5
3. Gray MW, Covello PS (1993) RNA editing in plant mitochondria and chloroplasts. FASEB J 7:64–71
4. Pring D, Brennicke A, Schuster W (1993) RNA editing gives a new meaning to the genetic information in mitochondria and chloroplasts. Plant Mol Biol 21:1163–1170
5. Araya A, Bégu D, Litvak S (1994) RNA editing in plants. Physiol Plant 91:543–550
6. Scott J (1995) A place in the world for RNA editing. Cell 81:833–836
7. Bonnard G, Gualberto JM, Lamattina L, Grienenberger JM (1992) RNA editing in plant mitochondria. Crit Rev Plant Sci 10:503–524
8. Hoch B, Maier RM, Appel K, Igloi GL, Kössel H (1991) Editing of a chloroplast mRNA by creation of an initiation codon. Nature 353:178–180
9. Benne R, Van den Burg J, Brakenhoff JP, Sloof P, Van Boom JH, Tromp MC (1986) Major transcript of the frameshifted *cox*II gene from trypanosome mitochondria contains four nucleotides that are not encoded in the DNA. Cell 46:819–826
10. Powell LM, Wallis SC, Pease RJ, Edwards YH, Knott TJ, Scott J (1987) A novel form of tissue-specific RNA processing produces apolipoprotein B-48 in intestine. Cell 50:831–840
11. Covello PS, Gray MW (1989) RNA editing in plant mitochondria. Nature 341:662–666
12. Gualberto JM, Lamattina L, Bonnard G, Weil JH, Grienenberger JM (1989) RNA editing in wheat mitochondria results in the conservation of protein sequences. Nature 341:660–662
13. Hiesel R, Wissinger B, Schuster W, Brennicke A (1989) RNA editing in plant mitochondria. Science 246:1632–1634
14. Kudla J, Igloi GL, Metzlaff M, Hagemann R, Kössel H (1992) RNA editing in tobacco chloroplasts leads to the formation of a translatable *psbL* mRNA by a C to U substitution within the initiation codon. EMBO J 11:1099–1103
15. Maier RM, Hoch B, Zeltz P, Kössel H (1992) Internal editing of the maize chloroplast *ndhA* transcript restores codons for conserved amino acids. Plant Cell 4:609–616
16. Maier RM, Neckermann K, Hoch B, Akhmedov NB, Kössel H (1992) Identification of editing positions in the *ndhB* transcript from maize chloroplasts reveals sequence similarities between editing sites of chloroplasts and plant mitochondria. Nucleic Acids Res 20:6189–6194
17. Mahendran R, Spottswood MR, Miller DL (1991) RNA editing by cytidine insertion in mitochondria of *Physarum polycephalum*. Nature 349:434–438
18. Cattaneo R (1991) Different types of messenger RNA editing. Annu Rev Genet 25:71–78

19. Walbot V (1991) RNA editing fixes problems in plant mitochondrial transcripts. Trends Genet 7:37–39

20. Wissinger B, Brennicke A, Schuster W (1992) Regenerating good sense: RNA editing and *trans*-splicing in plant mitochondria. Trends Genet 8:322–328

21. Hiesel R, Combettes B, Brennicke A (1994) Evidence for RNA editing in mitochondria of all major groups of land plants except the Bryophyta. Proc Natl Acad Sci USA 91:629–633

22. Yoshinaga K, Iinuma H, Masuzawa T, Ueda K (1996) Extensive RNA editing of U to C in addition to C to U substitution in the *rbcL* transcripts of hornwort chloroplasts and the origin of RNA editing in green plants. Nucleic Acids Res 24:1008–1014

23. Fox TD, Leaver CJ (1981) The *Zea mays* mitochondrial gene coding cytochrome oxidase subunit II has an intervening sequence and does not contain TGA codons. Cell 26:315–323

24. Bégu D, Graves PV, Domec C, Arselin G, Litvak S, Araya A (1990) RNA editing of wheat mitochondrial ATP synthase subunit 9: direct protein and cDNA sequencing. Plant Cell 2:1238–1290

25. Graves PV, Bequ D, Velours J, Neau E, Belloc F, Araya A (1990) Direct protein sequencing of wheat mitochondrial ATP synthase subunit 9 confirms RNA editing in plants. J Mol Biol 214:1–6

26. Dell'Orto P, Moenne A, Graves PV, Jordana X (1993) The potato mitochondrial ATP synthase subunit 9: gene structure, RNA editing and partial protein sequence. Plant Sci 88:45–53

27. Malek O, Lätting K, Hiesel R, Brennicke A, Koop V (1996) RNA editing in bryophytes and a molecular phylogeny of land plants. EMBO J 15:1403–1411

28. Schuster W, Ternes R, Knoop V, Hiesel R, Wissinger B, Brennicke A (1991) Distribution of RNA editing sites in *Oenothera* mitochondrial mRNAs and rRNAs. Curr Genet 20:397–404

29. Marechal-Drouard L, Ramamonjisoa D, Cosset A, Weil JH, Dietrich A (1993) Editing corrects mispairing in the acceptor stem of bean and potato mitochondrial phenylalanine transfer RNAs. Nucleic Acids Res 21:4909–4914

30. Knoop V, Schuster W, Wissinger B, Brennicke A (1991) *Trans* splicing integrates an exon of 22 nucleotides into the *nad5* mRNA in higher plant mitochondria. EMBO J 10:3483–3493

31. Binder S, Marchfelder A, Brennicke A, Wissinger B (1992) RNA editing in *trans*-splicing intron sequences of *nad2* mRNAs in *Oenothera* mitochondria. J Biol Chem 267:7615–7623

32. Schuster W, Unseld M, Wissinger B, Brennicke A (1990) Ribosomal protein S14 transcripts are edited in *Oenothera* mitochondria. Nucleic Acids Res 18:229–233

33. Schuster W, Brennicke A (1991) RNA editing makes mistakes in plant mitochondria: editing loses sense in transcripts of a *rps19* pseudogene and in creating stop codons in *coxI* and *rps3* mRNAs of *Oenothera*. Nucleic Acids Res 19:6923–6929

34. Brandt P, Unseld M, Eckert-Ossenkopp U, Brennicke A (1993) An *rps14* pseudogene is transcribed and edited in *Arabidopsis* mitochondria. Curr Genet 24:330–336

35. Kuntz M, Camara B, Weil J-H, Schantz R (1992) The *psbL* gene from bell pepper (*Capsicum annuum*): plastid RNA editing also occurs in non-photosynthetic chromoplasts. Plant Mol Biol 20:1185–1188

36. Bock R, Hagemann R, Kössel H, Kudla J (1993) Tissue- and stage-specific modulation of RNA editing of the *psbF* and *psbL* transcript from spinach plastids—a new regulatory mechanism? Mol Gen Genet 240:238–244

37. Neckermann K, Zeltz P, Igloi GL, Kössel H, Maier RM (1994) The role of RNA editing in conservation of start codons in chloroplast genomes. Gene (Amst) 146:177–182

38. Casey JL, Bergmann KF, Brown TL, Gerin JL (1992) Structural requirements for RNA editing in hepatitis virus: evidence for a uridine-to-cytidine editing mechanism. Proc Natl Acad Sci USA 89:7149–7153

39. Scott J (1990) Editing and modification of messenger RNA. In: Eckstein F, Lilley DMJ (eds) Nucleic acids and molecular biology, vol 4. Springer, Berlin Heidelberg New York, pp 258–273

40. Hernould M, Suharsono S, Litvak S, Araya A, Mouras A (1993) Male-sterility induction in transgenic tobacco plants with an unedited *atp9* mitochondrial gene from wheat. Proc Natl Acad Sci USA 90:2370–2374

41. Iwabuchi M, Kyozuka J, Shimamoto K (1993) Processing followed by compete editing of an altered mitochondrial *atp6* RNA restores fertility of cytoplasmic male sterile rice. EMBO J 12:1437–1446

42. Gualberto JM, Bonnard G, Lamattina L, Grienenberger JM (1991) Expression of the wheat mitochondrial *nad-rps12* transcription unit: correlation between editing and mRNA maturation. Plant Cell 3:1109–1120

43. Sutton CA, Conklin PL, Pruitt KD, Hanson MR (1991) Editing of pre-mRNAs can occur before *cis*- and *trans*-splicing in *Petunia* mitochondria. Mol Cell Biol 11:4274–4277

44. Yang AJ, Mulligan RM (1991) RNA editing intermediates of *cox2* transcripts in maize mitochondria. Mol Cell Biol 11:4278–4281

45. Freyer R, Hoch B, Neckermann K, Maier RM, Kössel H (1993) RNA editing in maize chloroplasts is a processing step independent of splicing, cleavage to monocistronic mRNAs. Plant J 4:621–629

46. Ruf S, Zeltz P, Kössel H (1994) Complete RNA editing of unspliced and dicistronic transcripts of the intron-containing reading frame IRF 170 from maize chloroplasts. Proc Natl Acad Sci USA 91:2295–2299

47. Wissinger B, Schuster W, Brennicke A (1991) Trans-splicing in *Oenothera* mitochondria: *nad1* mRNAs are edited in exon and trans-splicing group II intron sequences. Cell 65:473–482

48. Börner GV, Mörl M, Wissinger B, Brennike A, Schmelzer C (1995) RNA editing of a group II intron in *Oenothera* as a prerequisite for splicing. Mol Gen Genet 246:739–744

49. Phreaner CG, Williams MA, Mulligan M (1996) Incomplete editing of *rps12* transcripts results in the synthesis of polymorphic polypeptides in plant mitochondria. Pant Cell 8:107–117

50. Chapdelaine Y, Bonen L (1991) The wheat mitochondrial gene for subunit I of the NADH dehydrogenase complex: a *trans*-splicing model for this gene-in-pieces. Cell 65:465–472

51. Kadowaki K, Ozawa K, Kazama S, Kubo N, Akihama (1995) Creation of an initiation codon by RNA editing in the *cox1* transcript from tomato mitochondria. Curr Genet 28:415–422

52. Zeltz P, Hess WR, Neckermann K, Börner T, Kössel H (1993) Editing of the chloroplast *rpoB* transcript is independent of chloroplast translation and shows different patterns in barley and maize. EMBO J 12:4291–4296

53. Sutton CA, Conklin PL, Pruitt KD, Calfee AJ, Cobb AG, Hanson MR (1993) Editing of *rps3/rpl16* transcripts creates a premature truncation of the *rpl16* open reading frame. Curr Genet 23:472–476

54. Blum B, Bakalara N, Simpson L (1990) A model for RNA editing in kinetoplastid mitochondria: "guide" RNA molecules transcribed from maxicircle DNA provide the edited information. Cell 60:189–198

55. Chen SH, Habib G, Yang CY, Gu ZW, Lee BR, Weng SA, Silberman SR, Cai SJ, Deslypere JP, Rosseneu M, Gotto AM Jr, Li WH, Chan L (1987) Apolipoprotein B-48 is the product of a messenger RNA with an organ-specific in-frame stop codon. Science 238:363–366

56. Navaratnam N, Morrison JR, Battacharya S, Patel D, Funahashi T, Giannoni F, Teng B, Davidson ND, Scott J (1993) The p27 catalytic subunit of the apolipoprotein B mRNA editing enzyme is a cytidine deaminase. J Biol Chem 268:20709–20712

57. Araya A, Domec C, Bégu D, Litvak S (1992) An *in vitro* system for the editing of ATP synthase subunit 9 mRNA using wheat mitochondrial extract. Proc Natl Acad Sci USA 89:1040–1044

58. Yu W, Schuster W (1995) Evidence for a site-specific cytidine deamination reaction involved in C to U RNA editing of plant mitochondria. J Biol Chem 270:18227–18233

59. Lu B, Hanson MR (1992) A single nuclear gene specifies the abundance and extent of RNA editing of a plant mitochondrial transcript. Nucleic Acids Res 20:5699–5703

60. Sugita M, Ohta M, Sugiura M (1995) Structure and function of RNA-binding proteins in chloroplasts and cyanobacteria. In: Go M, Schimmel P (eds) Tracing biological evolution and gene structures. Elsevier, Amsterdam, pp 77–86

61. Gualberto JM, Weil J-H, Grienenberger J-M (1990) Editing of the wheat *coxIII* transcript: evidence for twelve C to U and one U to C conversions and for sequence similarities around editing sites. Nucleic Acids Res 18:3771–3776

62. Bock R, Maliga P (1995) In vivo testing of a tobacco plastid DNA segment for guide RNA function in *psbL* editing. Mol Gen Genet 247:439–443

63. Sutton CA, Zoubenko OV, Hanson MR, Maliga P (1995) A plant mitochondrial sequence transcribed in transgenic tobacco chloroplasts is not edited. Mol Cell Biol 15:1377–1381

64. Zeltz P, Kadowaki K, Kubo N, Maier RM, Hirai A, Kössel H (1996) A promiscuous chloroplast DNA fragment is transcribed in plant mitochondria but the encoded RNA is not edited. Plant Mol Biol 31:647–656

65. Chaudhuri S, Carrer H, Maliga P (1995) Site-specific factor involved in the editing of the *psbL* mRNA in tobacco plastids. EMBO J 14:2951–2957

66. Hirose T, Wakasugi T, Sugiura M, Kössel H (1994) RNA editing of tobacco *petB* mRNAs occurs both in chloroplasts and non-photosynthetic proplastids. Plant Mol Biol 26:509–513

67. Glaubitz JC, Carlson JE (1992) RNA editing in the mitochondria of a conifer. Curr Genet 22:163–165

68. Freyer R, Lopez C, Maier RM, Martin M, Sabater B, Kössel H (1995) Editing of the chloroplast *ndhB* encoded transcript shows divergence between closely related members of the grass family (Poaceae). Plant Mol Biol 29:679–684

69. Ohyama K, Fukuzawa H, Kohchi T, Shirai H, Sano T, Sano S, Umesono K, Shiki Y, Takeuchi M, Chang Z, Aota S-I, Inokuchi H, Ozeki H (1986) Chloroplast gene organization deduced from complete sequence of liverwort *Marchantia polymorpha* chloroplast DNA. Nature 322:572–574

70. Wakasugi T, Hirose T, Horihata M, Tsudzuki T, Kössel H, Sugiura M (1996) Creation of a novel protein-coding region at the RNA level in black pine chloroplasts: the pattern of RNA editing in the gymnosperm chloroplast is different from that in angiosperms. Proc Natl Acad Sci USA 93:8766–8770

71. Manhart JR (1994) Phylogenetic analysis of green plant *rbcL* sequences. Mol Phylogenet Evol 3:114–127
72. Manhart JR, Palmer JD (1990) The gain of two chloroplast tRNA introns makes the green algal ancestors of land plants. Nature 345:268–270
73. Ueda K, Yoshinaga K (1996) Can *rbcL* tell us the phylogeny of green plants? In: Nei M, Takahata N (eds) Current topics of molecular evolution. Pennsilvania State University Press, pp 97–103
74. Hirose T, Fan H, Suzuki JY, Wakasugi T, Tsudzuki T, Kössel H (1996) Occurrence of silent RNA editing in chloroplasts: its species specificity and the influence of environmental and developmental conditions. Plant Mol Biol 30:667–672

Phylogenetic Reconstruction of Some Conifer Families: Role and Significance of Permineralized Cone Records

TAKESHI ASAKAWA OHSAWA

4.1 Introduction

The quest for a better understanding of nature is a major human motivation in the advancement of natural science. In many fields of biological science, reconstruction of the entire course of evolution, that is, the phylogeny of organisms, has been and is still attracting those who attempt to understand the biodiversity of this living planet. Paleobotany is a study that has greatly contributed to our knowledge of plant history.

Paleobotanists have continued with the often frustrating process of establishing natural systems of classification because such systems must be objective. To avoid subjectivity, cladistics has recently been adopted in paleobotany as in other fields of plant systematics. This is one of the most important and rewarding practices in the history of plant systematics.

Another advance in plant systematics during this decade is the introduction of cladism and molecular data. Morphological characters that have been used to compare plants are limited in number, and their evolutionary importance has to be "evaluated" in relation to other characters, which often results in a circular argument about the "naturalness" or "objectivity" of the proposed system. Contemporary plant systematics incorporating molecular systematics and cladism together resolves the evolutionary nests into clades based on a large number of characters much more objectively than before. However, the recent increase in knowledge of the present has brought the importance of the past into perspective. Fossil records in combination with modern results of living plant phylogeny can more clearly answer the questions of plant history. For example, Rothwell and Serbet [1] assessed the phylogenetic relationships of extant and extinct lignophytes and showed that the extinct taxa provided data to clarify the order and age of character originations.

As for the studies of conifer relationships, traditional studies of relationships were based on one or a few morphological characters. However, it is difficult to

Department of Biology, Faculty of Science, Chiba University, 1-33 Yayoi-cho, Inage-ku Chiba 263, Japan determine whether a certain character reflects the phylogenetic relationship or exhibits homoplasy. To reduce this problem, as many morphological characters as possible are now used to resolve phylogenetic relationships. Hart [2] and Miller [3] used cladistic analysis and produced cladograms for all conifers. They added primitive fossil cordaites and conifers such as Voltziaceae and Lebachyaceae as outgroups, but did not include fossils of recent families. Eckenwalder [4] compared the degree of similarity using many morphological characters and showed a close relationship between the Cupressaceae s.s. and the Taxodiaceae.

Protein similarity has also been used to reconstruct phylogenetic relationships. Among the conifers, the immunological response of seed protein between pairs of taxa has been compared to demonstrate the relationships of genera in the Pinaceae [5] and in the Cupressaceae and Sciadopityaceae [6]. Comparison of DNA sequences has become a powerful tool for reconstructing phylogenetic relationships. Phylogenetic analyses of all the conifers [7], of the Cupressaceae s.l. [8], of the Cupressaceae s.s. [9], of *Pinus* [10], and of *Picea* [11] have been completed. Although these analyses have not led to definite conclusions because of insufficient data, these problems may be overcome by using more rapidly evolving genes [7,8].

After clarifying the phylogenetic relationships between living taxa, the following points become interesting problems. When and how did the investigated taxa diverge? Which genes are involved in the evolutionary changes? Why or how have living taxa survived in contrast to the many extinct fossil taxa? The fossil record will undoubtedly provide crucial data for resolving the first problem. This chapter addresses this question by resolving the history of some modern families of conifers (i.e., the Cupressaceae s.l. and the Pinaceae) using the exceptionally rich Cretaceous fossils recently described from Japan.

4.2 The Cupressaceae

4.2.1 Classification of the Cupressaceae

The Cupressaceae s.l. have been split into two families, the Cupressaceae s.s. and the Taxodiaceae since Pilger [12]. However, recent numerical and cladistic analyses of conifer families leads to the conclusion that both families should be merged into the Cupressaceae s.l. [2–4]. In these studies, Miller [3] concluded that *Metasequoia* and *Cryptomeria* are sister groups of the Cupressaceae s.s. in which *Taxodium* is included as an ingroup taxon. Hart [2] concluded that the Cupressaceae s.s. occupy one trichotomous branch with a *Glyptostrobus-Metasequoia-Taxodium* clade and a *Taiwania-Cryptomeria-Cunninghamia* clade.

Intergeneric relationships of the Cupressaceae s.l. inferred from immunological distance of seed proteins [6] indicate that *Cryptomeria*, *Glyptostrobus*, and *Taxodium* form a sister group of the Cupressaceae s.s., and that *Sequoia*, *Sequoiadendron* and *Metasequoia* are distantly related to each other. Phylogenetic relationships obtained from the *rbcL* gene sequences also show the monophyly of the Cupressaceae s.s. and the Taxodiaceae [8].

Recently, many permineralized ovulate cones of the Cupressaceae *s.l.* have been described from the Upper Cretaceous of Hokkaido, Japan. Based on these, Ohsawa [13] reconstructed the morphological evolution in the Cupressaceae *s.l.* This work was based on a few morphological characters that were considered to be useful for the recognition of fossil species and for considering their affinities. However, it is difficult to detect objectively whether these characters accurately reflect phylogenetic relationships. To help overcome this problem, as many morphological characters as possible should be used to resolve the phylogenetic relationships. I analyze here the phylogenetic relationships of all permineralized cones of the Cupressaceae *s.l.* from the Cretaceous of Japan and consider the phylogeny of the Cupressaceae *s.l.*

4.2.2 Phylogenetic Analysis of Ovulate Cones of the Cupressaceae

Phylogenetic analysis of the permineralized ovulate cones of the Cupressaceae *s.l.* using Wagner parsimony was performed on specimens of six fossil species; *Archicupressus nihongii* Ohsawa et al. [14], *Cunninghamiostrobus yubariensis* Stopes et Fujii [15], *Haborosequoia nakajimae* Ohsawa et al. [16], *Parataiwania nihongii* Nishida et al. [17], *Yezosequoia shimanukii* Nishida et al. [18], *Yubaristrobus nakajimae* Ohsawa et al. [19], and a specimen of a *Sequoia*-like cone reported from the Upper Cretaceous of Hokkaido [16] were used for character examinations. Microscopic slides of modern cones of *Athrotaxis cupressoides* Don, *A. selaginoides* Don, *Cryptomeria japonica* (L.) Don, *Cunninghamia lanceolata* (Lambert) Hook., *Chamaecyparis obtusa* (Sieb. et Zucc.) Endl., *Cupressus sempervirens* L., *Glyptostrobus pencilis* (Staunton) K. Koch, *Thuja orientalis* L., *Taxodium disticum* (L.) Richards, *Metasequoia glyptostroboides* Hu et Cheng, *Sequoia sempervirens* (Don) Endl. and *Sequoiadendron giganteum* (Lindl.) Buchholz were also examined and compared with the fossil cones. Some of these micropreparations were borrowed from Prof. C. N. Miller, University of Montana, U.S.A. Many anatomical studies on the seed cones of the family and the original description of *Mikasastrobus* Saiki et Kimura [20] were also referred to.

The cladogram was generated by PAUP Mac ver. 3.1.1. using a heuristic search with simple step-wise addition and tree bisection–reconnection (TBR) branch swapping options. It was rooted using multi-outgroup taxa: *Araucaria* of the Araucariaceae, *Picea* of the Pinaceae, *Podocarpus* of the Podocarpaceae, and *Sciadopitys* of the Sciadopityaceae, because the nearest sister group of the Cupressaceae *s.l.* has not been exactly determined. Since the rooting is based on multi-outgroups, the character polarity is not discussed. Ten characters, each exhibiting two to seven character states (Appendix 4.1, Table 4.1), were used for phylogenetic analysis. The character proximity was decided to determine the

TABLE 4.1. Characters used in phylogenetic analysis of petrified cones of the Cupressaceae

Characters

1   Arrangement of cone scale
     0: alternate   1: opposite

2   Complex number
     0: many   1: several

3   Bract-scale complex thickness
     0: tapered   1: inflated   2: peltate with apical depression

4   Separation of scale
     0: separate   1: separate at apex   2: fused   1B: dentate apex

5   Form of complex trace at origin
     0: dorsoventral   1: cylindrical

6   Arrangement of vascular strands in bract-scale complex
     0: no adaxial bundle   1: single adaxial bundle
     2: four to five adaxial bundles   3: rhomboidal arrangement
     4: reniform arrangement   5: arranged in ring with few medullary bundles
     3B: single abaxial bundle

7   Number of resin canals of bract-scale complex at origin
     0: several   1: one or less

8   Resin canal distribution in bract-scale complex
     0: abaxial   1: both sides

9   Seed orientation
     0: inverted   1: erect

10   Fibers in ground tissue
     0: rare   1: moderate   2: abundant

steps between the character states. The character states of each taxon are shown in Table 4.2.

A heuristic search found only one most parsimonious tree, 34 steps long with a consistency index (*CI*) of 0.559 and retention index (*RI*) of 0.815 (Fig. 4.1). The genera of the Cupressaceae s.s. formed a monophyletic group. However, the taxodiaceous genera became paraphyletic. This result supports the merger of the Cupressaceae s.s. and Taxodiaceae *sensu* Page [21]. The Cupressaceae *s.l.* was split into two major groups. The first group consists of *Cunninghamia, Taiwania, Cunninghamiostrobus, Mikasastrobus*, and *Parataiwania*. This group is called the *Cunninghamia* group here (Table 4.3). In *Cunninghamia* group, *Cunninghamiostrobus* diverges at the most basal position, and *Taiwania, Mikasastrobus* and *Parataiwania* form the ultimate monophyletic group (Fig. 4.1).

The other taxa form a larger group called the *Sequoia*-Cupressoid group here (Table 4.3). The division of *Yezosequoia* from other taxa is the first event in this group, then *Athrotaxis* clades diverge (Fig. 4.1). The next clades exhibit a trichotomy diverging into *Haborosequoia* clade and two monophyletic groups. One of the latter clades consists of *Sequoia, Sequoiadendron, Metasequoia*, and *Sequoia-*

TABLE 4.2. Character distribution in the seed cone of the Cupressaceae

	Taxa	1	2	3	4	5	6	7	8	9	10
Living taxa	*Araucaria*	0	0	0	1	0	2	0	1	0	1
	Picea	0	0	0	0	1	3B	0	1	0	1
	Podocarpus	0	1	0	1	0	3B	0	1	0	0
	Sciadopitys	0	0	0	0	1	3B	0	1	0	1
	Athrotaxis cupressoides	0	0	1	2	1	2	?	1	0	0
	Cunninghamia	0	0	0	1	0	1	1	0	0	1
	Taiwania	0	0	0	2	0	0	1	0	0	0
	Sequoia	0	0	2	2	1	3	1	0	0	1
	Sequoiadendron	0	0	2	2	1	3	1	1	0	1
	Metasequoia	1	0	2	2	1	3	1	0	0	1
	Taxodium	0	0	1	1	1	3	1	1	0	1
	Glyptostrobus	0	0	1	1B	1	3	1	1	0	1
	Cryptomeria	0	0	1	1B	1	3	1	1	0	1
	Chamaechyparis	1	1	1	2	1	5	1	1	1	1
	Cupressus	1	1	1	2	1	4	1	1	1	1
	Thuja	1	1	1	2	1	3B	1	1	1	1
	Juniperus	1	1	1	2	1	3	1	1	1	1
Fossil taxa	*Archicupressus*	1	0	2	2	1	4	1	1	1	1
	Cunninghamiostrobus	0	0	0	1	1	1	1	1	0	1
	Haborosequoia	0	0	1	2	1	3	1	1	0	1
	Mikasastrobus	0	0	0	1	0	0	1	0	0	0
	Parataiwania	0	0	0	1	0	0	1	0	0	0
	Seqoia-like cone	0	0	2	2	1	3	1	1	0	1
	Yezosequoia	0	0	1	2	0	2	1	1	0	2
	Yubaristrobus	0	0	1	2	1	5	1	1	1	1

See Table 4.1 for definition of characters.

like cone, and is called the *Sequoia* subgroup here (Table 4.3). The other clade consists of the Cupressaceae *s.s.*, *Cryptomeria*, *Glyptostrobus*, *Taxodium*, *Archicupressus*, and *Yubaristrobus*, where the separation of the Cupressaceae *s.s.* including *Archicupressus* and *Yubaristrobus* and three taxodiaceous genera occurs at the base (Fig. 4.1). In this chapter the clade of Cupressaceae *s.s.*, *Archicupressus*, and *Yubaristrobus* is called the Cupressoid subgroup, and the other clade is called the *Taxodium* subgroup (Table 4.3).

The present cladogram conforms well to the results of broad comparative or cladistic analyses of conifer [2,3,4] in concluding to merge the Cupressaceae *s.s.* and the Taxodiaceae with some inconsistency on the sister group of the Cupressaceae *s.s.* and the relationships of taxodiaceous genera. The present cladogram also fits the relationships between the genera of the Cupressaceae *s.l.* based on the immunological distance or *rbcL* gene sequence in that *Cryptomeria* and *Glyptostrobus* form a sister group of the Cupressaceae *s.s.* [6,8]. The inconsistency between immunological tree and the present result is seen in the position of *Athrotaxis* forming a monophyletic group with the *Sequoia* group. Although the bootstrap probability is very low, *Cryptomeria* and *Glyptostrobus* become the sister group of the Cupressaceae *s.s.*, and *Cunninghamia* diverges at the base of

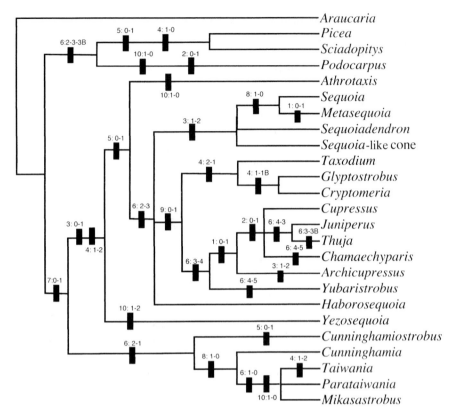

FIG. 4.1. Most parsimonious tree obtained from the seed cone characters. Character changes are rearranged on the most parsimonious tree. Numbers of character and character states should be referred to Table 4.1 and Appendix 4.1

the Cupressaceae *s.l.* in the molecular tree. The present study is consistent with the molecular tree on these points but differs in the positions of *Taiwania* and *Taxodium*. However, since the bootstrap probability is low, the relationships between the taxodiaceous genera are not highly reliable. Data from more rapidly evolving genes may help resolve these relationships.

Although some differences exist between the present cladogram and previous classifications, groupings, and phylogenetic relations, there are no questions about the position of the Cupressaceae *s.s.* consisting of an ultimate monophyletic group of the Taxodiaceae and that *Cryptomeria* and *Glyptostrobus* are the most closely related sister taxa of the Cupressaceae *s.s.* Although not supported by DNA sequence data, *Taxodium* is probably included in the same group of *Cryptomeria* and *Glyptostrobus*, because great similarity in seed proteins and cone structure between *Glyptostrobus* and *Taxodium* has been shown [6,22,23]. *Sequoia*, *Sequoiadendron*, and *Metasequoia* form a monophyletic group which is a sister group of the Cupressaceae *s.s.* and *Taxodium* group. The positions of

TABLE 4.3. Grouping of the Cupressaceae *s.l.* used in this study

Group	Members
Cunninghamia group	*Cunninghamia* Brown 1826
	Taiwania Hayata 1906
	Cunninghamiostrobus Stopes et Fujii 1910
	Mikasastrobus Saiki et Kimura 1993
	Parataiwania Nishida et al. 1992
Sequoia-Cupressoid group	*Yezosequoia* Nishida et al. 1991
	Athrotaxis Don 1841
	Haborosequoia Ohsawa et al. 1992
Sequoia sub-group	*Sequoia* Endlicher 1847
	Metasequoia Hu et Cheng 1848
	Sequoiadendron Buchholz 1939
	Sequoia-like cone (Ohsawa et al. 1992)
Taxodium sub-group	*Cryptomeria* Don 1841
	Glyptostrobus Endlicher 1847
	Taxodium Richard 1810
Cupressoid sub-group	Cupressaceae *s.s.*
	Archicupressus Ohsawa et al. 1992
	Yubaristrobus Ohsawa et al. 1993

Taiwania and *Athrotaxis* are not clear but the basal division of the *Cunninghamia* clade is highly reliable.

4.2.3 Character Evolution in the Cupressaceae s.l.

The character changes were plotted on the most parsimonious tree (Fig. 4.1). All character changes are consistent in both ACCTRAN and DELTRAN optimizations. The characteristics at the basal node of the family are many spirally arranged bract-scale complexes with tapered apex (character no. 3 = 0), the scale separated from the bract only at the apical region (character no. 4 = 1), a dorsiventral vascular trace to the bract-scale complex which splits into a larger set of abaxial bundles and several adaxial bundles (character no. 6 = 2), a single resin canal which becomes subdivided into both abaxial and adaxial canals (character nos. 7 = 1, 8 = 1), inverted seeds (character no. 9 = 0), and fundamental tissue containing sclerenchymatous cells (character no. 10 = 1). Subsequent character changes in Fig. 4.1 can be referred to in Table 4.1.

The seed cone synapomorphy of the Cupressaceae s.l. is the reduction of the number of the scale resin canals at origin (character no. 7 = 1). The *Cunninghamia* group is defined by one synapomorphy: decrease in the number of adaxial vascular bundles to one (character no. 6 = 1). On the other hand, the *Sequoia*-Cupressoid group share two synapomorphies: thick bract-scale complexes and completely-fused bract and scale (character nos. 3 = 1, 4 = 2).

In the *Cunninghamia* group, *Cunninghamia*, *Taiwania*, *Mikasastrobus*, and *Parataiwania* form a monophyletic group with the synapomorphy of the loss of adaxial resin canals in the bract-scale complex. Moreover, *Taiwania*, *Mikasastrobus*, and *Parataiwania* are most closely related in sharing two synapomorphies: loss of sclerenchyma cells and adaxial vascular bundles in the bract-scale complex. *Cunninghamiostrobus* exhibits a transitional form between the hypothetical ancestral form and living genera of the *Cunninghamia* group in having adaxial resin canals in the complex, but is specialized in diverging a cylindrical strand to the complex. *Mikasastrobus* and *Parataiwania* exhibit the same character states and have no autoapomorphy nor synapomorphies representing a transition between *Cunninghamia* and *Taiwania*. *Taiwania* could have evolved from a *Cunninghamia*-like ancestor through a *Parataiwania*-like precursor primarily by a loss of the adaxial vascular bundles and sclerenchyma cells in the bract-scale complex, and subsequently by a reduction of the external projection of the scale.

In the *Sequoia*-Cupressoid clade, all genera except *Yezosequoia* form a monophyletic group supported by one synapomorphy: a cylindrical complex trace at its origin. *Yezosequoia* exhibits an intermediate form between this group and the hypothetical ancestor, but is specialized in having highly sclerenchymatous bract-scale complex. The group consisting of the *Sequoia* subgroup, *Taxodium* subgroup, the Cupressoid subgroup, and *Haborosequoia* is monophyletic defined by well-developed abaxial and adaxial vascular bundles. The *Sequoia* group is defined by the apical depression of peltate bract-scale complex. The *Taxodium* subgroup is characterized by the apical separation of the bract and scale. The reniform arrangement of vascular bundles in the bract-scale complex is a synapomorphy of the Cupressoid subgroup.

The *Taxodium* subgroup and the Cupressoid subgroup exhibit erect seeds, which is a synapomorphy. The extant Cupressaceae genera *s.s.* are defined by the synapomorphies of (1) decussate or trimerous whorled phyllotaxis, and (2) a small cone consisting of a small number of bract-scale complexes. *Yubaristrobus* is intermediate in having spirally arranged bract-scale complexes, and vascular arrangement derived from the reniform arrangement seen in some species of the Cupressaceae *s.s.* *Archicupressus* exhibits a cylindrical cone consisting of many whorls of bract-scale complexes.

The Cupressaceae *s.s.* probably evolved by acquiring (1) the reniform arrangement of the complex trace, (2) decussate or whorled phyllotaxis, and (3) a small and spherical cone from an ancestor common to the *Taxodium* and Cupressoid group, which had erect seed and spirally arranged bract-scale complexes.

4.2.4 Minimum Estimated Age of Diversification Suggested from Mesozoic Fossil Records

Among the fossil taxa analyzed in this study, *Cunninghamiostrobus* is the earliest record of the Cupressaceae *s.l.* and is known from the Lower Cretaceous (Fig.

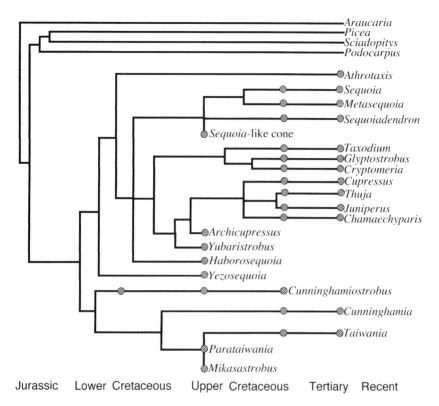

FIG. 4.2. Distribution of each taxon in time shown on the cladogram obtained in this study. *Shaded circles*, occurrence of each taxon

4.2). This suggests that the *Cunninghamia* group and the *Sequoia*-Cupressoid group had diverged no later than the early Cretaceous.

In the *Sequoia*-Cupressoid group, no taxa examined in this study are known from the Lower Cretaceous (Fig. 4.2). The divergence of the *Sequoia* subgroup, *Taxodium* subgroup, and the Cupressoid subgroup probably occurred by the Middle Cretaceous, because fossil taxa of each group are known from the Upper Cretaceous.

4.3 Phylogenetic Analysis of Ovulate Pinaceous Cones

4.3.1 *Taxonomy of the Pinaceae*

The Pinaceae are characterized by having two seeds on each cone-scale complex consisting of a large ovuliferous scale and a free bract. However, Pinaceous genera exhibit considerable diversification of vegetative and reproductive morphology. Many intergeneric relationships among the pinaceous genera have been

TABLE 4.4. Previous grouping of pinaceous genera

	Author			
Grouping	Van Tieghem	Pilger (1926)	Melcher and Wedermann (1954)	Frankis (1989)
	Abietoid group (Mtélocéles ou Cédrée)	subfam. Pinoideae	subfam. Pinoideae	subfam. Pinoideae
	Abies	*Pinus*	*Pinus*	*Pinus*
	Keteleeria	subfam. Abietoideae	subfam. Lauricoideae	subfam. Piceoideae
	Cedrus	*Pseudotsuga*	*Larix*	*Picea*
	Pseudolarix	*Picea*	*Cedrus*	subfam. Lauricoideae
	Tsuga	*Latix*	*Pseudolarix*	*Larix*
	Pinoid group (Epixylocéles ou Pinée)	*Abies*	subfam. Abietoideae	*Cathaya*
	Pseudotsuga	*Keteleeria*	*Pseudotsuga*	*Pseudotsuga*
	Picea	*Cedrus*	*Picea*	subfam. Abietoideae
	Larix	*Pseudolarix*	*Abies*	*Cedrus*
	Pinus	*Tsuga*	*Keteleeria*	*Pseudolarix*
			Tsuga	*Abies*
				Keteleeria
				Nothotsuga
				Tsuga

Subfam., subfamily.

discussed based on some of these varied features. Van Tieghem [24] recognized two groups in the family based on the number and position of resin canals in the vascular cylinder of a young tap root (Table 4.4). His Abietoid group (Cédrées), characterized by single central resin canal, consists of *Abies*, *Cedrus*, *Keteleeria*, *Pseudolarix*, and *Tsuga*. The Pinoid group (Pinée) is characterized by two resin canals adjacent to the protoxylem pole. This grouping has been supported by many authors based on the anatomy of wood and seed coat, pollination type, and immunological analysis [5,25,26].

Vierhapper [27] divided the Pinaceae, which he classified as the subfamily Abietoideae, into two tribes based on the occurrence of dimorphic shoots and foliage. The tribe Pineae includes only *Pinus*, and other genera belong to the tribe Sapineae. Pilger [12] ranked the subfamily Abietoideae to the family Pinaceae, and each tribe to the subfamilies, Pinoideae and Abietoideae (Table 4.4). Melchior and Werdermann [28] improved Pilger's classification and divided the family Pinaceae into three subfamilies (Table 4.4). Their classification is accepted in many conifer textbooks. Each subfamily is defined by the shoot morphology and the strobili attachment. The subfamily Pinoideae consists only of *Pinus* and is characterized by a strongly dimorphic shoot and foliage. All leaves on long shoots are reduced to scale leaves without chloroplasts and all photosynthetic leaves are on dwarf shoots subtended by scale leaves on long shoots. The strobili occur on long shoots. The subfamily Lauricoideae consists of *Cedrus*, *Larix*, and *Pseudolarix*. Their shoots are dimorphic consisting of long and short shoots. However, the leaves are monomorphic and linear or needlelike. All strobili occur on short shoots. The subfamily Abietoideae consists of *Abies*, *Cathaya*, *Keteleeria*, *Picea*, *Pseudotsuga*, and *Tsuga* characterized by monomorphic or weakly dimorphic shoots, monomorphic foliage, and strobili on long shoots. The significant feature in this classification is the morphology of vegetative organs, which is, in Farjon's opinion [29], frequently influenced by ecological conditions.

Since the 1980s, physiological characters have been used to resolve infrafamiliar relationships in the Pinaceae. Immunological analysis by Price et al. [5] identified two infrafamilial groups that were comparable to Van Tieghem's two subfamilies. Chase et al. [7] proposed phylogenetic relationships based on *rbcL* gene sequences. The results indicated monophyly of the Pinoid group, but the Abietoid group was paraphyletic. However, the suggested topology showed inconsistency in the root position and more data are needed for better resolution. Frankis [30] used many morphological characters and proposed an infrafamilial classification (Table 4.4). This classification was supported by Farjon [29] based on cladistic analysis using 25 characters. Their results were basically consistent with Van Tieghem's grouping, except that they added two monotypic subfamilies Pinoideae and Piceoideae, for the very distinct genera *Pinus* and *Picea*.

Most earlier studies on infrafamilial relationships in the Pinaceae have reached a consensus recognizing two groups. Moreover, immunological, molecular, and cladistic methods suggested close relationships between *Pinus* and *Picea*, and between *Pseudotsuga* and *Larix*. This chapter adopts the classification of Frankis

[30] and recognizes four subfamilies: Pinoideae, Piceoideae, Lauricoideae, and Abietoideae. After the relationships between living taxa have been resolved, evolution of the four subfamilies presents an interesting problem, particularly the evolution of the Pinoideae, which exhibit unique cone structure.

4.3.2 Permineralized Cones of the Pinaceae

The study of fossil pinaceous cones started in the 1830s with some descriptions of the external morphology of Creto-Tertiary materials in Great Britain [31,32]. *Pinus* and *Picea* are extant genera that are known from petrified cone fossils. The earliest record is *Pinus bergica* from the Neocomian of Belgium [33]. About ten more species of *Pinus* are known from the Upper Cretaceous [34] and the Tertiary [35–43]. In contrast, the earliest *Picea* occur much later. Two species of *Picea*, *P. diettertiana* and *P. eichhornii*, are known from the Oligocene [44,45], and one species, *P. wolfei*, is known from the Miocene [46].

Most anatomically preserved cones of the Pinaceae that are not included in extant genera have been described under the name *Pityostrobus* (Nathorst) Dutt [47]; about two dozen species have been described [33,47–60]. *Pityostrobus* is considered to be such a varied group that each species in this genus could represent an independent genus as compared to extant genera [61]. The affinities of *Pityostrobus* species are difficult to determine, because most species exhibit a combination of characters of some extant genera [61].

Two more extinct genera are known from permineralized ovulate cones: *Pseudoaraucaria* (Fliche) Alvin [62] and *Obirastrobus* Ohsawa et al. [63]. Six species are recognized in *Pseudoaraucaria*, all of which exhibit many stable characters at the genus level [61,62,64,65]. *Pseudoaraucaria* is considered to be a distinct and monophyletic genus. Miller [58] suggested a basal position of *Pseudoaraucaria* in the family. This hypothesis was supported by the degree of similarity between two species of *Pseudoaraucaria* and extant genera [66], but its relationship to extant genera has not been resolved. Two species have been recognized in *Obirastrobus* [63]. Based on the resin canals, they are believed to be related to *Keteleeria* and *Pseudolarix* [63]. However, the phylogenetic relationships of these species is not clear, because they exhibit character combinations of some extant genera and also apomorphies.

Phylogenetic analysis of the Pinaceae including the main fossil cones should help resolve these relationships. Relationships between fossil taxa and extant genera will also help identify when each line diverged and morphological changes over time. This chapter analyzes phylogenetic relationships of permineralized ovulate Pinaceae cones worldwide using cladistics, and discusses evolution within the family.

4.3.3 Phylogenetic Analysis of Permineralized Cones of the Pinaceae

Seed cones of the Pinaceae reported from the Cretaceous and Tertiary worldwide were used in this study. Specimens of *Pityostrobus matsubarae* Ohsawa et al.

[59], *Obirastrobus kokubunii* Ohsawa et al. [63], and *O. nihongii* Ohsawa et al. [63] were examined and used for character determination. Characters of other fossil species were taken from original descriptions. However, some species of *Pityostrobus* were excluded from the analysis, because of a lack of character information. Inclusion of these species would relax global parsimony by increasing the number of the most parsimonious trees. Some species of *Pinus* and *Pseudoaraucaria* were also excluded because of poor character information. Characters of living taxa were determined from anatomical studies on the seed cones of the families. Additional characters were recorded from microscopic preparations of modern cones of *Pinus banksiana* Lambert, *P. parviflora* Sieb. et Zucc., *P. pumilla* (Pallas) Regel, *Picea abies* (L.) Karsten, *P. jezoensis* (Sieb. et Zucc.) Carriere, *Tsuga diversifolia* (Maximowicz) Masters, *Keteleeria davidiana* (Bertrand) Beissner, *Pseudotsuga macrocarpa* Mayr, *Cedrus atlantica* (Endl.) Carriere, *Abies mariesii* Masters, and *Larix kaempferi* (Lambert) Carriere. Since the anatomical seed cone features of *Pseudolarix*, *Cathaya*, and *Nothotsuga*, the genera endemic to China, have not been examined, these taxa were not included in the analysis.

Thirty four seed-cone characters were used for the phylogenetic analysis (Appendix 4.2, Table 4.5). All characters were binary. Character states in the outgroup were scored as "0", and apomorphic states as "1". The missing and multistate characters were scored as "?" (Table 4.6).

The cladogram was generated by PAUP Mac ver. 3.1.1, using a heuristic search with simple step-wise addition and TBR branch swapping options. It was rooted using *Sciadopitys* of the Sciadopityaceae as the outgroup, because its bract character states most closely resemble those of the Pinaceae (both have bracts free from ovuliferous scales and single vascular strand). For consistency with the previous phylogenetic relationships proposed for living taxa, an enforced topological constraint option was used with the backbone constraint tree [67]. The constraint tree was based on the phylogenetic relationships obtained from the morphological characters [29,30] and immunological distance [5].

A heuristic search without enforcing topological constraint found 1205 most-parsimonious trees, each 162 steps long with a consistency index (*CI*) of 0.210 and retention index (*RI*) of 0.633. A strict consensus tree was generated based on the 1205 most-parsimonious trees (Fig. 4.3).

In this consensus tree, *Pityostrobus matsubarae* divides at the base and is the sister taxon of the other species. The next branch was not resolved and divided into 15 clades as polytomy (Fig. 4.3). In these clades, two large monophyletic groups were recognized: one consisting of *Pinus* and six *Pityostrobus* species, and the other consisting of *Larix* and *Pseudotsuga* of the Lauricoideae, *Pseudoaraucaria*, and two *Pityostrobus* species. *Pinus* became paraphyletic containing four *Pityostrobus* species at the terminal of some clades. Occurrence of characters found in *Pinus* has been pointed out in *Pityostrobus* [61]. The result of the present cladogram supports the close relationship between *Pinus* and some *Pityostrobus* species. The position of *Picea* species was not resolved, because each is arranged in polytomy. The Abietoideae genera were arranged in two monophyletic

TABLE 4.5. Characters used in phylogenetic analysis of petrified cones of the Pinaceae

Character

Pith of one axis
1   Sclerenchyma;   0-absent, 1-present

Vascular cylinder
2   Dissection;   0-complete, 1-dissected
3   Growth ring;   0-absent, 1-present
4   Resin canal;   0-absent, 1-present
5   Resin canals' ring;   0-absent, 1-present

Cortex
6   Sclerenchyma;   0-present, 1-absent
7   Outer thick-walled layer;   0-absent, 1-present

Bract-scale complex
8   Sclerenchyma;   0-present, 1-absent
9   Resin canal number of scale at origin;   0-more than 2, 1-2
10   Bract trace divergence;   0-separated from scale trace,
                                                   1-connected with scale trace
11   Scale trace diverge as;   0-two strands, 1-one strand

Bract
12   Sclerenchyma;   0-present, 1-absent
13   Separation of bract;   0-laterally first, 1-medially first
14   Resin canal;   0-present, 1-absent
15   Resin canal;   0-more than two, 1-two
16   Bract trace;   0-present, 1-terminate before entering bract

Scale base
17   Sclerenchyma (fiber) on abaxial side;   0-present, 1-absent
18   Sclerenchyma (fiber) on adaxial side;   0-present, 1-absent
19   Hypodermal sclerenchyma;   0-present, 1-absent
20   Lateral sclerenchyma (fiber);   0-absent, 1-present
21   Interseminal ridge;   0-absent, 1-present
22   Abaxial resin canal;   0-present, 1-absent
23   Adaxial resin canal;   0-present, 1-absent
24   Adaxial resin canal;   0-several, 1-one
25   Abaxial resin canal in part opposite seed;   0-present, 1-absent

Scale middle
26   Adaxial curvature of scale trace;   0-absent, 1-present
27   Sclerenchyma;   0-present, 1-absent
28   Abaxial canal;   0-present, 1-absent
29   Adaxial canal;   0-present, 1-absent
30   Interfascicular canal;   0-present, 1-absent
31   Inflated apex;   0-absent, 1-present
32   Papery thin apex;   0-absent, 1-present
33   Lacunae in the scale;   0-absent 1-present

Seed
34   Resin cavities;   0-absent, 1-present

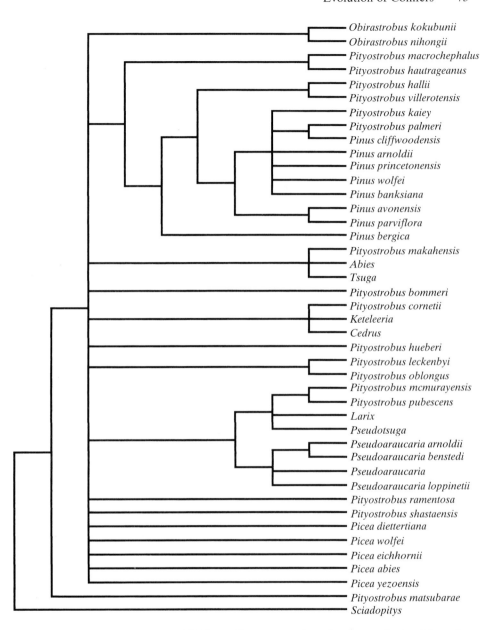

Fig. 4.3. Strict consensus tree of 1205 equally most-parsimonious trees obtained from the anatomical features of cones of the Pinaceae

TABLE 4.6. Character distribution in the seed cone of the Pinaceae

	Species	Character no.																																	
		1	2	3	4	5	6	7	8	9	10	11	12	13	14	15	16	17	18	19	20	21	22	23	24	25	26	27	28	29	30	31	32	33	34
Fossil taxa	*Obirastrobus kokubunii*	0	1	0	1	0	0	1	0	0	0	?	0	1	0	0	0	0	1	1	1	1	0	0	0	0	0	0	0	0	0	0	0	0	0
	Obirastrobus nihongii	0	0	0	1	0	0	1	0	0	0	1	0	1	0	0	0	0	1	1	1	1	0	0	0	0	0	0	0	0	0	0	0	0	0
	Pityostrobus macrochephalus	0	1	0	0	0	1	0	1	?	?	0	1	0	0	0	1	1	1	0	0	0	0	1	1	0	0	0	0	0	0	0	0	0	0
	Pityostrobus makahensis	1	0	0	1	1	0	1	0	1	0	0	1	?	?	1	1	0	1	1	0	0	1	0	1	1	?	0	0	0	0	1	0	0	1
	Pityostrobus bommeri	1	0	0	0	1	0	0	0	0	1	0	0	0	1	1	1	0	1	1	0	0	0	1	0	0	1	0	0	0	0	0	0	0	0
	Pityostrobus cornetii	0	0	0	0	0	0	1	1	0	0	1	?	0	1	1	1	1	1	1	0	0	0	0	1	1	0	0	1	0	0	0	0	0	1
	Pityostrobus hallii	0	0	0	0	0	0	0	0	1	1	1	1	0	0	1	1	1	1	1	0	1	0	1	1	0	0	1	1	0	0	0	1	0	0
	Pityostrobus hautrageanus	1	1	1	0	0	0	0	0	0	0	0	0	?	0	1	1	0	1	1	1	0	0	0	1	0	1	1	0	0	0	1	0	0	0
	Pityostrobus hueberi	1	0	0	0	0	0	0	1	0	1	1	1	1	1	1	1	1	1	1	?	0	0	1	1	1	1	0	0	1	0	0	0	0	0
	Pityostrobus kaiey	0	0	0	1	1	0	0	0	2	0	1	1	1	1	0	0	0	1	1	0	0	1	0	1	0	0	0	0	0	0	1	0	0	1
	Pityostrobus leckenbyi	0	1	0	0	1	?	?	?	1	1	1	1	0	0	1	1	0	0	1	1	1	0	1	0	0	0	0	1	0	0	1	0	0	1
	Pityostrobus matsubarae	0	0	0	0	0	1	0	0	0	0	0	0	0	0	1	0	0	1	1	1	1	1	0	1	0	0	0	0	0	0	0	0	0	0
	Pityostrobus mcmurayensis	1	1	1	0	0	0	?	?	1	0	1	1	0	0	0	1	0	0	1	1	1	1	1	0	0	?	?	?	1	?	0	0	0	1
	Pityostrobus oblongus	0	1	0	1	0	?	0	?	0	1	1	1	0	0	0	1	0	1	1	0	1	0	0	1	1	0	0	0	0	0	0	0	0	1
	Pityostrobus palmeri	0	0	0	1	1	1	0	1	1	1	1	1	0	0	0	1	1	1	1	0	0	1	1	0	0	1	1	0	1	0	0	1	0	0
	Pityostrobus pubescens	0	0	0	0	0	1	0	1	?	0	0	0	0	1	1	1	0	1	1	1	1	0	0	0	1	?	0	1	1	0	1	0	0	0
	Pityostrobus ramentosa	0	0	0	0	0	0	1	0	0	0	1	0	0	0	0	0	1	1	1	1	1	0	1	0	0	?	1	0	0	0	0	0	0	0
	Pityostrobus shastaensis	1	0	0	1	1	1	0	0	1	1	0	?	1	0	0	1	0	1	1	0	1	1	0	1	1	1	0	0	1	1	?	?	0	?
	Pityostrobus villerotensis	0	1	0	0	0	0	1	0	1	0	0	0	0	1	1	0	0	1	0	1	1	1	0	1	1	1	0	0	0	1	0	0	0	0
	Pseudoaraucaria arnoldii	0	1	0	0	0	0	1	0	0	0	0	0	0	1	0	1	0	?	0	?	1	1	0	0	0	0	0	0	0	0	0	0	1	1

76

Taxon																													
Pseudoaraucaria benstedi	0	1	0	0	0	1	0	0	0	1	0	0	1	0	0	0	0	1	1	1	1	0	0	0	0	0	1	0	0	1
Pseudoaraucaria major	0	1	0	0	0	1	0	0	0	1	0	0	1	0	0	0	0	1	1	1	1	0	0	0	0	0	1	0	0	0
Pseudoaraucaria loppinetii	0	1	0	0	0	1	0	0	0	1	0	0	1	0	0	0	0	1	1	1	0	0	0	0	0	0	1	0	0	0
Pinus arnoldii	0	0	0	1	1	0	1	1	1	0	1	0	1	0	1	1	1	0	0	0	1	1	1	1	1	1	0	1	1	0
Pinus avonensis	0	0	0	?	?	?	0	1	1	1	1	0	?	?	1	0	0	0	0	0	1	1	1	1	1	1	0	1	1	0
Pinus bergica	1	0	1	0	0	0	0	1	1	1	1	1	1	?	1	1	1	1	1	1	1	1	1	1	1	1	0	1	1	0
Pinus cliffwoodensis	0	1	1	1	0	1	0	1	1	1	1	1	1	1	1	1	1	1	1	1	1	1	1	1	1	1	0	1	1	0
Pinus princetonensis	0	0	0	0	1	1	1	1	1	0	1	1	1	1	1	1	1	0	0	0	1	1	1	1	1	1	0	1	1	0
Pinus wolfei	0	0	0	1	?	1	1	1	0	1	1	0	1	0	0	0	0	1	1	0	1	1	0	0	0	0	0	1	1	0
Picea dietertiana	1	0	0	0	0	1	0	0	0	1	0	0	1	1	1	0	1	0	0	?	0	0	0	0	0	0	0	0	1	0
Picea wolfei	1	0	0	0	0	1	0	0	0	1	0	1	0	1	0	1	1	0	?	?	0	0	0	0	0	0	0	0	1	0
Picea eichhornii	0	0	0	1	0	1	0	0	1	1	0	1	1	1	1	0	1	0	1	1	1	1	0	0	0	1	0	0	1	0

Living taxa																														
Keteleeria	0	0	0	0	0	0	1	0	1	0	1	0	0	1	0	0	0	0	0	0	1	0	0	0	0	0	0	0	0	1
Cedrus	1	0	0	0	0	1	0	0	0	1	0	0	0	1	0	1	1	0	1	1	1	0	1	0	0	0	0	0	0	1
Abies	1	0	0	?	0	0	0	0	0	0	0	0	0	0	1	0	0	0	0	0	0	0	0	0	0	0	0	0	0	1
Tsuga	1	0	0	0	0	0	0	0	0	1	0	0	1	1	0	0	1	0	0	0	1	0	0	0	0	0	0	0	0	1
Larix	1	0	0	?	0	0	0	0	1	0	1	0	1	1	0	1	0	1	0	0	1	0	0	?	0	0	0	0	0	1
Pseudotsuga	1	0	0	0	0	1	0	1	0	1	0	0	1	1	1	0	1	0	1	0	1	0	1	1	0	1	0	0	0	1
Picea abies	1	0	0	?	0	0	0	0	0	1	0	0	1	1	0	0	1	0	0	0	1	0	0	0	0	0	0	0	1	1
Picea yezoensis	1	0	0	0	0	1	0	0	0	1	0	0	1	1	0	0	1	0	0	0	1	0	0	0	0	0	0	0	1	1
Pinus parviflora	0	0	0	0	1	1	0	1	0	0	1	1	0	1	1	1	1	1	1	1	1	1	1	1	1	1	0	1	1	0
Pinus banksiana	0	0	1	1	0	1	1	0	0	1	1	0	0	1	1	0	0	1	1	1	1	1	1	1	1	1	0	1	1	0
Sciadopitys	0	0	0	0	0	0	0	0	0	0	0	0	0	0	0	0	0	0	0	0	0	0	0	0	0	0	0	0	0	0

77

groups: one consisting of *Abies*, *Tsuga*, and *Pityostrobus makahensis*, and the other consisting of *Cedrus*, *Keteleeria*, and *Pityostrobus cornetii*.

Both *Pseudoaraucaria* and *Obirastrobus* are monophyletic in the present cladogram. Each of them has been considered as a compact natural group because of their uniform characteristics in all species [48,61,63]. This result was supported by the present analysis.

These results are concordant with a classification proposed by Frankis [30] in recognizing two of the four subfamilies. However, the monophyly of his Abietoideae and Piceoideae was not supported by the present analysis. Genera of the Abietoideae *sensu* Pilger [12] occur at a large polytomy and do not form a monophyletic group. Either of the Abietoideae and Lauricoideae *sensu* Melchior and Wedermann [28] are polyphyletic in the current results, because *Larix* and *Cedrus* of his Lauricoideae form monophyletic groups with *Pseudotsuga* and *Keteleeria* of Abietoideae, respectively.

Two groups proposed by Van Tieghem [24] do not occur in the present cladogram. They are recognized by a number of features such as the number of resin canals in young tap root [24], the presence or absence of resin cavity in the seed coat [30,61], the distribution of resin canals in wood [26,68], the pollination type [25], and the immunological distance [5]. Considered from the uniformity of these characters, the two groups proposed by Van Tieghem seem to reflect true phylogenetic relationships. The above-mentioned characteristics are vegetative characters, except for resin cavities in the seed coat. The present cladogram is based only on seed-cone characters, which may be insufficient for reconstructing phylogenetic relationships.

Recent phylogenetic analyses based on immunology and *rbcL* gene sequences confirmed the monophyly of *Pinus* and *Picea* [5,7], which was not supported in the present cladogram. A morphological cladogram proposed by Farjon [29] also indicates the monophyly of *Pinus* and *Picea*. Their synapomorphies are the large number of cotyledons and the micropylar fluid in ovulate cones. Neither character can be recognized in fossil cones.

The cladogram obtained by the standard method was not consistent with all former phylogenetic arguments inferred from several different phylogenetic analyses. I, therefore, tried to make a cladogram of fossil pinaceous cones consistent with a topology obtained from phylogenetic analyses of living taxa. To obtain a tree that is consistent with the results of studies on living taxa, a topological constraint was enforced using the "backbone" constraint tree option of PAUP Mac ver. 3.1.1. This option adds the analyzed taxa at any point on the backbone tree, as long as the backbone is not violated [67]. The backbone constraint tree was based on the consensus of the phylogenetic relationships obtained from immunological distance and morphological data [5,29,30], and enforced a topology maintaining the monophyly of both Abietoid and Pinoid groups of Van Tieghem [24], and that of both *Pinus* and *Picea*, and *Larix* and *Pseudotsuga* (Fig. 4.4).

A heuristic search enforced by the topological constraint tree found 11 609 most-parsimonious trees, each 163 steps long with a *CI* of 0.209 and *RI* of 0.630.

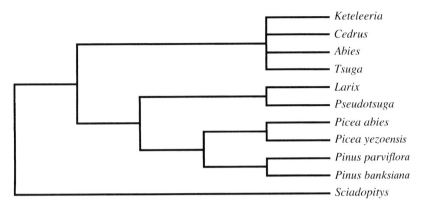

FIG. 4.4. Backbone constraint tree based on the consensus of the immunological tree [5] and morphological cladogram [29]

A strict consensus tree was generated based on 11 609 most-parsimonious trees (Fig. 4.5).

The enforced cladogram resolved a polytomy that occurred in the cladogram without any constraint enforcement. *Pityostrobus matsubarae* was arranged at the basal position of the family as it was in the most-parsimonious trees. The next branching also exhibited a polytomy but recognized only four clades diverging in *Obirastrobus*, *Pityostrobus ramentosa*, and Abietoideae clades and a clade consisting of other taxa. This polytomy is caused by the unstable position of *Obirastrobus* and *Pityostrobus ramentosa*, which are located in either of the bases of the family except *P. matsubarae*, the base of the Abietoideae clade, or the base of the clade consisting of the other taxa. The Abietoideae clade comprises *Keteleeria*, *Cedrus*, *Abies*, *Tsuga*, *Pityostrobus cornetii*, *P. leckenbyi*, *P. makahensis*, and *P. oblongus*. In this clade three monophyletic groups are recognized. In the largest monophyletic group composed of the rest of the taxa, two major monophyletic groups were recognized. One of them is the same group that is recognized in the strict consensus tree of most-parsimonious trees and comprises *Larix*, *Pseudotsuga*, *Pseudoaraucaria*, *Pityostrobus makahensis*, *P. mcmurrayensis*, and *P. pubescens*. The other group contains *Picea*, *Pinus*, and its related taxa. The topology of this group is imbalanced, consisting of several clades such as *Picea* species, *Pityostrobus shastaensis*, *P. bommeri*, and *P. hueberi* that are attached to the stem clade in a pectinate arrangement. The next branching defines two monophyletic groups. One consists of two *Pityostrobus* species, and another consists of all species of *Pinus* and four *Pityostrobus* species.

Although the cladograms generated by the search enforced by backbone topological constraint are less parsimonious, they are only one step longer than the most-parsimonious trees. Compared with living taxa, the fossil taxa have a limited number of characters that are useful for phylogenetic analysis. This is because the fossils are usually found as separate organs forcing one to perform the

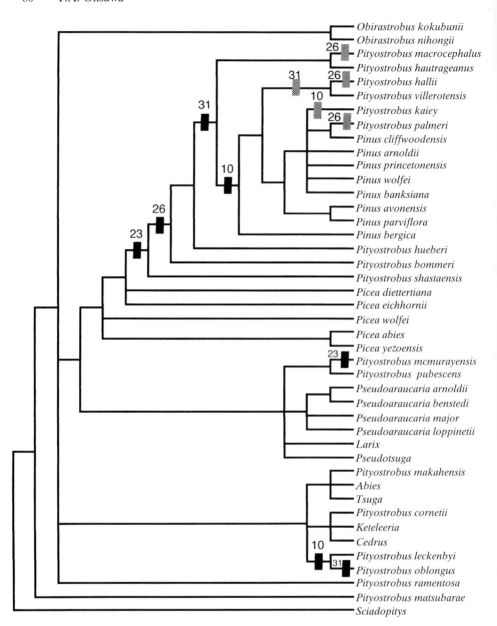

Fig. 4.5. Strict consensus tree of 11 609 equally most-parsimonious trees enforced to maintain topology of backbone constraint tree. *Pinus* cone characters are rearranged on tree. *Black bars*, character change from state absent to present; *shaded bars*, character change from state present to absent; *10*, bract trace divergence; *23*, occurrence of resin canals in adaxial tissue in scale base; *26*, strong abaxial curvature of scale trace; *31*, inflated scale apices

analysis based on a single organ. In discussing the evolution of pinaceous cones, the enforced cladograms seem more reliable than the most-parsimonious cladograms, because they exhibit topology supported by phylogenetic analysis of living taxa using more characters than in the analysis of fossils. Consequently the enforced cladograms were adopted as the reference tree in the following discussions.

4.3.4 Character Evolution in the Pinus Clade

Pinus cones exhibit four characteristics that do not occur in other living genera: (1) the bract-scale complex trace forming a ring at its origin; (2) the resin canals restricted to the abaxial side in scale base (=loss of resin canals in the adaxial tissue of scale base); (3) the strong abaxial curvature of the scale trace; and (4) the inflated scale apices [61]. To demonstrate the evolution of *Pinus* cones, I rearranged these four characters. For rearrangement of the characters on the tree, both ACCTRAN and DELTRAN optimization were applied, and the four characters mentioned above were unambiguously arranged (Fig. 4.5). The fusion of the bract and the scale traces is acquired at the base of the *Pinus* clade consisting of all *Pinus* species and four *Pityostrobus* species that are an ingroup of *Pinus* (Fig. 4.5, character no. 10). Reversion occurs once at the *Pityostrobus kaieyi* clade, and parallel evolution occurs at the clade consisting of *Pityostrobus oblongus* and *P. leckenbyi* in the Abietoideae. The adaxial resin canals in the scale base are lost at the most basal position of the Pinoideae clade including *Pityostrobus shastaensis* (Fig. 4.5, character no. 23). Parallel evolution occurs at the clade of *P. mcmurrayensis*. The abaxial curvature of the scale trace basically defines the monophyletic group comprising *Pityostrobus bommeri, P. hueberi, P. macrocephalus, P. hautrageanus*, and the *Pinus* clade (Fig. 4.5, character no. 26). However, the reversions occur in three *Pityostrobus* species ingroups of *Pinus*. The inflated scale apices is acquired at the base of monophyletic group comprising *Pityostrobus macrocephalus, P. hautrageanus*, and *Pinus* clade. Reversion and parallel evolution occur once (Fig. 4.5, character no. 31). The order in which the four characters were acquired in the course of *Pinus* evolution could be considered. The first acquired character is the loss of the adaxial resin canal at the scale base, the second is the abaxial curvature of the scale trace, the third is the inflated apices of the scale, and the evolution of the fusion of bract and scale traces is the last event.

4.3.5 Minimum Estimated Age of Subfamily Divergence

Inclusion of fossil taxa in the phylogenetic analysis can help suggest the minimum age of appearance and development of each clade. Figure 4.6 is the cladogram in which the end of each branch indicates the age at which each taxon occurs. The Abietoideae clade was the most compact group during the Cretaceous containing only a few taxa in the early Cretaceous (Fig. 4.6). No fossil taxa forming a monophyletic group with Abietoideae are known from the Upper Cretaceous. In

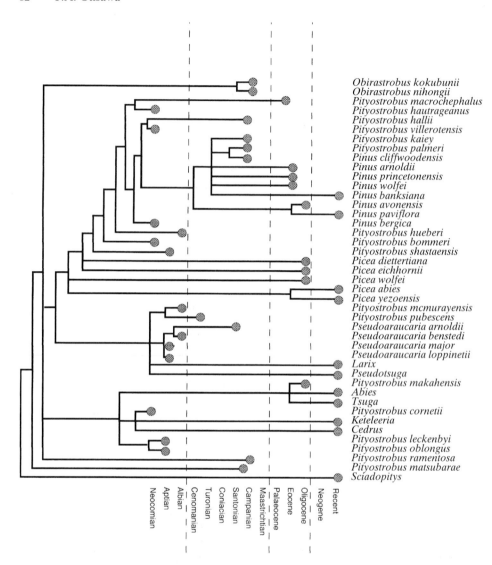

FIG. 4.6. Distribution of extant and petrified cone records of each taxon in time. *Shaded circles*, occurrence of each taxon

the Lauricoideae clade, only two species of *Pityostrobus* are known from the Albian and Cenomanian (Fig. 4.6). However, the diversification of the Lauri-coideae clade is demonstrated by *Pseudoaraucaria*, in which six species are known from the Cretaceous worldwide. The extant genera of the Abietoideae and Lauricoideae are known from as early as the Oligocene or Eocene [69]. However, the result of the present cladogram indicates that the *Keteleeria* and *Cedrus*, the *Abies* and *Tsuga*, the *Larix*, and the *Pseudotsuga* lineages diverged before the early Cretaceous. As the Upper Cretaceous taxa monophyletic with these extant genera are scarce, the precise pattern of divergence of these extant taxa is difficult to demonstrate. The Piceoideae is inconsistent in the present cladogram because *Picea* becomes paraphyletic with each fossil species arising independently in different clades during the early Cretaceous or earlier.

The lineage of Pinoideae, which comprises only one extant genus, *Pinus*, shows much diversification during the Cretaceous (Fig. 4.6). The earliest record of *Pinus* is from the Neocomian and is one of the earliest known permineralized cones of the family [33]. Except for *Pityostrobus macrocephalus*, all sister taxa near the base of *Pinus* are known as early as *Pinus bergica*. The Pinoideae have shown more diversification during the Cretaceous than at present; four *Pityostrobus* and one *Pinus* species in the Lower Cretaceous and three *Pityostrobus* and one *Pinus* species in the Upper Cretaceous belong to this group. The diversification of *Pinus* during the Upper Cretaceous is shown by the clade comprising *Pinus* species except *Pinus bergica*, and *Pityostrobus kaieyi* and *P. palmeri* (Fig. 4.6).

In contrast to the long fossil record of *Pinus*, other living genera are known from the Eocene or Oligocene [69]. However, each of three subfamilies, the Pinoideae, Lauricoideae, and Abietoideae, contains some early taxa from the Neocomian or Aptian (Fig. 4.6), indicating that these subfamilies had already appeared by the early Cretaceous. Since *Picea* is the sister taxon of *Pinus*, the Piceoideae should have diverged before the early Cretaceous, despite the lack of fossil records. To demonstrate the divergence of subfamilies, research on earlier fossils is necessary to clarify relationships between *Picea* species and their ancestral taxa.

4.3.6 Phylogentic Implications for Fossil Taxa

Among the fossil taxa, both *Obirastrobus* and *Pseudoaraucaria* form monophyletic groups, indicating the validity of the two genera. Some authors suggested the basal position of *Pseudoaraucaria* in the family [61,65,66]. In the present cladogram, they form a sister group of *Larix*, *Pseudotsuga*, and related *Pityostrobus* species. The relationship between *Obirastrobus* and *Keteleeria* was pointed out [63]. However, *Obirastrobus* exhibits no close relationship with any living taxa in this analysis.

According to Miller [61], *Pityostrobus* is a very varied genus, no two species of which could be interpreted as representatives of the same natural genus. Each species of *Pityostrobus* also varies in degree of similarity to seed cones of recent genera [61]. The result of the present study largely supports Miller's opinion, because *Pityostrobus* species are distributed through most parts of the cladogram. The affinities of *Pityostrobus* species are difficult to determine, because of the mixture of features of plural extant genera. For example, species such as *P. andraei* (Coemans) Alvin [48], *P. argonensis* (Fliche) Creber [70], and *P. kayei* Miller et Robison [65], exhibit three of four characteristics of *Pinus*, but it is not certain whether they are ancestors or descendants of *Pinus*. *Pityostrobus pubescens* Miller [58] was believed to be an intermediate between *Pseudoaraucaria* and some extant genera, but the most closely related extant genera were not specified. However, based on phylogenetic analysis of the seed cones, most *Pityostrobus* species form monophyletic groups with one of the subfamilies, except *P. matsubarae*, which is located in the basal position of the family, and phylogenetic relationships between *Pityostrobus* species and extant genera are suggested.

Nine species of *Pityostrobus* are included in a monophyletic group containing *Pinus* but excluding *Picea*. Of these, *Pityostrobus shastaensis*, *P. bommeri*, *P. hueberi*, *P. macrocephalus*, and *P. hautrageanus* are located in the outgroup of *Pinus*, and the others are in the ingroup of *Pinus*. *Pityostrobus cornetii*, *P. leckenbyi*, *P. oblongus*, and *P. makahensis* form a monophyletic group with Abietoideae. *Pityostrobus pubescens* and *P. mcmurrayensis* are sister taxa of *Larix* and *Pseudotsuga*. Miller [58] described *Pityostrobus pubescens* as exhibiting characteristics of *Pseudoaraucaria* and some extant taxa. The result of this study supports his opinion. The position of *Pityostrobus ramentosa* is not detected by the present cladogram, because it is arranged in the basal polytomy.

Pityostrobus is polyphyletic as shown in the cladogram (Fig. 4.6), suggesting the need of taxonomic division of the genus. At least four groups are recognized. However, the present study only used some of the *Pityostrobus* species and the position of the other species is not known. Thus, I leave the taxonomic emendation until further studies have been carried out to elucidate the position of all *Pityostrobus* species.

4.4 Conclusions

4.4.1 Application of the Constraint Tree in Phylogenetic Analysis

A limitation of fossil information in phylogenetic analysis is that fossils are found as separate organs and the phylogenetic analysis is forced to be performed based on a single organ and limited information. The organ used in this study was the seed cone, which provides information more useful than does a vegetative organ. The phylogenetic reconstruction of the Cupressaceae *s.l.* proposed in this study mostly agreed with the topology referred from some recent phylogenetic analyses of living taxa. For example, the Cupressaceae *s.s.* forms a monophyletic group arranged in an ultimate branch in the Taxodiaceae. This result supports the validity of merging Cupressaceae *s.s.* and Taxodiaceae into one family Cupressaceae *s.l.*, which was proposed by many authors using living materials. However, the phylogenetic reconstruction of the Pinaceae was considerably inconsistent with the relationships inferred from the analysis of living taxa. In this paper, a backbone topological constraint was applied to resolve such inconsistency. This method can help determine the position of fossil taxa in the reference tree obtained from phylogenetic analysis of living taxa. If a reliable cladogram is proposed based on living taxa and if the cladogram based on fossil taxa conflicts with the former, the cladogram enforced by backbone constraint tree analysis seems more reliable than the non-constraint cladograms to discuss the evolution of morphological features and the phylogenetic relationships of fossil taxa. The application of the constraint tree to analyzing the relations of fossil pinaceous cones is the test case of this method. The more reliable phylogenetic relationships of living taxa that are obtained, the more useful will be this method for reconstructing morphological evolution.

4.4.2 Taxonomical Implications of Fossil Taxa Using Phylogenetic Analysis

Determining the exact affinities of fossil pinaceous cones is quite difficult, because they exhibit combinations of features characterizing extant taxa and unique in fossil taxa [58]. The affinities of fossil taxa have traditionally been discussed rather subjectively, which sometimes provides a misleading conclusion as shown in the example of *Obirastrobus*. However, if a fossil taxon is included in the phylogenetic analysis, its affinities are defined without any subjective bias. Based on phylogenetic analysis, the position of *Obirastrobus*, *Pseudoaraucaria* and of each species of *Pityostrobus* were defined. *Pseudoaraucaria* became a sister group of extant Lauricoideae genera *Larix* and *Pseudotsuga*. *Pityostrobus* could be divided into at least four groups: species becoming monophyletic with the Pinoideae, species becoming monophyletic with Lauricoideae, species becoming monophyletic with Abietoideae, and species at the basal position of the family. The position of *Obirastrobus* was not detected exactly, but it was located at the basal position either of the family Abietoideae, or of the Pinoideae, Piceoideae, and Lauricoideae.

4.4.3 Advantages of Fossil Information in Reconstructing the Phylogenetic Relationships

Fossils provide information that is unavailable from living plants. Extinct taxa sometimes exhibit evolutionary intermediate forms, which are significant in reconstructing the evolution of conifers [3,71–73]. Recent phylogenetic analyses have shown that *Pinus* and *Picea* are closely related forming a monophyletic group at the top of one clade. However, *Pinus* cones exhibit four autoapomorphic features and their evolutionary course cannot be resolved by the information from living taxa. The present phylogenetic analysis including the fossil taxa clarified morphological evolution in the *Pinus* clade. Similar alignment of character evolution was clarified in the evolution of *Taiwania*. Morphological trends in the evolution of Cupressaceae *s.s.* was also shown successfully by including *Yubaristrobus* and *Archicupressus* as fossil intermediate forms. These examples show that in certain cases the fossil data give significant information in considering morphological evolution.

The second advantage of fossils is their use as chronological evidence. The minimum estimated age of appearance of each family or genus has been discussed using fossil taxa by some palaeobotanists [69,71]. In the present Pinaceae cladogram, three of the four extant subfamilies were recognized in early Cretaceous taxa, suggesting pre-Cretaceous separations of the four subfamilies. Similar evidence was shown in the Cupressaceae, in which two major groups had separated before the early Cretaceous. The Cupressaceae *s.s.* probably appeared in the Middle Cretaceous. As discussed by Brunsfeld et al. [8], the reliable time of appearance of each group can be applied to assessing the molecular evolution rate.

As shown above, fossil information plays a significant role in the phylogenetic reconstruction of some coniferous taxa, even if imperfect and fragmentary. If more molecular data will infer more precise phylogenetic relationships between living taxa, fossil information will be more important for the reconstruction of the history of target taxa.

Acknowledgments. The author wishes to express his deep gratitude to Dr. Makoto Nishida, Prof. Emeritus, Chiba University, and to Harufumi Nishida of International Budo University for their guidance and valuable suggestions in completing this study, and for critical reading of the manuscript. Gratitude is also due to Dr. Charles N. Miller of the University of Montana for loan of some slides of extant species of the Cupressaceae. Research on which this review is based was funded by Grants-in-Aid for Scientific Research Nos. 05740517, 07740652, and 07304057 from the Ministry of Education, Science, and Culture of Japan.

Appendices

Appendix 4.1 Characters Used in the Phylogenetic Analysis of the Cupressaceae s.l.

1. Arrangement of bract-scale complex. 0: alternate; 1: opposite. The arrangement of the bract-scale complex is divided into two types in the Cupressaceae. The genera of the Cupressaceae s.s. is characterized by decussate or whorled phyllotaxis [3]. On the other hand, except for *Metasequoia*, the Taxodiaceae genera exhibit spiral phyllotaxy.

2. Number of bract-scale complexes. 0: many; 1: several. The seed cone of the Cupressaceae s.s. is usually smaller than that of the Taxodiaceae and consists of a small number of bract-scale complexes [21,74]. The seed cone of the Taxodiaceae is composed of many bract-scale complexes.

3. Thickness of bract-scale complex. 0: tapered, 1: inflated, 2: peltate with apical depression. Two types are recognized in the thickness of the bract-scale complex. The bract-scale complex is thickened in most genera of the Cupressaceae [21,22,74,75]. However, in taxodiaceous genera, *Cunninghamia*, *Taiwania*, *Cunninghamiostrobus*, *Mikasastrobus*, and *Parataiwania* have thin and

tapered complexes [15,17,21,75], while *Sequoia, Sequoiadendron, Metasequoia,* and *Archicupressus* exhibit thick, peltate complexes with apical depression [16]. They were recognized as ordered characters in this analysis.

4. Separation of scale. 0: separate; 1: separate at apex; 2: fused; 1B: dentate apex. In the Cupressaceae, the bract and scale form a fused bract-scale complex [21,74,75]. They are completely fused and the scale cannot be recognized externally in most genera, but *Cunninghamia, Cryptomeria, Glyptostrobus, Taxodium, Cunninghamiostrobus, Mikasastrobus,* and *Parataiwania* have externally discernible scale tips as minute lobes [15,17,20,21]. Moreover the scale tips of *Cryptomeria* and *Glyptostrobus* develop dissected and dentate apices. In *Picea* and *Sciadopitys* of the outgroup, the bract and ovuliferous scale are separate from each other. Of these four character states, separate type, apically separate type, and completely fused type are arranged linearly. However the dentate scale apex seems to be a specialized form of the apically separate type.

5. Form of bract-scale complex trace at origin. 0: dorsoventral; 1: cylindrical. In most extant genera of the Cupressaceae, the vascular supply to the bract-scale complex is cylindrical in cross section, while in *Cunninghamia* and *Taiwania,* which have a tapered bract-scale complex, there is massive dorsoventral vascular supply [75,76]. In the fossil taxa, *Mikasastrobus, Parataiwania,* and *Yezosequoia* have a dorsoventral or adaxially opened and C-shaped vascular bundle in cross section [17,18,20]. *Cunninghamiostrobus, Haborosequoia, Yubaristrobus,* and *Archicupressus* have a cylindrical vascular bundle [14,16,19,77].

6. Arrangement of vascular bundles in bract-scale complex. 0: no adaxial bundle; 1: single adaxial bundle; 2: four to five adaxial bundles; 3: Rhomboidal arrangement; 4: reniform arrangement; 5: arranged in ring with few medullary bundles; 3B: single abaxial bundle. This character is very diversified and seven types were recognized. *Taiwania, Mikasastrobus,* and *Parataiwania* have abaxial vascular bundles only. *Cunninghamia* usually has a single adaxial vascular bundle [13,78–80]. *Cunninghamiostrobus* exhibits this type [15,77,81,82]. Four to five adaxial bundles occur consistently in *Athrotaxis* [78,83]. *Yezosequoia* is also this type [18]. In most other genera, adaxial as well as abaxial vasculature develops [75,76,78,79,84,85]. Some Cupressaceae genera such as *Thuja orientalis* have well-developed adaxial vasculature and a single abaxial vascular bundle [13,84,85]. This type could be derived from reduction of *Sequoia*-type abaxial bundles. A reniform pattern, which could be derived from a rhomboidal pattern, occurs in some genera of the Cupressaceae s.s. [13,84,85]. In *Chamaechyparis obtusa* and *Yubaristrobus,* vascular bundles are arranged in a ring with a few medullary bundles. This is believed to be derived by modification of reniform outline [13,19].

7. Number of resin canals in bract-scale complex at origin. 0: several; 1: one. Only one resin canal supplies the bract-scale complex in the Cupressaceae genera [13,78,80,86], instead of the two or more resin canals found in the outgroup taxa [13,78,80].

8. Resin canal distribution in bract-scale complex. 0: abaxial; 1: both sides. The resin canals are restricted to the abaxial side of the bract-scale complex in

Cunninghamia, Taiwania, Sequoia, Metasequoia, Mikasastrobus, and *Parataiwania* [13,76,78,86]. In other genera, they occur on both abaxial and adaxial sides [13,78].

9. Seed orientation. 0: Inverted; 1: erect. The seed of *Cryptomeria, Glyptostrobus, Taxodium* and the Cupressaceae s.s. is oriented erectly and the micropyle faces the apex of the bract-scale complex [22,23,73,83,87,88]. In other taxa, it is inverted with a micropyle facing the cone axis [22,73,83,87].

10. Fibers in fundamental tissue of bract-scale complex. 0: absent; 1: present; 2: abundant. In *Athrotaxis, Taiwania, Mikasastrobus,* and *Parataiwania,* the fundamental tissue of the bract-scale complex consists mostly of parenchymatous cells [78,89]. In other genera, the fundamental tissue consists of parenchyma and sclerenchyma. Of these, *Yezosequoia* has exceptionally sclerized tissue.

Appendix 4.2   Characters Used in the Phylogenetic Analysis of the Pinaceae

Pith of cone axis

1. Occurrence of sclerenchyma cells in pith. 0: absent; 1: present. Living Pinaceae usually have sclerenchyma cells in the pith of the cone axis, but *Keteleeria* and *Pinus* lack sclerenchyma. Among the fossil genera, *Obirastrobus* and *Pseudoaraucaria* lack sclerenchyma cells in the pith, but about half the species of *Pityostrobus* have them.

Vascular cylinder

2. Dissection of vascular cylinder. 0: complete; 1: dissected. The vascular cylinder of the cone axis is usually complete, but *Pseudolarix* has a highly dissected vascular cylinder [61]. *Pseudoaraucaria* is characterized by a highly dissected vascular cylinder like *Pseudolarix* [62]. The vascular cylinder of some species of *Pityostrobus, Obirastrobus,* and fossil *Pinus* is somewhat dissected.

3. Growth rings. 0: absent; 1: present. Growth rings in the wood of the vascular cylinder are characteristic of some species of *Pinus* [61]. However, they were not found in the species examined in this study. Growth rings occur in some species of fossil *Pinus* and *Pityostrobus buchananii.*

4. Resin canals. 0: absent; 1: present. Resin canals in the wood are characteristic of *Pinus* and some species of *Picea* [61]. I also found resin canals in the wood of *Pseudotsuga.* In both extant and fossil *Pinus,* all species except *P. bergica* have resin canals. The resin canal is also a characteristic of *Obirastrobus* and some species of *Pityostrobus.*

5. Resin canals arranged in ring. 0: absent; 1: present. In both extant and fossil species of *Pinus,* resin canals in the wood are arranged in a ring [61].

Cortex

6. Sclerenchyma cells in cortex of cone axis. 0: present; 1: absent. *Pinus, Tsuga Keteleeria,* and some species of *Picea* do not have scattered sclerenchyma cells in

the cortex of the cone axis. They are also absent in some species of *Pinus* and *Pityostrobus* among fossil taxa.

7. Outer thick-walled cell layer of cortex. 0: absent; 1: present. The cortex of *Keteleeria, Cedrus, Abies, Tsuga*, and some species of *Pinus* consists of uniform tissue, but that of the other species consists of an inner thin-walled layer and an outer thick-walled layer. Among the fossil genera, *Obirastrobus* and *Pseudoarau-caria* have an outer thick-walled cortex, and *Pityostrobus* has both types.

Bract-scale complex

8. Sclerenchyma cells in the bract-scale complex. 0: present; 1: absent. Funda-mental tissue of the bract-scale complex of *Keteleeria* and *Cedrus* consists of parenchymatous cells. However, that of the other genera contains thick-walled and elongate sclerenchyma or fibers [78]. The bract-scale complex of *Obirastro-bus* and *Pseudoaraucaria* also contains sclerenchymatous cells, and that of *Pity-ostrobus* show both type depending on the species.

9. Number of resin canals in scale at origin. 0: more than two canals; 1: two canals. In *Tsuga, Pseudotsuga*, and *Pinus*, two resin canals supply the scale [78]. In *Keteleeria, Pseudolarix, Cedrus, Abies, Larix*, and *Picea*, more than two resin canals supply the scale [78]. In the fossil genera, *Pseudoaraucaria* and most species of *Pityostrobus*, two resin canals supply the scale at the origin. *Obirastro-bus* and a few species of *Pityostrobus* have more than two resin canals supplying the scale at the origin.

10. Type of bract trace divergence. 0: separated from scale trace; 1: connected with scale trace. The divergence of the bract trace is a characteristic that sepa-rates *Pinus* from other genera [61]. In *Pinus*, the bract trace is connected with the scale trace at the origin to form a cylindrical unit. In the other extant genera, the bract trace is separated from the scale trace at the origin. The connected type occurs in some *Pityostrobus* species.

11. Type of scale trace divergence. 0: two strands; 1: one strand. The divergence of the scale trace is also divided into two types. In *Pseudoaraucaria* and some species of *Pityostrobus*, the scale trace diverges as two independent strands.

Bract

12. Occurrence of sclerenchyma cells. 0: present; 1: absent. Sclerenchyma cells are absent in the fundamental tissue of the bract in *Keteleeria, Cedrus, Abies, Tsuga, Picea*, and in some species of *Pinus*. Among the fossil taxa, sclerenchyma cells occur in *Obirastrobus* and *Pseudoaraucaria*, and about half the species of *Pityostrobus*.

13. Separation of bract. 0: laterally first; 1: medially first. Two types are recog-nized in the separation of the bract [61]. In *Abies, Tsuga* and *Larix*, the bract separates from the scale medially first. In the other genera, it separates laterally first. *Obirastrobus* and a few species of *Pityostrobus* show the former type.

14. Resin canals in bract. 0: present; 1: absent. Resin canals are usually present in the bract. However, there are no resin canals in the bract of *Cedrus* and a few *Pityostrobus* species.

15. Resin canal number. 0: more than two; 1: two. Two resin canals are usually present in the bract. However, *Obirastrobus* and some species of *Pinus* and *Pseudoaraucaria* have more than two resin canals.

16. Bract trace behavior. 0: present in bract; 1: terminate before entering bract. A single vascular strand runs through the bract in most taxa. However, the bract trace of *Keteleeria*, some species of fossil *Picea*, and some species of *Pityostrobus* and *Pseudoaraucaria* terminates before entering the bract.

Scale base

17, 18, 19, 20. Distribution of sclerenchyma cells at scale base. 17: Sclerenchyma on abaxial side. 0: present; 1: absent. 18: Sclerenchyma on adaxial side. 0: present; 1: absent. 19: Hypodermal sclerenchyma. 0: present; 1: absent. 20: Lateral sclerenchyma forming wing-like structure. 0: absent; 1: present. In most genera, sclerenchyma cells are present in the abaxial side and absent in the adaxial side. However, *Larix, Pseudotsuga, Pseudoaraucaria*, and some species of *Pityostrobus* have sclerenchyma cells in both sides. *Keteleeria, Cedrus*, and some species of *Pityostrobus* lack them. Hypodermal sclerenchyma occur on the adaxial surface of the scale base in *Larix, Pseudotsuga, Pseudoaraucaria*, and in a few species of *Pityostrobus*. The lateral masses of sclerenchyma cells forming wing-like structures occur in *Abies, Picea, Tsuga, Obirastrobus, Pseudoaraucaria*, and in some species of *Pityostrobus*.

21. Interseminal ridge. 0: absent; 1: present. An interseminal ridge is conspicuous in *Tsuga, Larix, Picea*, and *Pinus* among living genera [78]. Among the fossil taxa, *Pseudoaraucaria, Obirastrobus* and some species of *Pityostrobus* have a conspicuous interseminal ridge.

22, 23, 24, 25. Resin canals in scale base. 22: Occurrence of abaxial resin canals. 0: present; 1: absent. 23: Occurrence of adaxial resin canals. 0: present; 1: absent. 24: Number of adaxial resin canals. 0: several; 1: one. 25: Occurrence of abaxial resin canal in part opposite seed. 0: present; 1: absent. Among living genera, *Pinus* is characterized by having resin canals restricted to the abaxial side [61,78]. In *Abies, Tsuga, Larix*, and *Pseudotsuga*, resin canals are restricted to the adaxial side [61,78]. The resin canals are present in both abaxial and adaxial sides in *Cedrus, Keteleeria*, and *Pseudolarix*. Of these, only one resin canal occurs in the adaxial side in *Cedrus* and more than one canal occurs in *Keteleeria* and *Pseudolarix* [61,78]. *Picea* has an intermediate form with resin canals on both sides and an adaxially restricted type. In this case, the resin canals are restricted to the adaxial side at the base, but enter the abaxial side at the part opposite the seed [61,78]. In the fossil taxa, all of these types are *exhibited*.

Scale middle

26. Adaxial curvature of the scale trace. 0: absent; 1: present. The scale vascular trace of *Pinus* shows prominent adaxial curvature [61,78]. This character also occurs in some species of *Pityostrobus*.

27. Occurrence of sclerenchyma cells. 0: present; 1: absent. Sclerenchyma cells are present in the fundamental tissue of the middle part of the ovuliferous scale

in most taxa [78], but they do not occur in a few *Pityostrobus* species nor in *Pinus bergica*.

28, 29, 30. Resin canal distribution at middle of scale. 28: Abaxial canals. 0: present; 1: absent. 29: Adaxial canals. 0: present; 1: absent. 30: Interfascicular canals. 0: present; 1: absent. Resin canals usually occur in the abaxial and adaxial sides and interfascicular regions of the scale. However, *Pseudotsuga* have only an adaxial canal; *Larix*, *Tsuga*, and some *Picea* species lack the abaxial canal. Consequently, *Cedrus* and some *Pinus* species lack the adaxial canal [78]. The same tendency is found in the fossil taxa and some species lack resin canals of some part.

31. Inflated scale apex. 0: absent; 1: present. The scale apex is inflated in *Pinus* and tapered in other taxa [61,78]. In fossil taxa, *Pinus* and some species of *Pityostrobus* exhibit the inflated apex.

32. Papery thin apex. 0: absent; 1: present. The scale apex is papery thin in some species of *Picea* [61,78]. In fossil taxa, some *Picea* species and one *Pityostrobus* exhibit the papery thin apex.

33. Lacunae in scale tissue. 0: absent; 1: present. The lacunae characteristically appear in the scale tissue of *Picea* [61,78].

Seed

34. Resin cavity in seed coat. 0: absent; 1: present. This is characteristic of the subfamily Abietoideae [30,61], and is found in some species of *Pityostrobus* and *Pseudoaraucaria*.

References

1. Rothwell GW, Serbet R (1994) Lignophyte phylogeny and the evolution of spermatophytes. Syst Bot 19:443–482
2. Hart JA (1987) A cladistic analysis of conifers: preliminary reports. J Arnold Arb 68:269–307
3. Miller CN (1988) Origin of modern conifer families. In: Beck CB (ed) Origin and evolution of gymnosperms. Columbia University Press, New York, pp 448–486
4. Eckenwalder JE (1976) Re-evaluation of Cupressaceae and Taxodiaceae: a proposed merger. Madroño 23:237–300
5. Price RA, Olsen-Stojkovich J, Lowenstein JM (1987) Relationships among the genera of Pinaceae: an immunological comparison. Syst Bot 12:91–97
6. Price RA, Lowenstein JM (1989) An immunological comparison of the Sciadopityaceae, Taxodiaceae, and Cupressaceae. Syst Bot 14:141–149
7. Chase MW, Soltis DE, Olmstead RG, Margan D, Les DH, Mishler BD, Duvall MR, Price RA, Hill HG, Qiu Y, Kron KA, Rettig JH, Conti E, Palmer JH, Manhart JR, Sysma KJ, Michaels HJ, Kress WJ, Karol KG, Clark WD, Hedrén M, Gaut BS, Jansen RK, Kim K, Wimpee CF, Smith JF, Furnier GR, Strauss SH, Xiang Q, Plunkett GM, Soltis PS, Swensen SM, Williams SE, Gadek PA, Quinn CJ, Eguiarte LE, Golenberg E, Learn GH, Graham SW, Barrett SCH, Dayanandan S, Albert VA (1993) Phylogenetics of seed plants: an analysis of nucleotide sequences from the plastid gene *rbc*L. Ann Misouri Bot Gard 80:528–580

8. Brunsfeld SJ, Soltis PS, Soltis DE, Gadek PA, Quinn CJ, Strenge DD, Ranker TA (1994) Phylogenetic relationships among the genera of Taxodiaceae and Cupressaceae: evidence from *rbc*L sequences. Syst Bot 19:253–262

9. Gadeck PA, Quinn CJ (1993) A preliminary analysis of relationships within the Cupressaceae *sensu stricto* based on *rbc*L sequences. Ann Miss Bot Gard 80:581–586

10. Wang X, Szmidt AE (1993) Chloroplast DNA-based phylogeny of Asian *Pinus* species (Pinaceae). Pl Syst Evol 188:197–211

11. Smith DE, Klein AS (1994) Phylogenetic inferences on the relationship of North American and European *Picea* species based on nuclear ribosomal 18S sequences and the internal transcribed spacer 1 region. Molec Phyl Evol 3:17–26

12. Pilger R (1926) Coniferae. In: Engler A, Prantl K (eds) Die Naturlichen Pflanzenfamilien. 2nd edition. Wilhelm Engelmann, Leipzig, pp 121–166

13. Ohsawa T (1994) Anatomy and relationships of petrified seed cones of the Cupressaceae, Taxodiaceae, and Sciadopityaceae. J Plant Res 107:503–512

14. Ohsawa T, Nishida H, Nishida M (1992) Structure and affinities of the petrified plants from the Cretaceous of northern Japan and Saghalien, XI. A cupressoid seed cone from the Upper Cretaceous of Hokkaido. Bot Mag Tokyo 105:125–133

15. Stopes MC, Fujii K (1910) Studies on the structure and affinities of Cretaceous plants. Philos Trans R Soc London Ser B 210:1–90

16. Ohsawa T, Nishida M, Nishida H (1992) Structure and affinities of the petrified plants from the Cretaceous of northern Japan and Saghalien, X. Two *Sequoia*-like cones from the Upper Cretaceous of Hokkaido. J Jap Bot 67:72–82

17. Nishida M, Ohsawa T, Nishida H (1992) Structure and affinities of petrified plants from the Cretaceous of northern Japan and Saghalien VIII. *Parataiwania nihonghii* gen. et sp. nov. a taxodiaceous cone from the Upper Cretaceous of Hokkaido. J Jpn Bot 67:1–9

18. Nishida M, Nishida H, Ohsawa T (1991) Structure and affinities of petrified plants from the Cretaceous of northern Japan and Saghalien VI. *Yezosequoia shimanukii* gen. et sp. nov. a petrified taxodiaceous cone from Hokkaido. J Jpn Bot 66:280–291

19. Ohsawa T, Nishida H, Nishida M (1993) Structure and affinities of the petrified plants from the Cretaceous of northern Japan and Saghalien. XIII. *Yubaristrobus* gen. nov., A new taxodiaceous cone from the Upper Cretaceous of Hokkaido. J Plant Res 106:1–9

20. Saiki K, Kimura T (1993) Permineralized taxodiaceous seed cone from the Upper Cretaceous of Hokkaido, Japan. Rev Palaeobot Palynol 76:83–96

21. Page CN (1990) Pinatae. In: Kramer KU, Green PS (eds) Pteridophytes and gymnosperms. Springer, Berlin Heidelberg New York, pp 290–236

22. Hida M (1962) The systematic position of *Metasequoia*. Bot Mag Tokyo 73:316–323

23. Takaso T, Tomlinson PB (1990) Cone and ovule onogeny in *Taxodium* and *Glyptostrobus* (Taxodiaceae-Coniferales). Am J Bot 77:1209–1221

24. Van Tieghem P (1891) Structure et affinites des *Abies* et des generes les plus voisins. Bull Soc Bot France 38:406–415

25. Doyle JC (1945) Developmental lines in pollination mechanisms in Coniferales. Sci Proc R Dublin Soc 24:43–62

26. Jeffrey EC (1905) Comparative anatomy and phylogeny of conifers. Pt.2-Abietineae. Mem Boston Soc Nat Hist 6:1–37

27. Vierhapper F (1910) Enteurf eines neuen Systems der Coniferen. Abh KK Zool-Bot Ges Wien 5:1–56

28. Melchior H, Wedermann E (1954) A. Englers Syllabus der Pflanzen familien I. Allg. Teil. Bakterien bis Gymnosperm. Vol 12. Aufl, Berlin

29. Farjon A (1990) Pinaceae. Drawing and description of the genera *Abies*, *Cedrus*, *Pseudolarix*, *Keteleeria*, *Nothotsuga*, *Tsuga*, *Cathaya*, *Pseudotsuga*, *Larix* and *Picea*. Koeltz Sci Books, Königstein

30. Frankis MP (1989) Generic inter-relationships in Pinaceae. Notes R Bot Gard Edinburgh 45:527–548

31. Seward AC (1919) Fossil plants vol 4. Cambridge University Press, Cambridge

32. Stopes MC (1915) Catalogue of the Mesozoic plants in the British Museum. The Cretaceous flora. Part 2 Lower Greensand (Aptian) plants of Britain. British Museum, London

33. Alvin KL (1960) Further conifers of the Pinaceae from the Wealden Formation of Belgium. Mém Inst Roy Sci Nat Belg 146:1–39

34. Miller CN, Marinky CA (1986) Seed cones of *Pinus* from the Late Cretaceous of New Jersey. Rev Palaeobot Palynol 46:257–272

35. Miller CN (1969) *Pinus avonensis*, a new species of petrified cones from the Oligocene of western Montana. Am J Bot 56:972–978

36. Miller CN (1973) Silicified cones and vegetative remains of *Pinus* from the Eocene of British Colombia. Cont Univ Mich Mus Paleontol 24:101–118

37. Miller CN (1974) *Pinus wolfei*, a new petrified cone from the Eocene of Washington. Am J Bot 61:772–777

38. Miller CN (1978) *Pinus burtii*, a new species of petrified cones from the Miocene of Martha's Vineyard. Bul Torrey Bot Club 105:93–97

39. Underwood JC, Miller CN (1980) *Pinus buchananii* a new species based on a petrified cone from the Oligocene of Washington. Amer J Bot 67:1132–1135

40. Stockey RA (1983) *Pinus driftwoodensis* sp. n. from the Early Tertiary of British Colombia. Bot Gaz 144:148–156

41. Stockey RA (1984) Middle Eocene *Pinus* remains from British Colombia. Bot Gaz 145:262–274

42. Miller CN (1992) Structurally preserved cones of *Pinus* from the Neogene of Idaho and Oregon. Int J Plant Sci 153:147–154

43. Banks HP, Ortiz-Stomaor A, Hartmanb CM (1981) *Pinus escalantensis* sp. nov. a new permineralized cone from the Oligocene of British Colombia. Bot Gaz 142:286–293

44. Miller CN (1970) *Picea diettertiana*, a new species of petrified cones from the Oligocene of Western Montana. Am J Bot 57:579–585

45. Miller CN (1989) A new species of *Picea* based on silicified seed cones from the Oligocene of Washington. Am J Bot 76:749–754

46. Crabtree DR (1983) *Picea wolfei*, a new species of petrified cone from the Miocene of northwestern Nevada. Am J Bot 70:1356–1364

47. Dutt CP (1916) *Pityostrobus macrochephalus* L. and H. A Tertiary cone showing ovular structures. Ann Bot 30:529–549

48. Alvin KL (1953) Three Abietaceous cones from the Wealden of Belgium. Mém Inst R Sci Nat Belg 125:1–42

49. Crabtree DR, Miller CN (1989) *Pityostrobus makaensis*, a new species of silicified pinaceous seed cone from the Middle Tertiary of Washington. Am J Bot 76:176–184

50. Creber GT (1956) A new species of abietaceous cone from the Lower Greensand of the Isle of Wight. Ann Bot, NS 20:375–383

51. Creber GT (1960) On *Pityostrobus leckenbyi* (Carruthers) Seward and *Pityostrobus oblongus* (Lindley and Hutton) Seward fossil abietaceous cones from the Cretaceous. J Linn Soc (Bot) 56:421–429

52. Louvel C (1960) Contribution l'étude de *Pityostrobus oblongus* (Fliche sp.) appareil femelle d'un conifère Albien de l'Argonne. Mém Soc Géol Françias NS 90:1–26

53. Miller CN (1972) *Pityostrobus palmeri*, a new species of structurally preserved conifer cones from the Late Cretaceous of New Jersey. Am J Bot 59:352–358

54. Miller CN (1974) *Pityostrobus hallii*, a new species of structurally preserved conifer cones from the Late Cretaceous of Maryland. Am J Bot 61:798–804

55. Miller CN (1976) Two new pinaceous cone from the Early Cretaceous of California. J Paleontol 50:821–832

56. Miller CN (1977) *Pityostrobus lynni* (Berry) comb. nov., a pinaceous seed cone from the Palaeocene of Virginia. Bull Torrey Bot Club 104:5–9

57. Miller CN (1978) *Pityostrobus cliffwoodensis* (Berry) comb. nov., a pinaceous seed cone from the Late Cretaceous of New Jersey. Bot Gaz 139:284–287

58. Miller CN (1985) *Pityostrobus pubescens*, a new species of pinaceous cones from the Late Cretaceous of New Jersey. Am J Bot 76:133–142

59. Ohsawa T, Nishida H, Nishida M (1991) Structure and affinities of petrified plants from the Cretaceous of northern Japan and Saghalien VII. Petrified pinaceous cone from the Upper Cretaceous of Hokkaido. J Jpn Bot 66:356–368

60. Stockey RA (1981) *Pityostrobus mcmurrayensis* sp. nov., a permineralized pinaceous cone from the Cretaceous of Alberta. Can J Bot 59:75–82

61. Miller CN (1976) Early evolution in the Pinaceae. Rev Paleobot Palynl 21:101–117

62. Alvin KL (1957) On *Pseudoaraucaria* Fliche emend. a new genus of fossil pinaceous cones. Ann Bot 21:33–51

63. Ohsawa T, Nishida M, Nishida H (1992) Structure and affinities of the petrified plants from the Cretaceous of northern Japan and Saghalien. XII. *Obirastrobus* gen. nov., petrified pinaceous cones from the Upper Cretaceous of Hokkaido. Bot Mag Tokyo 105:461–484

64. Alvin KL (1957) On the two cones *Pseudoaraucaria heeri* (Coemans) nov. comb. and *Pityostrobus villerotenensis* nov. sp. from the Wealden of Belgium. Mém Inst Roy Sci Nat Belg 135:1–27

65. Miller CN, Robison CR (1975) Two new species of structural preserved pinaceous cones from the Late Cretaceous of Martha's Vineyard Island Massachusetts. J Paleontol 49:138–150

66. Alvin KL (1988) On a new specimen of *Pseudoaraucaria major* Fliche (Pinaceae) from the Cretaceous of Isle of Wight. Bot J Linn Soc 97:159–170

67. Swofford DL (1993) PAUP, phylogenetic analysis using parsimony. Version 3.1. Computer program distributed by the Illinois Natural History Survey. Champaign, Illinois, pp 257

68. Greguss P (1972) Xylotomy of the living conifers. Akad Kiado, Budapest

69. Florin R (1963) The distribution of conifer and taxad genera in time and space. Acta Horti Bergiani 20:121–312

70. Creber GT (1967) Notes on some petrified cones of the Pinaceae from the Cretaceous. Linn Soc Lond Proc 178:147–152

71. Miller CN (1977) Mesozoic conifers. Bot Gaz 43:217–280

72. Stockey RA (1981) Some comments on the origin and evolution of conifers. Can J Bot 59:1932–1940

73. Florin R (1951) Evolution of cordaites and conifers. Acta Horti Bergiani 17:259–388
74. Hart JA, Price RA (1990) The genera of Cupressaceae (including Taxodiaceae) in the southeastern United States. J Arnold Arbor 71:275–322
75. LaPasha CA, Miller CN (1981) New taxodiaceous cones from the Upper Cretaceous of New Jersey. Am J Bot 68:1374–1382
76. Lemoine-Sebastian C (1968) La vascularisation du complexe bractée-écaille chez le Taxodiacées. Trav Lab Forest Toulouse 7:1–22
77. Ogura Y (1930) On the structure and affinities of some Cretaceous plants from Hokkaido. J Fac Sci Imp Univ Tokyo Sect III (Bot) 2:381–412
78. Radais M (1894) Contribution l'étude de l'anatomie comparée du fruit des conifères. Ann Sci Nat Bot Sér7 19:165–368
79. Satake Y (1934) On the systematic importance of the Japanese Taxodiaceae. Bot Mag Tokyo 48:186–205
80. Aase HC (1915) Vascular anatomy of the megasporophyllus of conifers. Bot Gaz 60:277–313
81. Miller CN (1975) Petrified cones and needle-bearing twings of a new taxodiaceous conifer from the Early Cretaceous of California. Am J Bot 62:706–713
82. Miller CN, Crabtree DR (1989) A new taxodiaceous seed cone from the Oligocene of Washington. Am J Bot 76:133–142
83. Hirmer M (1936) Die Blüten der Coniferen. I. Entwicklungsgeschichte und vergleichende Morphologie des weiblichen Blütenzapfens der Coniferen. Bibliot Bot 114:1–100
84. Lemoine-Sebastian C (1969) La vascularisation du complexe bractée-écaille dans le cône femelle des Cupressacées. Botanica Rhedonica, Sér A 7:3–27
85. Lemoine-Sebastian C (1972) Étude comparative de la vascularisation et du complexe séminal chez les Cupressacées. Phytomorphology 22:246–260
86. Liu T, Su H (1983) Biosystematic studies on *Taiwania* and numerical evaluations on the systematics of Taxodiaceae. Taiwan Museum Special Publication Series. The Taiwan Museum, Taipei, pp 1–113
87. Hida M (1957) The comparative study of Taxodiaceae from the standpoint of development of the cone scale. Bot Mag Tokyo 70:44–51
88. Takaso T, Tomlinson PB (1989) Aspects of cone and ovule ontogeny in *Cryptomeria* (Taxodiaceae). Am J Bot 76:692–705
89. Hayata B (1907) On *Taiwania* and its affinity to other genera. Bot Mag Tokyo 21:21–27

Origin and Diversification of Angiosperms

Evolutionary Biology of Flowers: Prospects for the Next Century

PETER K. ENDRESS

5.1 Introduction

There are perhaps no other things as deeply ingrained in the needs of mankind as flowers. The Romans summed it up two millennia ago with their slogan "panem et circenses" or "food and pleasure." Our main sources of food are flowers in the form of their final stage, the fruits and seeds, such as wheat, rice, or corn, but flowers are also an unending source of pleasure with their exciting forms, colors and scents in an often ephemeral display. The manifold forms of flowers and their various interactions with animals are also the epitome of biodiversity on our planet. The short lifespan of flowers and their vanishing diversity strike emotional chords. Both the food and pleasure aspects of flowers are heavily and increasingly involved in industrial activities including biotechnology. On all these grounds the development, and, in a broader context, the evolution of flowers in their diversity and complexity will inevitably remain one of the central fields of biology.

5.2 Approaches Toward the Understanding of Flower Evolution

In the past decade, very different fields have contributed to the understanding of flower evolution: (1) comparative development and functional morphology of a diversity of flowers (floral organization and architecture) and (2) studies on floral biology (floral mode) from population to higher levels continued to be the backbone of studies in angiosperm evolution. In addition, (3) a wealth of uncovered fossil flowers greatly enhanced the interpretation of early diversification of the angiosperms. With the advent of new molecular techniques two new fields opened up in an almost explosive development: (4) molecular developmental genetics of flowers, and (5) molecular systematics of flowering plants. (6) A

Institute of Systematic Botany, University of Zürich, Zollikerstrasse 107, 8008 Zürich Switzerland

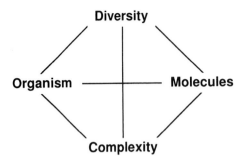

FIG. 5.1. Major dimensions of life

refinement of analytical techniques, such as the further development of cladistics, went hand in hand with these new developments, which, in turn, gave new impulses to the classical fields (1) and (2).

The elucidation of the phylogeny of life on earth is an unending task. At present, the enlarged methodology is able, more than ever before, to enhance this task by rapidly moving fronts and by continuous reciprocal illumination. All these fields cover different parts of the spectrum of evolutionary biology, between diversity and complexity as well as molecules and organisms (Fig. 5.1).

5.3 Floral Diversity—the Colorful Phylogenetic Tree

With the new electronic techniques to study large databases the generation of explicit phylogenetic hypotheses has become important in systematics. This is expressed with the fashionable catchword "tree-thinking" [1,2]. However, the comparison should also be made in a fuller sense of the tree as an organism: it should not only be analyzed in its branching system but in its entirety with leaves, flowers and fruits. The branching system of a tree is only the supporting structure of its shoots where the colorful scenes of life are displayed. Of the two facets of phylogeny, anagenesis and cladogenesis, cladogenesis often receives more attention. However, both are complementary aspects of diversification and should be given equal consideration.

Comparative studies should be conducted with a fully "biological," i.e., circumspect, scope: large scale, using larger monophyletic groups, such as genera, families, or even orders, or parts of larger genera. Such studies may be extremely difficult if the species of a group are scattered over the world, or if they grow at remote places that are not easily accessible. There are exemplary studies in comparative floral biology on *Adansonia* (Bombacaceae) [3], *Dalechampia* (Euphorbiaceae) [4,5], *Lapeirousia* (Iridaceae) [6,7], and *Parkia* (Mimosaceae) [8]. Comparative floral developmental studies were done on legumes and on paleoherbs [9,10], on *Loasaceae* [11], and on *Besseya* (Scrophulariaceae) [12]. Such studies are also invaluable from the point of view of conservation, especially for endangered groups such as the cited example *Adansonia* [3].

It is no less important to focus on a particular biological phenomenon and to do large-scale comparative research. Examples are the discovery of and comparative work on oil flowers [13], the extensive comparative studies of floral ultraviolet reflection conducted in more than 8000 angiosperm species [14–16], the comparative studies on floral color change during anthesis [17], on secondary pollen presentation [18], and the developmental studies of patterns of sympetaly [19] and of androecia [20,21].

Particularly fascinating are studies of coevolution due to mutualism of plants and pollinating insects in a coherent phylogenetic context beyond the limits of species, such as in the genera *Yucca* and *Trollius* [22–28]. Attempts to collate and synthesize a large database on floral features over a broad range of angiosperm diversity will be of great help for many kinds of studies in angiosperm evolution [29,30].

5.4 Origin and Early Evolution

One of the exciting recent results of paleobotany is that small flowers with few organs were most likely the starting point of angiosperm flower evolution and diversification [31–35]. There is renewed discussion as to whether large flowers evolved by congestion of several small flowers [36,37], or by an increase in organ number per flower during a phase of flexibility of organ numbers caused by the lack of synorganization constraints [38,39]. As long as there is no counterevidence from developmental studies of extant critical groups, such as paleoherbs [10], the second scenario seems more parsimonious. However, this problem should be addressed anew using more technical methods, e.g., those employed by molecular developmental geneticists. Further, studies on fossil and extant basal angiosperms make it probable that early flowers were perianthless or had a simple perianth, floral bud protection was by subtending bracts and stamen connectives, pollinator attraction was exerted by stamens, and floral rewards included stigmatic and other secretions; the stamens had a thick connective, often associated with valvate dehiscence; flowers were protogynous (review in [39,39]).

Although there are no unequivocal angiosperm fossils older than from the earliest Cretaceous [33], the most likely scenario from an overall paleobotanical evaluation of seed plants is that angiophytes (the stem group of the angiosperms) originated as early as in the late Triassic [40]. All the major insect groups that are considered to be early flower pollinators [41,42] were already present by the late Triassic, such as Coleoptera, Diptera, and Thysanoptera [43], or early Jurassic, such as Lepidoptera [22]. It will be exciting to search for Jurassic angiosperm fossils and to further study some still enigmatic Jurassic finds [44].

Thus, the details of the origin of the flower are still a great mystery. What was the sequence of innovations that eventually led to conventional flowers? When did these steps happen? What were the evolutionary events between the origin of ovules/seeds and the origin of carpels/fruits? [41]. Still too many pieces are missing. It is not to be expected that a nice chain of evolutionary events may be reconstructed by available fossils.

The extraordinary diversity of the angiosperms has led to the question of why and when their diversification rate was elevated (as compared with other seed plants). The advent of flowers seems to have played a key role in this respect. Among the various hypotheses discussed, features of flowers are prominent, such as (1) angiospermy and concomitant pollen tube selection [45]; (2) rapid development (also of flowers and seeds) [46–48]; (3) insect pollination [22,49–51]; and (4) varied dispersal modes (as a consequence of angiospermy) [49,52,53]. These questions have to be readdressed continuously in the light of new evidence, with the availability of new methods, and with broad comparative studies [22,48].

5.5 Booming Fields in the Study of Flowers—the Paradox of the Ovules

Of the fields mentioned above that contribute to our understanding of flower evolution, molecular developmental genetics of flowers and molecular systematics of flowering plants have been especially prominent in the last few years. Landmark studies such as the establishment of a cascade of regulatory genes in flower development that are homologous in *Arabidopsis* and *Antirrhinum* [54] and the breakthrough of molecular systematics at the macroevolutionary level [55] have greatly stimulated the study of angiosperm phylogeny and flower evolution. Both these molecular fields have much funding support and have exploded also in terms of positions for researchers [56–58].

After the establishment and elaboration of the "ABC model" of floral development [54,59–61], which encompasses the four basic organ categories that are commonly present in the eudicots (sepals, petals, stamens, carpels), molecular developmental genetics has concentrated on the development of the ovules. The ovules are the last floral parts to be formed in flower development, and they are only subunits of the carpels, which are the last formed floral organs. The paradoxical situation is that ovules are evolutionarily much older than angiosperms and even older than flowers, if flowers are seen as constitutional parts of the anthophytes (the group including angiosperms, Bennettitales, Gnetales, Caytoniales, and Pentoxylales) [62]. It may therefore be expected that ovules are more difficult to elucidate in the evolution of their developmental mechanisms than the other floral parts. Yet ovules are crucial floral parts since they become the seeds. One direction of research would be tracing genes that play a role in flower and ovule formation in a variety of angiosperms and in seed plants other than angiosperms [63–65].

The comparative investigation of ovules has mainly been carried out with microtome sections in embryological studies, which led to a large body of knowledge during the 20th century. Only recently has the three-dimensional structure come into focus with scanning electron microscopy (SEM) studies of ovules of some basal angiosperm groups [66]. Ovule development has also been described in detail in model species, such as *Arabidopsis* [67,68]. *Arabidopsis*, however, has tenuinucellar, campylotropous ovules with long funiculi, which do not represent the evolutionarily basal state in angiosperms. Attempts to interpret the evolutionary history of ovules with the variety observed in *Arabidopsis* ovules [69] are therefore not satisfactory. Complementary studies of crassinucellar ovules would be useful. However, molecular genetic studies are beginning to give insight into developmental idiosyncrasies of ovules in model plants [70]. Colombo et al. [71] found a new MADS box gene in *Petunia* that produces ovule-like structures ectopically on sepals and petals when overexpressed, while in *Arabidopsis* formation of ovules outside the carpels has hardly been observed [72]. However, distinctive parts of the ovule may be suppressed or replaced by other differentiations in mutants. *Bell* mutants of *Arabidopsis* suppress the formation of two integuments but lead to a single somewhat amorphous integument-like structure, and the ovule remains orthotropous, while strong expressions of *bell* produce distorted carpel-like structures instead of an integument-like structure [73,74]. In strong *ant* mutants integuments are lacking [75,76]. In *sup* mutants the inner integument becomes hood-like, and not cylindrical [77].

5.6 Key Innovations in Flower Evolution

Key innovations are novel characters that, once acquired, do not disappear again but become stabilized in the clade where they originated. They are successful traits and lead to evolutionary radiations. The result is that they become characteristics of large clades in the course of evolution (see, e.g., [48,78–82]). They can

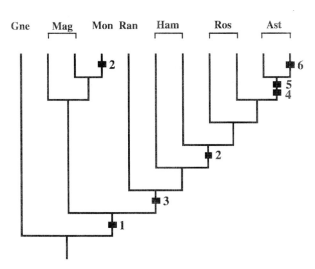

Fig. 5.2. Key innovations in flower evolution (simplified cladogram of angiosperms from topology of 2B tree in Chase et al. [55], and Nickrent and Soltis [83]). *Gne*, gnetophytes; *Mag*, magnoliids; *Mon*, monocots; *Ran*, ranunculids; *Ham*, hamamelidids; *Ros*, rosids; *Ast*, asterids

1 **postgenital fusion, carpels**
2 **syncarpy**
3 **triaperturate pollen**
4 **tenuinucellar, unitegmic ovules**
5 **sympetaly**
6 **fusion of stamens and petals**

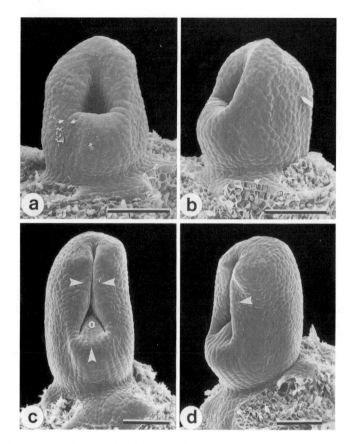

FIG. 5.3a–d. *Laurus nobilis* L. (Endress 2655 and 2669) (Lauraceae). Development of carpel closure. **a, b** young open carpel. **a** Ventral view, **b** Side view. **c, d** Carpel with the lateral flanks and the cross-zone (from below) closing (*arrows*). **c** Ventral view. **b** Side view. *o*, ovule; *scale bars* = 0.1 mm

only be recognized a posteriori, i.e., when they have become a constitutive feature of a large, i.e., evolutionarily successful, group. Angiosperm flowers are extremely diverse. But although number and kind of evolutionary changes are immense, there are relatively few key innovations [38].

The following are key innovations in flowers, as it appears from character plotting on cladograms of major angiosperm clades (Fig. 5.2).

(1) postgenital fusion (carpel) (all angiosperms) (Fig. 5.3)
(2) syncarpy (eudicots and monocots) (Fig. 5.4)
(3) triaperturate pollen (eudicots)
(4) tenuinucellar, unitegmic ovules (asterids)
(5) sympetaly (asterids) (Fig. 5.5)
(6) Fusion of stamens and petals (higher asterids)

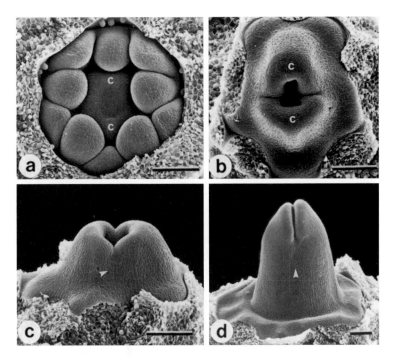

FIG. 5.4a–d. *Campsis × tagliabuana* Rehder (Endress 7884) (Bignoniaceae). Development of syncarpy. **a** Two young carpels forming shallow mound (*c*, carpel). **b, c** Slightly older gynoecium with congenitally united base (*arrowhead*). **b** From above, **c** from the side. **d** Older gynoecium, the congenitally united base elongated by intercalary growth (*arrowhead*). *Scale bars* = 0.1 mm

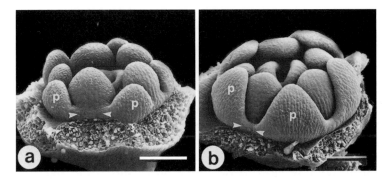

FIG. 5.5. *Penstemon parryi* Gray (Endress 9624) (Scrophulariaceae). Development of sympetaly. **a** Floral bud from the side with beginning sympetaly (*arrowheads*). **b** Slightly older floral bud from the side showing sympetalous base of corolla (*arrowheads*). *p*, petals; *scale bars* = 0.1 mm

5.7 The Evolutionary Biology of Postgenital Fusion— First Key Innovation in Angiosperm Flowers

The most comprehensive key innovation in angiosperm phylogeny was the advent of carpels in which the ovules are enclosed. Closely associated with the evolutionary origin of carpels is the ontogenetic process of postgenital fusion of the carpellary flanks (Fig. 5.3). Postgenital fusion completely secludes the ovules from the outside and, in addition, forms the site of pollen tube growth and guidance from the stigma to the ovary. In some angiosperms seclusion may also come about by a secretion instead of postgenital fusion of the carpel surface. Seclusion as such of originally exposed parts is not a novelty in angiosperms. It also occurs by thickening growth in the integument of Gnetales [84] and other gymnosperms [85] and cone scales of some conifers (Tomlinson, pers. comm.) or folding of cupular walls [86]. Even transient loose intermeshing of contiguous organ surfaces occurs in sporophylls of *Equisetum* [87]. However, the real innovation is postgenital fusion with tight adhesion of epidermal cells of two contiguous surfaces, which seems to be restricted to angiosperms. Most probably the evolutionary advent of postgenital fusion and carpels went hand in hand. The complete and early postgenital fusion of carpels also led to the novelty of inner morphological surfaces, which is otherwise restricted to animals where it arises by other mechanisms. Congenital fusion, which also plays an important role in flowers, is much less revolutionary. It also occurs in lower plants, e.g., in the vegetative sheathing structures of *Equisetum* and *Gnetum*, and it does not produce inner morphological surfaces.

The site of postgenital fusion and pollen tube guidance is a crucial meeting point of different fields in botany, not only because it was a most successful evolutionary step as the first key innovation in angiosperms, but also because the experimental manipulation of pollen tubes is of interest from a practical point of view in plant breeding. Postgenital fusion constitutes a fascinating meeting point of different fields as will be explained in the following sections.

5.7.1 Structure and Development of Postgenital Fusion and Pollen Tube Transmitting Tissue

Postgenital fusion may begin at different times of carpel ontogeny. If it occurs early, fusion by intermeshing accompanied by periclinal divisions of the epidermal cells of the contiguous carpel flanks is so tight that at anthesis the former epidermal surfaces are no longer apparent. The cuticles may also have vanished. In postgenitally fusing apices of the two carpels in Apocynaceae [88] even plasmodesmata form de novo between the fusing surfaces of the two carpels [89]. Circumstantial observations on different groups indicate that the carpel flanks may not close simultaneously throughout their length. The parts that close late may exhibit unordered growth that gives the impression of stuffing holes.

Parts of the pollen tube guiding surface in the style and especially in the ovary may be unfused. While the fusing zones are restricted to the epidermis, the pollen tube guiding zone may, in addition, encompass several cell layers below the epidermis, especially in plant groups with small inner carpel surfaces and with numerous ovules, thus forming a massive transmitting tissue (for diversity, see below). Plasmodesmata are commonly found in the transverse walls of the pollen tube transmitting tissue; they have now also been found in longitudinal walls in *Tibouchina* with a massive transmitting tissue [90].

5.7.2 Diversity of Pollen Tube Transmitting Tracts

While stigma diversity has found much attention (e.g. [91,92], the diversity of pollen tube transmitting tracts has less often been addressed. A rough classification may encompass three different kinds of tract differentiation: (1) pollen tube transmission at the epidermal surface; (2) pollen tube transmission in massive transmitting tissue (below epidermal surface, see above); and (3) pollen tube transmission free in extensive mucilage that may fill the entire stylar canal and ovarial cavity (outside the epidermal surface) [38]. The third kind is less common than the first two and has not been studied in detail as to ultrastructure, physiology and chemistry; it occurs, e.g., in some paleoherbs and aroids [39,93] and may be associated with unusual gynoecium architectures exhibiting an extragynoecial compitum [94]. Since in this third kind the transmitting medium is secreted by the carpel surface, the relationship to the epidermis is still present. However, a few strange cases are known where pollen tube transmission is not related to the morphological carpel surface. A few groups have carpels without an inner morphological surface in the style, e.g., *Alchemilla* [95], but still with internal pollen tube transmission. In some cleistogamous flowers pollen tubes grow inside the tissue through the floral base or even through the inflorescence axis [38]. Rare as they may be, these cases show that the potential for pollen tube transmission is not completely related to active organ surfaces.

5.7.3 Biochemistry and Biophysics of Pollen Tube Guidance

It has long been known that the fusing surfaces of the carpel and the pollen tube guiding zones down to the embryo sac are secretory [96] and that arabinogalactan-proteins are a major matrix substance for pollen tube growth in stigma and style. In *Nicotiana* arabinogalactan-proteins have now also been found at the ovarial (placental) guiding surfaces between style and ovules where postgenital fusion is absent [97]. Arabinogalactan-proteins are high-molecular-weight protoglycans with a high water-holding capacity that seem to contribute to the physical characteristics of the viscous matrix that provides support for the pollen tubes [97]. It was also shown that transmitting tissue-specific (TTS) proteins from the style (a particular kind of arabinogalactan-proteins) attract and stimulate pollen tube growth in vitro [98]. In turn, pollination affects quantity and quality of TTS proteins produced by the styles [99]. Pollen tube growth in the carpel can be seen as a kind of cell movement as it occurs in animals but not otherwise in plants [100,101]. In *Lilium* with pollen tube growth at the epidermal surface pollen tubes grow beneath the epidermal cuticle; a fibrous wall material covers the tip of the pollen tube cell wall and the surface of the epidermal cells where the two adhere [102]. How are pollen tubes directed to the ovules? One of the mechanisms postulated long ago is chemotropic attraction from the ovules [103]. This was corroborated by Hülskamp et al. [104], who could show in four female-sterile mutants of *Arabidopsis* that the less the ovules were developed, the less pollen tubes on the placental surface exhibited directed growth toward the ovules. A similar effect was also found in female-sterile plants of an *Oenothera* mutant [105].

5.7.4 Genetics of Postgenital Fusion and Functioning of Pollen Tube Transmitting Surface

Lolle et al. [106] found a mutant (*fiddlehead*) in *Arabidopsis* that produces postgenital fusion not only in the gynoecium but among different organs of the shoot. Organ proximity is of course a precondition. Thus, it seems that postgenital fusion is a genetically distinct program superimposed on other developmental events. Lolle and Cheung [107] also showed that pollen grains germinate on the surfaces of a broad range of shoot organs of this mutant. However, during development organs of these mutants first became fusion competent and only later allowed pollen tube growth, while mature organs could not fuse any longer but still allowed pollen tube growth. The process of postgenital fusion is now thought to be governed by at least four genes [108]. Likewise, Alvarez et al. [109] defined a gene (*spatula*) in *Arabidopsis*, which, in deficient form, hinders carpel fusion and precludes differentiation of pollen tube transmitting tissue. Pollen tube growth was also observed on the epidermis of young floral organs other than carpels [110]. These finds point to an intimate relationship but not complete congruence between postgenital fusion and pollen tube transmitting surface from a genetical point of view.

5.7.5 Self-incompatibility and Pollen Tube Competition in the Transmitting Tract

The evolutionary origin of a specialized pollen tube transmitting tract in the angiosperms was most successful, because it provided a filter to allow screening for pollen tube quality. This has long been known for the phenomenon of self-incompatibility, where male gametophytes that are genetically too similar to the carpel are not allowed access to the female gametophytes; they are arrested somewhere between the stigma and the egg cell. A vast literature has been assembled on the biology of the different types of self-incompatibility [111–113].

Although it is still unclear whether self-incompatibility or self-compatibility was ancestral in angiosperms, Weller et al. [114] in mapping the distribution of the different types on different phylogenetic scenarios found slightly more evidence in favor of self-compatibility. From this conclusion it would follow that self-incompatibility had multiple origins arising from self-compatibility in angiosperms (see also [115]). It would be worthwhile to trace these origins on refined phylogenetic hypotheses of larger angiosperm clades. Lloyd and Wells [41] hypothesized that the site of the incompatibility reaction was in the nucellus in early angiosperms (as, e.g., in Winteraceae). From there it was protracted to the style and stigma. This would also imply that the screening activity of the carpel in terms of incompatibility did not occur at the beginning of angiosperm evolution but was a somewhat later innovation. Another, more general way of male gametophyte selection, is by pollen tube competition among a population of pollen tubes growing simultaneously through the carpel. The majority of genes expressed in the sporophyte are also expressed in the microgametophyte and are, therefore, also subjected to natural selection while growing through the transmission tract [116]. It has been shown that the most rapidly growing pollen tubes lead to fertilization and they also produce the most vigorous offspring [116,117]. This may have been one of the most significant innovative traits in early angiosperm evolution [45]. Walsh and Charlesworth [118] point to difficulties in the assessment of the occurrence and the effects of pollen tube competition.

5.8 Sympetaly and the Evolutionary Establishment of Key Innovations

Sympetaly, congenital fusion of petals into a coherent corolla, is a prominent example of synorganization of originally independent organs. In this case synorganization comes about developmentally by a basal circular meristem that effects intercalary growth (Fig. 5.5). Sympetaly is a prominent feature of the Asteridae. Its dominance implies that it is a key innovation for this group. Sympetaly enables great evolutionary plasticity in floral architecture, while, conversely, variation in organ number becomes restricted [119,120]. The intercalary meristems of the corollas can bring about floral tubes of various lengths and widths that are extensively explored in adaptations for different pollinator groups. Evolutionary changes of the floral architecture and pollination mode are easily achieved by subtle differential growth of the intercalary meristems (review by Endress [121]).

Comparative developmental studies by Erbar (1991) on a wide range of asterids showed two kinds of sympetaly: either the petals appeared as distinct primordia and the fused part appeared only later ("late sympetaly") or the fused part appeared first as a ring meristem on which the individual petals appeared later ("early sympetaly"). Erbar [19] showed that these two kinds of sympetaly to some extent coincide with two major groups of asterids that now also appear in molecular systematic works. The asterid I group in Chase et al. [55] is largely characterized by late sympetaly, and the asterid II group by early sympetaly (see also Olmstead et al. [122]). In addition to the question of the systematic distribution of these traits, it is worthwhile to ask about correlations with other floral traits. There is a tendency of early sympetaly to be associated with inferior ovaries and with reduced calyces. Both lines of inquiry, the systematical and the structural, should be integrated into an evolutionary framework of approach.

In contrast to the evolutionary advent of carpels, for which there is no fossil record for reconstruction, the advent of sympetaly can be evaluated in some detail. This has not been done as yet, but the progressively more detailed molecular systematic research of large clades of the asterids provides the basis for pursuing this question [122,123–126]. A first look indicates, as may be expected for a key innovation, that sympetaly did not arise in fully fledged shape but started with trials. Key innovations start unnoticed as minor changes and become prominent and rooted in the organization much later when they are established in the genetic structure of the group, enhanced by their evolutionary success. The details of the process are encompassed by the fields of population genetics, developmental genetics, ecology, and systematics. In the basal clades of asterids (asterid III group in Chase et al. [55]) sympetaly is in general less constant than in the higher asterids. It either occurs only in parts of a family (e.g., Loasaceae [127]; Pittosporaceae, Cornaceae [128]), or there are apparent reversals to choripetaly (e.g., some Ericaceae [123]; Empetraceae [124]). It may also be weakly expressed or be difficult to recognize in mature stages (e.g., *Aralia*, *Hedera*, *Pittosporum* [129,130]). Sympetaly also occurs sporadically outside the asterids but has nowhere else led to a great diversity. It may be hypothesized that sympetaly is not genetically deeply rooted in the floral organization in the lower asterids so that it could easily disappear or come back again. In the higher asterids it is genetically more stabilized so that it is more difficult for choripetaly to return. Such an exceptional example is (partial) secondary choripetaly in *Besseya* (Scrophulariaceae), which most likely arose via a phase with high corolla reduction, as documented in detail by Hufford [12].

The more a favorable feature becomes genetically fixed, the less it can be discarded again. The stabilized feature becomes a genetic burden [79] or epigenetic trap [131]. But this burden is more than compensated for by enhanced potentials of evolutionary plasticity at a new level (cf. [120]).

The systematic structure of the asterids is in accordance with this general hypothesis. The most successful groups in the asterids, as measured in species richness, such as Asterales, Scrophulariales, Gentianales, and Rubiales, are all groups with stabilized sympetaly. It is also apparent that sympetaly allows a broad array of floral architectures and pollination biological modes, arising especially by their tubular base, including a high diversity of monosymmetric flowers [38]. In fact, some choripetalous groups have only now been suggested to belong to the asterids [55]. Likewise, only recently some groups with polyandrous flowers have been recognized as representatives of basal asterids. Polyandry hardly occurs any longer in those asterids with well-established sympetaly. The elaborated floral architectures of sympetalous flowers leave little room for polyandry; synorganization between stamens of a fixed low number has become a foremost trait. Such examples of basal asterids with partly or predominantly polyandrous flowers are Araliaceae, Hydrangeaceae, Loasaceae, Fouquieriaceae, and Lecythidaceae [11,55,122,125,128]. An expression of the great evolutionary plasticity in sympetalous asterids is the potential of rapid evolutionary changes at the population and species level from the genetic diversity available at that level. Sensible changes may occur in one or few generations when there are shifts in the frequency of certain pollinators. This was shown, e.g., for *Polemonium viscosum* (Polemoniaceae) [132] and various *Mimulus* species (Scrophulariaceae) [133–135].

5.9 Looking into the Future

The science of flowers was born two centuries ago with two pioneering works by J.W. von Goethe [136] on comparative morphology and by C.K. Sprengel [137] on comparative biology, both of which became highly influential for the developments up to the present. Biology today is characterized by a rapidly increasing sophistication of specialized and narrower approaches. This tendency requires a counterbalance by endeavors that try to transcend conventional borderlines and synthesize aspects of more than one field. Joint efforts between different fields become ever more important. In the era of e-mail and limited resources collaboration is made easier and is increasingly a necessity.

Such combined efforts may be seen in the elaboration of phylogenetic methods not just by an improvement of the algorithms but with the inclusion of biological aspects in the analysis and combination of classical and molecular data sets [138–144]. Reproductive ecology and evolution, not only at population level but also at higher levels, are increasingly seen in concert [22,51,145,146]. There is a wealth of information in comparative developmental studies on flowers, old and new, that should serve as a source of inspiration for molecular developmental geneticists. Conversely, it is surprising how little is known about the systematics and phylogeny of model organisms, such as *Arabidopsis* [147,148]. It is therefore important to bring model organisms into evolutionary perspective [58,59,149,150].

In summary, what we need is to further interdisciplinary cooperation to identify new hot-spot areas where research from different fields can be concentrated; to keep a balance between the molecular and organismic and between the diversity and complexity sides of our approach; more marriages between macro- and microevolutionary studies, between molecular and classical phylogenetic studies, and between paleo- and neobotanical studies; evaluation of floral evolution in concert with molecular macrosystematic studies that consider more taxa and more genes in particular groups; and expansion of molecular developmental genetics of flowers to more basal angiosperms.

The diversity and complexity of life are immense and even with all our optimism the most we can hope to achieve in the next century will only be a small move into the unknown world. There will not be many great breakthroughs (in the same way as there are not many key innovations in evolution!), but the most promising way to achieve them is to continuously make connections between the various lines of research. The threat to diversity is epitomized in flowers where plants and pollinating animals meet and extinctions therefore may have cumulative effects [151–153].

I will conclude with a quote from Wilson [154]: "The best of science doesn't consist of mathematical models and experiments, as textbooks make it seem. Those come later. It springs fresh from a more primitive mode of thought, wherein the hunter's mind weaves ideas from old facts and fresh metaphors and the scrambled crazy images of things recently seen. To move forward is to concoct new patterns of thought, which in turn dictate the design of the models and experiments. Easy to say. Difficult to achieve."

References

1. O'Hara RJ (1988) Homage to Clio, or, toward an historical philosophy for evolutionary biology. Syst Zool 37:142–155
2. Donoghue MJ, Sanderson MJ (1992) The suitability of molecular and morphological evidence in reconstructing plant phylogeny. In: Soltis PS, Soltis DE, Doyle JJ (eds) Molecular systematics of plants. Chapman & Hall, New York, pp 340–368
3. Baum DA (1995) The comparative pollination and floral biology of baobabs (*Adansonia*-Bombacaceae). Ann Missouri Bot Gard 82:322–348
4. Armbruster WS (1994) Evolution of plant pollination systems: Hypotheses and tests with the neotropical vine *Dalechampia*. Evolution 47:1480–1505
5. Armbruster WS (1996) Evolution of floral morphology and function: an integrative approach to adaptation, constraint, and compromise in *Dalechampia* (Euphorbiaceae). In: Lloyd DG, Barrett SCH (eds) Floral biology. Studies on floral evolution in animal-pollinated plants. Chapman & Hall, New York, pp 241–272
6. Goldblatt P, Manning JC, Bernhardt P (1995) Pollination biology of *Lapeirousia* subgenus *Lapeirousia* (Iridaceae) in southern Africa; floral divergence and adaptation for long-tongued fly pollination. Ann Missouri Bot Gard 82:517–534
7. Manning JC, Goldblatt P (1996) The *Prosoeca peringueyi* (Diptera: Nemestrinidae) pollination guild in southern Africa: long-tongued flies and their tubular flowers. Ann Missouri Bot Gard 83:67–86
8. Luckow M, Hopkins HCF (1995) A cladistic analysis of *Parkia* (Leguminosae: Mimosoideae). Am J Bot 82:1300–1320
9. Tucker SC, Douglas AW (1994) Ontogenetic evidence and phylogenetic relationships among basal taxa of legumes. In: Ferguson IK, Tucker SC (eds) Advances in legume systematics 6: structural botany. Royal Botanic Gardens, Kew, pp 11–32
10. Tucker SC, Douglas AW (1996) Floral structure, development, and relationships of paleoherbs: *Saruma, Cabomba, Lactoris*, and selected Piperales. In: Taylor DW, Hickey LJ (eds) Flowering plant origin, evolution and phylogeny. Chapman & Hall, New York, pp 141–175
11. Hufford L (1990) Androecial development and the problem of monophyly of Loasaceae. Can J Bot 68: 402–419
12. Hufford L (1995) Patterns of ontogenetic evolution in perianth diversification of *Besseya* (Scrophulariaceae). Am J Bot 82:655–680

13. Vogel S (1989) Fettes Oel als Lockmittel. Erforschung der ölbietenden Blumen und ihrer Bestäuber. Akademie der Wissenschaften und der Literatur 1949–1989, 113–130. Steiner, Stuttgart

14. Biedinger N, Barthlott W (1993) Untersuchungen zur Ultraviolettreflexion von Angiospermenblüten I. Monocotyledoneae. Trop Subtrop Pflanzenwelt 86:1–122

15. Burr B, Barthlott W (1993) Untersuchungen zur Ultraviolettreflexion von Angiospermenblüten II. Magnoliidae, Ranunculidae, Hamamelididae, Caryophyllidae, Rosidae. Trop Subtrop Pflanzenwelt 87:1–193

16. Burr B, Rosen D, Barthlott W (1995) Untersuchungen zur Ultraviolettreflexion von Angiospermenblüten III. Dilleniidae und Asteridae *s.l.* Trop Subtrop Pflanzenwelt 93:1–185

17. Weiss MR (1995) Floral color change: a widespread functional convergence. Am J Bot 82:167–185

18. Yeo PF (1993) Secondary pollen presentation. Pl Syst Evol, Suppl 6, 1–268

19. Erbar C (1991) Sympetaly—a systematic character? Bot Jahrb Syst 112:417–451

20. Ronse Decraene LP, Smets EF (1993) The distribution and systematic relevance of the androecial character polymery. Bot J Linn Soc 113:285–350

21. Ronse Decraene LP, Smets EF (1995) The distribution and systematic relevance of the androecial character oligomery. Bot J Linn Soc 118:193–247

22. Pellmyr O (1992a) Evolution of insect pollination and angiosperm diversification. Trends Ecol Evol 7:46–49

23. Pellmyr O (1992b) The phylogeny of a mutualism: evolution and co-adaptation between Trollius and its seed-parasitic pollinators. Biol J Linn Soc 47:337–365

24. Pellmyr O, Thompson JN (1992) Multiple occurrences of mutualism in the yucca moth lineage. Proc Natl Acad Sci USA 89:2927–2929

25. Pellmyr O, Huth CJ (1994) Evolutionary stability of mutualism between yuccas and yucca moths. Nature 372:257–260

26. Pellmyr O, Leebens-Mack J, Huth, CJ (1996) Non-mutualistic yucca moths and their evolutionary consequences. Nature 380:155–156

27. Thompson JN (1994) The coevolutionary process. University of Chicago Press, Chicago

28. Bogler DJ, Neff JL, Simpson BB (1995) Multiple origins of the yucca-yucca moth association. Proc Natl Acad Sci USA 92:6864–6867

29. Halevy AH (ed) (1985–1989) CRC handbook of flowering, vol I–VI. CRC Press, Boca Raton

30. Kubitzki K (ed) (1993) The families and genera of vascular plants, Vol. II. Springer, Berlin

31. Taylor DW, Hickey LJ (1992) Phylogenetic evidence for the herbaceous origin of angiosperms. Pl Syst Evol 180:137–156

32. Crane PR, Friis EM, Pedersen KR (1994) Palaeobotanical evidence on the early radiation of magnoliid angiosperms. Pl Syst Evol, Suppl 8, 51–72

33. Crane PR, Friis EM, Pedersen KR (1995) The origin and early diversification of angiosperms. Nature 374:27–33

34. Friis EM, Pedersen KR, Crane PR (1994) Angiosperm floral structures from the Early Cretaceous of Portugal. Pl Syst Evol, Suppl 8, 31–49

35. Friis EM, Pedersen KR, Crane PR (1995) *Appomattoxia ancistrophora* gen. et sp. nov., a new Early Cretaceous plant with similarities to Circaeaster and extant Magnoliidae. Am J Bot 82:933–943

36. Leroy J-F (1993) Origine et evolution des plantes a fleurs. Masson, Paris
37. Hickey LJ, Taylor DW (1996) Origin of the angiosperm flower. In: Taylor DW, Hickey LJ (ed) Flowering plant origin, evolution and phylogeny. Chapman & Hall, New York, pp 176–231
38. Endress PK (1994a) Diversity and evolutionary biology of tropical flowers. Cambridge University Press, Cambridge
39. Endress PK (1994b) Floral structure and evolution of primitive angiosperms: recent advances. Pl Syst Evol 192:79–97
40. Doyle JA, Donoghue MJ (1993) Phylogenies and angiosperm diversification. Paleobiology 19:141–167
41. Lloyd DG, Wells MS (1992) Reproductive biology of a primitive angiosperm, *Pseudowintera colorata* (Winteraceae), and the evolution of pollination systems in the Anthophyta. Pl Syst Evol 181:77–95
42. Kato M, Inoue T, Nagamitsu T (1995) Pollination biology of Gnetum (Gnetaceae) in a lowland mixed dipterocarp forest in Sarawak. Am J Bot 82:862–868
43. Fraser NC, Grimaldi DA, Olsen PE, Axsmith B (1996) A Triassic Lagerstätte from eastern North America. Nature 380:615– 619
44. Cornet B (1993) Dicot-like leaf and flowers from the Late Triassic tropical Newark Supergroup rift zone, U.S.A. Mod Geol 19:81–99
45. Mulcahy DL (1979) The rise of the angiosperms: a genecological factor. Science 206:20–23
46. Takhtajan A (1976) Neoteny and the origin of flowering plants. In: Beck CB (ed) Origin and early evolution of angiosperms. Columbia University Press, New York, pp 207–219
47. Doyle JA, Hickey LJ (1976) Pollen and leaves from the mid- Cretaceous Potomac Group and their bearing on early angiosperm evolution. In: Beck CB (ed) Origin and early evolution of angiosperms. Columbia University Press, New York, pp 139–206
48. Sanderson MJ, Donoghue MJ (1994) Shifts in diversification rate with the origin of angiosperms. Science 264:1590–1593
49. Regal PJ (1977) Ecology and evolution of flowering plant dominance. Science 196:622–629
50. Ericksson O, Bremer B (1992) Pollination systems, dispersal modes, life forms, and diversification rates in angiosperm families. Evolution 46:258–266
51. Bawa KS (1995) Pollination, seed dispersal and diversification of angiosperms. Trends Ecol Evol 10:311–312
52. Ricklefs RE, Renner SS (1994) Species richness within families of flowering plants. Evolution 48:1619–1636
53. Tiffney BH, Mazer SJ (1995) Angiosperm growth habit, dispersal and diversification reconsidered. Evol Ecol 9:93–117
54. Coen ES, Meyerowitz EM (1991) The war of the whorls: genetic interactions controlling flower development. Nature 353:31–37
55. Chase MW, Soltis DE, Olmstead RG, Morgan D, Les DH, Mishler BD, Duvall MR, Price RA, Hills HG, Qiu Y-L, Kron KA, Rettig JH, Conti E, Palmer JD, Manhart JR, Sytsma KJ, Michaels HJ, Kress WJ, Karol KG, Clark WD, Hedren M, Gaut BS, Jansen RK, Kim K-J, Wimpee CF, Smith JF, Furnier GR, Strauss SH, Xiang Q-Y, Plunkett GM, Soltis PS, Swensen SM, Williams SE, Gadek PA, Quinn CJ, Eguiarte LE, Golenberg E, Learn GH Jr, Graham SW, Barrett SCH, Dayanandan S, Albert VA (1993) Phylogenetics of seed plants: an analysis of nucleotide sequences from the plastid gene *rbc*L. Ann Missouri Bot Gard 80:528–580

56. Doyle JJ (1993) DNA, phylogeny, and the flowering of plant systematics. BioScience 43:380–389

57. Sytsma KJ, Hahn WJ (1994) Molecular systematics: 1991–1993. Progr Bot 55:307–333

58. Meyerowitz EM (1995) The molecular genetics of pattern formation in flower development: a perspective after ten years of the Flowering Newsletter. Flow Newsl 20:4–12

59. Meyerowitz EM (1994a) The genetics of flower development. Sci Am 271(5):40–47

60. Meyerowitz EM (1994b) Flower development and evolution: new answers and new questions. Proc Natl Acad Sci USA 91:5735–5737

61. Yanofsky MF (1995) Floral meristems to floral organs: genes controlling early events in *Arabidopsis* flower development. Ann Rev Pl Physiol Pl Mol Biol 46:167–188

62. Doyle JA (1994) Origin of the angiosperm flower: a phylogenetic perspective. Pl Syst Evol, Suppl 8, 7–29

63. Doyle JJ (1994) Evolution of a plant homeotic multigene family: towards connecting molecular systematics and molecular developmental genetics. Syst Biol 43:307–328

64. Purugganan MD, Rounsley SD, Schmidt RJ, Yanofsky MF (1995) Molecular evolution of flower development: diversification of the plant MADS-box regulatory gene family. Genetics 140:345–356

65. Tandre K, Albert VA, Sundas A, Engström P (1995) Conifer homologues to genes that control floral development in angiosperms. Pl Mol Biol 27:69–78

66. Imaichi R, Kato M, Okada H (1995) Morphology of the outer integument in three primitive angiosperm families. Can J Bot 73:1242–1249

67. Robinson-Beers K, Pruitt RE, Gasser CS (1992) Ovule development in wild-type *Arabidopsis* and two female-sterile mutants. Pl Cell 4:1237–1249

68. Schneitz K, Hülskamp M, Pruitt RE (1995) Wild-type ovule development in *Arabidopsis thaliana*: a light microscope study of cleared whole-mount tissue. Pl J 7:731–749

69. Herr JM Jr (1995) The origin of the ovules. Am J Bot 82:547–564

70. Weigel D (1995) The genetics of flower development: from floral induction to ovule morphogenesis. Ann Rev Gen 29:19–39

71. Colombo L, Franken J, Koetje E, van Went J, Dons HJM, Angenent GC, van Tunen AJ (1995) The Petunia MADS box gene FBP11 determines ovule identity. Pl Cell 7:1859–1868

72. Chasan R (1995) Ovule origins. Pl Cell 7:1735–1737

73. Ray A, Robinson-Beers K, Ray S, Baker SC, Lang JD, Preuss D, Milligan SB, Gasser CS (1994) Arabidopsis floral homeotic gene BELL (BEL1) controls ovule development through negative regulation of AGAMOUS gene (AG). Proc Natl Acad Sci USA 91:5761–5765

74. Modrusan Z, Reiser L, Feldmann KA, Fischer RL, Haughn GW (1994) Homeotic transformation of ovules into carpel-like structures in Arabidopsis. Pl Cell 6:333–349

75. Klucher KM, Chow H, Reiser L, Fischer RL (1996) The AINTEGUMENTA gene of *Arabidopsis* required for ovule and female gametophyte development is related to the floral homeotic gene APETALA2. Pl Cell 8:137–153

76. Elliott RC, Betzner AS, Huttner E, Oakes MP, Tucker WQJ, Gerentes D, Perez P, Smyth DR (1996) AINTEGUMENTA, an APETALA2-like gene of *Arabidopsis* with pleiotropic roles in ovule development and floral organ growth. Pl Cell 8:155–168

77. Gaiser JC, Robinson-Beers K, Gasser CS (1995) The *Arabidopsis* SUPERMAN gene mediates asymmetric growth of the outer integument of ovules. Pl Cell 7:333–345

78. Stebbins GL (1971) Relationships between adaptive radiation, speciation and major evolutionary trends. Taxon 20:3–16

79. Riedl R (1978) Order in living organisms. A systems analysis of evolution. Wiley, Chichester

80. Muller GB, Wagner GP (1991) Novelty in evolution: restructuring the concept. Ann Rev Ecol Syst 22:229–256

81. Baum DA, Larson A (1991) Adaptation reviewed: a phylogenetic methodology for studying character macroevolution. Syst Zool 40:1–18

82. Ghiselin MT (1995) Perspective:Darwin, progress, and economic principles. Evolution 49:1029–1037

83. Nickrent DL, Soltis DE (1995) A comparison of angiosperm phylogenies from nuclear 18S rDNA and rbcL sequences. Ann Missouri Bot Gard 82:208–234

84. Takaso T, Bouman F (1986) Ovule and seed ontogeny in *Gnetum gnemon* L. Bot Mag Tokyo 99:241–266

85. Serbet R, Rothwell GW (1995) Functional morphology and homologies of gymnospermous ovules: evidence from a new species of *Stephanospermum* (Medullosales). Can J Bot 73:650–661

86. Taylor TN, Del Fueyo GM, Taylor EL (1994) Permineralized seed fern cupules from the Triassic of Antarctica: implications for cupule and carpel evolution. Am J Bot 81:666–677

87. Endress PK (1975) Nachbarliche Formbeziehungen mit Hüllfunktion im Infloreszenz- und Blütenbereich. Bot Jahrb Syst 96:1–44

88. Verbeke JA (1992) Fusion events during floral morphogenesis. Ann Rev Pl Physiol Mol Biol 43:583–598

89. van der Schoot C, Dietrich MA, Storms M, Verbeke JA, Lucas WJ (1995) Establishment of a cell-to-cell communication pathway between separate carpels during gynoecium development. Planta 195:450–455

90. Ciampolini F, Faleri C, Cresti M (1995) Structural and cytochemical analysis of the stigma and style in *Tibouchina semidecandra* Cogn. (Melastomataceae). Ann Bot 76:421–427

91. Heslop-Harrison Y (1981) Stigma characteristics and angiosperm taxonomy. Nord J Bot 1:401–420

92. Schill R, Baumm A, Wolter M (1985) Vergleichende Mikromorphologie der Narbenoberflächen bei den Angiospermen; Zusammenhange mit Pollenoberflächen bei heterostylen Sippen. Pl Syst Evol 148:185–214

93. Buzgó M (1994) Inflorescence development of *Pistia stratiotes* (Araceae). Bot Jahrb Syst 115:557–570

94. Endress PK (1982) Syncarpy and alternative modes of escaping disadvantages of apocarpy in primitive angiosperms. Taxon 31:48–52

95. van Heel WA (1983) The ascidiform early development of free carpels, a S.E.M.-investigation. Blumea 28:231–270

96. Sage TL, Williams EG (1995) Structure, ultrastructure, and histochemistry of the pollen tube pathway in the milkweed *Asclepias exaltata* L. Sex Pl Reprod 8:257–265

97. Gane AM, Clarke AE, Bacic A (1995) Localisation and expression of arabinogalactan-proteins in the ovaries of *Nicotiana alata* Link and Otto. Sex Pl Reprod 8:278–282

98. Cheung AY, Wang H, Wu H-M (1995) A floral transmitting tissue-specific glycoprotein attracts pollen tubes and stimulates their growth. Cell 82:383–393

99. Wang H, Wu H-M, Cheung AY (1993) Development and pollination regulated accumulation and glycosylation of a stylar transmitting tissue-specific proline-rich protein. Pl Cell 5:1639–1650

100. Lord EM, Sanders LC (1992) Roles for the extracellular matrix in plant development and pollination: a special case of cell movement in plants. Develop Biol 153:16–28

101. Sanders LC, Lord EM (1992) A dynamic role for the stylar matrix in pollen tube extension. Int Rev Cytol 140:297–318

102. Jauh GY, Lord EM (1995) Movement of the tube cell in the lily style in the absence of the pollen grain and the spent pollen tube. Sex Pl Reprod 8:168–172

103. Mascarenhas JP (1993) Molecular mechanisms of pollen tube growth and differentiation. Pl Cell 5:1303–1314

104. Hülskamp M, Schneitz K, Pruitt RE (1995) Genetic evidence for a long range activity that directs pollen tube guidance in *Arabidopsis*. Pl Cell 7:57–64

105. Sniezko R, Winiarczyk K (1995) Pollen tube growth in pistils of female-sterile plants of *Oenothera* mut. *brevistylis*. Protoplasma 187:31–38

106. Lolle SJ, Cheung AY, Sussex IM (1992) Fiddlehead: an *Arabidopsis* mutant constitutively expressing an organ fusion program that involves interactions between epidermal cells. Develop Biol 152:383–392

107. Lolle SJ, Cheung AY (1993) Promiscuous germination and growth of wildtype pollen from Arabidopsis and related species on the shoot of the *Arabidopsis* mutant, fiddlehead. Develop Biol 155:250–258

108. Lolle SJ, Hsu W, Kopczak S, Pruitt RE (1996) Genetic analysis of ontogenetic fusion in *Arabidopsis*. In: Knox RB, Singh MB (eds) Plant reproduction '96. School of Botany, University of Melbourne, p 28

109. Alvarez J, Heisler MGB, Atkinson A, Smyth DR (1996) Spatula: a gene involved in carpel fusion and the development of transmitting tract tissue in *Arabidopsis thaliana*. In: Knox RB, Singh MB (eds) Plant reproduction '96. School of Botany, University of Melbourne, p 86

110. Kandasamy MK, Nasrallah JB, Nasrallah ME (1994) Pollen-pistil interactions and developmental regulation of pollen tube growth in *Arabidopsis*. Development 120:3405–3418

111. Nasrallah JB, Nasrallah ME (1993) Pollen-stigma signaling in the sporophytic self-incompatibility response. Pl Cell 5:1325–1335

112. Clarke AE, Newbigin E (1993) Molecular aspects of self-incompatibility in flowering plants. Ann Rev Genet 27:257–279

113. Cheung AY (1995) Pollen-pistil interactions in compatible pollination. Proc Natl Acad Sci USA 92:3077–3080

114. Weller SG, Donoghue MJ, Charlesworth D (1995) The evolution of self-incompatibility in flowering plants: a phylogenetic approach. Monogr Syst Bot Missouri Bot Gard 53:355–382

115. Olmstead RG (1989) The origin and function of self-incompatibility in flowering plants. Sex Pl Reprod 2:127–136

116. Mulcahy DL, Mulcahy GB, Searcy KB (1992) Evolutionary genetics of pollen competition. In: Wyatt R (ed) Ecology and evolution of plant reproduction. Chapman & Hall, New York, pp 25–36

117. Mulcahy DL, Mulcahy GB (1987) The effects of pollen tube competition. Am Sci 75:44–50

118. Walsh NE, Charlesworth D (1992) Evolutionary interpretation of differences in pollen tube growth rates. Quart Rev Biol 67:19–37
119. Endress PK (1987) Floral phyllotaxis and floral evolution. Bot Jahrb Syst 108:417–438
120. Endress PK (1990) Patterns of floral construction in ontogeny and phylogeny. Biol J Linn Soc 39:153–175
121. Endress PK (1996) Homoplasy in flowers. In: Sanderson MJ, Hufford L (eds) Homoplasy and the evolutionary process. Academic, Orlando, pp 303–325
122. Olmstead RG, Bremer B, Scott KM, Palmer JD (1993) A parsimony analysis of the Asteridae sensu lato based on *rbcL* sequences. Ann Missouri Bot Gard 80:700–722
123. Kron KA, Chase MW (1993) Systematics of the Ericaceae, Empetraceae, Epacridaceae and related taxa based upon *rbcL* sequence data. Ann Missouri Bot Gard 80:735–741
124. Anderberg AA (1994) Phylogeny of the Empetraceae, with special emphasis on character evolution in the genus Empetrum. Syst Bot 19:35–46
125. Hempel AL, Reeves PA, Olmstead RG, Jansen RK (1995) Implications of *rbcL* sequence data for higher order relationships of the Loasaceae and the anomalous aquatic plant *Hydrostachys* (Hydrostachyaceae). Pl Syst Evol 194:25–37
126. Gustafsson MHG, Backlund A, Bremer B (1996) Phylogeny of the Asterales *sensu lato* based on *rbcL* sequences with particular reference to the Goodeniaceae. Pl Syst Evol 199:217–242
127. Hufford L (1988) The evolution of floral morphological diversity in *Eucnide* (Loasaceae): The implications of modes and timing of ontogenetic changes on phylogenetic diversification. In: Leins P, Tucker SC, Endress PK (eds) Aspects of floral development. Cramer, Berlin, pp 103–119
128. Hufford L (1992) Rosidae and their relationships to other nonmagnoliid dicotyledons: a phylogenetic analysis using morphological and chemical data. Ann Missouri Bot Gard 79:218–248
129. Erbar C, Leins P (1988) Blütenentwicklungsgeschichtliche Studien an *Aralia* und *Hedera* (Araliaceae). Flora 180:391–406
130. Erbar C, Leins P (1996) An analysis of the early floral development of *Pittosporum tobira* (Thunb.) Aiton and some remarks on the systematic position of the family Pittosporaceae. Fedde Rep 106:463–473
131. Wagner G (1989) The origin of morphological characters and the biological basis of homology. Evolution 43:1157–1171
132. Galen C (1996) Rates of floral evolution: adaptation to bumblebee pollination in an alpine wildflower, *Polemonium viscosum*. Evolution 50:120–125
133. Carr DE, Fenster CB (1994) Levels of genetic variation and covariation for Mimulus (Scrophulariaceae) floral traits. Heredity 72:606–618
134. Fenster CB, Ritland K (1994) Evidence for natural selection on mating system in *Mimulus* (Scrophulariaceae). Int J Pl Sci 155:588–596
135. Bradshaw HD Jr, Wilbert SM, Otto KG, Schemske DW (1995) Genetic mapping of floral traits associated with reproductive isolation in monkeyflowers (*Mimulus*). Nature 376:762–765
136. von Goethe JW (1790) Versuch die Metamorphose der Pflanzen zu erklären. Ettinger, Gotha
137. Sprengel CK (1793) Das entdeckte Geheimniss der Natur im Bau und in der Befruchtung der Blumen. Vieweg, Berlin

138. Donoghue MJ, Sanderson MJ (1992) The suitability of molecular and morphological evidence in reconstructing plant phylogeny. In: Soltis PS, Soltis DE, Doyle JJ (eds) Molecular systematics of plants. Chapman & Hall, New York, pp 340–368

139. Albert VA, Backlund A, Bremer K, Chase MW, Manhart JR, Mishler BD, Nixon KC (1994) Functional constraints and *rbcL* evidence for land plant phylogeny. Ann Missouri Bot Gard 81:534–567

140. Bremer K (1994) Branch support and tree stability. Cladistics 10:295–304

141. Donoghue MJ (1994) Progress and prospects in reconstructing plant phylogeny. Ann Missouri Bot Gard 81:405–418

142. Doyle JA, Donoghue MJ, Zimmer EA (1994) Integration of morphological and ribosomal RNA data on the origin of angiosperms. Ann Missouri Bot Gard 81:419–450

143. de Queiroz A, Donoghue MJ, Kim J (1995) Separate versus combined analysis of phylogenetic evidence. Ann Rev Ecol Syst 26:657–681

144. Sytsma KJ, Baum DA (1996) Molecular phylogenies and the diversification of angiosperms. In: Taylor DW, Hickey LJ (eds) Flowering plant origin, evolution and phylogeny. Chapman & Hall, New York, pp 314–340

145. Renner SS, Ricklefs RE (1995) Dioecy and its correlates in the flowering plants. Am J Bot 82:596–606

146. Barrett SCH, Harder LD (1996) Ecology and evolution of plant mating. Trends Ecol Evol 11:73–79

147. Endress PK (1992) Evolution and floral diversity: the phylogenetic surroundings of *Arabidopsis* and *Antirrhinum*. Int J Pl Sci 153:S106–S122

148. Price RA, Palmer JD, Al-Shehbaz IA (1994) Systematic relationships of *Arabidopsis*: a molecular perspective. In: Meyerowitz EM, Somerville CR (eds) Arabidopsis. Cold Spring Harbor Laboratory, New York, pp 7–19

149. Kellogg EA, Shaffer HB (1993) Model organisms in evolutionary studies. Syst Biol 42:409–414

150. Zimmer EA (1994) Perspectives on future applications of experimental biology to evolution. In: Schierwater B, Streit B, Wagner GP, DeSalle R (eds) Molecular ecology and evolution: approaches and applications. Birkhäuser, Basel, pp 607–616

151. Frankie GW, Vinson SB, Newstrom LE, Barthell JF, Haber WA, Frankie JK (1990) Plant phenology, pollination ecology, pollinator behaviour and conservation of pollinators in Neotropical dry forest. In: Bawa KS, Hadley M (eds) Reproductive ecology of tropical forest plants. UNESCO, Paris, pp 37–47

152. Prance GT (1991) Rates of loss of biological diversity: a global view. In: Spellerberg IF, Goldsmith FB, Morris MG (eds) The scientific management of temperate communities for conservation. Blackwell, London, pp 27–44

153. Vogel S, Westerkamp C (1991) Pollination: an integrating factor of biocenoses. In: Seitz A, Loeschke V (eds) Species conservation: a population-biological approach. Birkhäuser, Basel, pp 159–170

154. Wilson EO (1992) The diversity of life. Harvard University Press, Cambridge

Chapter 6

Fossil History of Magnoliid Angiosperms

Else Marie Friis[1], Peter R. Crane[2], and Kaj Raunsgaard Pedersen[3]

6.1 Introduction

The exploration of the origin and diversification of angiosperms has entered an exciting new era. Fresh insights and new results are rapidly accumulating from paleobotany and a wide spectrum of botanical disciplines, and an increasing number of phylogenetic models for seed plant and angiosperm relationships are being developed. Current phylogenetic hypotheses broadly support previous views that the most basal angiosperm taxa fall within a grade of organization corresponding to the subclass Magnoliidae; however, there are divergent views on the resolution of relationships within the magnoliid grade. In part, discrepancies among the results from different analyses reflect difficulties in the polarization of critical reproductive characters, which arise because of the substantial morphological gap between angiosperms and other seed plants and the absence of important angiosperm features (e.g., carpel) in their closest seed plant relatives. These problems are also compounded by the extreme floral diversity among extant Magnoliidae, which ranges from the minute and naked, unisexual, unistaminate/unicarpellate flowers of *Hedyosmum* (Chloranthaceae) to the large bisexual and multipartite flowers of *Magnolia* (Magnoliaceae).

Despite uncertainties in resolving relationships at the base of the angiosperm phylogenetic tree some consensus on patterns of character evolution is beginning to emerge. In particular there are strong indications from the study of fossil material that small, simple flowers may be basic in angiosperms, rather than larger multipartite floral structures similar to those of the woody Magnoliaceae [1–4]. This result is broadly congruent with several recent phylogenetic analyses of extant angiosperms based on comparative morphology [5–7], as well as phylogenetic analyses of molecular data [8–10]. Several authors now support the

[1] Department of Palaeobotany, Swedish Museum of Natural History, P.O. Box 50007 S-104 05 Stockholm, Sweden
[2] Department of Geology, The Field Museum, Roosevelt Road at Lake Shore Drive Chicago, IL 60605, USA
[3] Department of Geology, University of Aarhus, DK-8000, Århus C, Denmark

hypothesis that "ancestral" angiosperms were herbaceous or semiherbaceous rather than large woody plants [5,8,9–11], and this hypothesis is also embodied in the concept that the angiosperm tree may be "rooted" among an assemblage of magnoliid families termed the paleoherbs [12,13].

The paleoherb concept was introduced by Donoghue and Doyle [12,13] for a putative clade of herbaceous magnoliids (Aristolochiales, Lactoridales, Nymphaeales, Piperales) and monocotyledons. Subsequently the term has been widely, but generally informally, used in the phylogenetic literature. Even though the paleoherbs do not form a monophyletic group in most phylogenetic analyses [7–10], and their circumscription varies from work to work, the concept nevertheless remains useful in focusing attention on the potential phylogenetic importance of non-woody magnoliids. Two Early Cretaceous fossils have also been highlighted in support of an herbaceous model of early angiosperm evolution. These include an Aptian impression fossil from Australia with characters perhaps indicating relationships with extant Piperales and Chloranthaceae [14], and a Hauterivian impression fossil from southern England [15]. Unfortunately, both fossils are too poorly preserved to establish their systematic affinities with certainty.

Important advances in understanding the origin and diversification of angiosperms have been generated by paleobotanical discoveries of extremely well preserved flowers from the early phases of angiosperm evolution. These flowers, discovered through sieving of unconsolidated Cretaceous sediments, often have their original three-dimensional form intact with all floral parts present and pollen preserved in situ in the stamens [16–31]. Numerous Cretaceous localities yielding such flowers and other angiosperm reproductive organs have now been discovered, particularly in Europe and North America, and these provide powerful data for testing hypotheses on floral evolution against the timing of appearance of floral features [1–4,32,33].

To illustrate the extent and significance of the information now available from the fossil record, in this chapter we summarize current knowledge of early fossil angiosperms at the magnoliid grade, focusing particularly on flowers and other reproductive structures from Cretaceous strata. In a few cases we also consider vegetative remains and other fossils from the Tertiary, particularly where there is no Cretaceous floral record, but it is not our intention here to provide a detailed review of the Cenozoic record. Because relationships within the Magnoliidae are still unresolved taxa are simply arranged in alphabetic sequence after orders.

6.2 Fossil Evidence on the Diversification of Magnoliid Angiosperms

6.2.1 Aristolochiales

The order Aristolochiales includes a single family, the Aristolochiaceae. Walker and Walker [34] placed it close to the Illiciaceae, Schisandraceae, and Winteraceae, while Takhtajan [35] and Cronquist [36] suggested that it may have been derived directly from the Magnoliales. Recent cladistic analyses indicate a position more remote from the Magnoliales. In the morphology-based analyses of Donoghue and Doyle [12,13] the family is placed among the paleoherbs together with Piperales, monocots, Nymphaeales, and Lactoridaceae. In analyses based on combined molecular and morphological data [10], or molecular data alone [8,9], the Aristolochiaceae typically group with the Piperales or monocots.

Aristolochiaceae. The Aristolochiaceae comprise about 12 genera and about 275 species distributed mainly in the tropics with some taxa in temperate regions [37]. They are mainly vines or more rarely shrubs or herbs with entomophilous flowers. The flowers are typically trimerous, but exhibit considerable architectural variation, particularly in the genus *Aristolochia* where the flowers may be highly modified and up to 1 m long [38].

The fossil record of the Aristolochiaceae is sparse and there is no reliable pollen record of the family [39]. The megafossil record includes wood from the Deccan Intertrappean Beds (Late Cretaceous/Early Tertiary?) of India (*Aristolochioxylon prakashii* [40]) and leaf fossils assigned to *Aristolochia* from the Late Tertiary of Abkhasia (*A. africanii*, *A. colchica* [41,42]), Ukraine [43] and Poland [44], as well as the Early to Late Tertiary of North America [45–47].

6.2.2 Ceratophyllales

The order Ceratophyllales comprises a single family, the Ceratophyllaceae. The family has generally been placed in the Nymphaeales and its simple vegetative and reproductive structure has often been interpreted as advanced among magnoliids [35,36]. Several recent phylogenetic analyses, however, have resolved the Ceratophyllaceae as potentially the most basal angiosperm lineage and the sister group to all other angiosperm families [5,8,9].

Ceratophyllaceae. The Ceratophyllaceae are a cosmopolitan family of submerged aquatic plants with small, simple, unisexual water-pollinated flowers [48]. The single genus *Ceratophyllum* contains six species.

There are only two records of putative Ceratophyllaceae from the Cretaceous [49], but these specimens remain to be fully documented. The fruits of *Ceratophyllum* are woody, and combined with the aquatic habit this should result in good fossilization potential. The horned or spiny fruits characteristic of the extant genus are also distinctive and easily identified. Fossil remains of *Ceratophyllum* are common in Tertiary and Quaternary strata, and thus their apparent absence from the Cretaceous needs to be addressed by future paleobotanical studies. In Europe and western Asia the genus is known from the Oligocene onward and is represented by at least 12 species of fruits [50–53]. In North America the genus is known from the Paleocene onward where it is represented by a variety of fruits. In the Eocene, the fruits occur in association with distinctive *Ceratophyllum* axes and leaves [5,54]. The pollen wall of extant *Ceratophyllum* pollen is extremely thin and no dispersed fossil pollen has been recorded.

6.2.3 Illiciales

The Illiciales comprise two families, the Illiciaceae and the Schisandraceae. In early phylogenetic analyses based on morphological data [12,13] the two families formed a monophyletic group that was placed with the Winteraceae as the "winteroid clade." However, in several molecular-based analyses, as well as combined molecular/morphological analyses, the Illiciaceae and the Schisandraceae are associated with the Austrobaileyaceae, while the Winteraceae are associated with the Magnoliales [8–10].

Illiciaceae. The Illiciaceae consist of a single genus, *Illicium*, and 47 species of shrubs and small trees distributed in southeast Asia and southeastern North America [55]. Flowers are bisexual and multipartite with a variable number of parts.

In the Late Cretaceous the presence of the family is suggested by *Illicium*-type pollen from the Maastrichtian of California [56]. Fossil leaves of *Longstrethia varidentata* from the Cenomanian of Nebraska exhibit several illiciaceous features, but also show similarities to extant Trimeniaceae. The more precise systematic affinity of these fossils has not been established [57]. The Tertiary record of the Illiciaceae is more extensive with several records of leaves and seeds from Europe, Asia, and North America, which range in age from the Paleocene to the Miocene [53,58].

Schisandraceae. The Schisandraceae comprise two genera of woody climbers: *Schisandra* (about 25 spp.) and *Kadsura* (about 24 spp.). Species of both genera are distributed in tropical and temperate eastern Asia with the exception of *Schisandra glabra* which occurs in southeastern North America [59]. Flowers are unisexual and multipartite.

The fossil record of the family is poor and the only report from the Cretaceous is based on dispersed pollen from the Maastrichtian of California comparable to that of modern *Schisandra* [39,56]. The Tertiary record is more extensive with fossil leaves and seeds of *Schisandra* reported from the Eocene and Miocene–Pliocene of Europe, from the Pliocene of western Asia, from the Eocene and Miocene–Pleistocene of eastern Asia, and from the Eocene of North America [53,60–62]. Fossil leaves of *Kadsura* are also reported from the Late Tertiary of Europe [53].

6.2.4 Lactoridales

The order Lactoridales includes one family, the Lactoridaceae, of uncertain affinity. The family is sometimes included in the Laurales [35], in the Magnoliales [36], or in the Piperales [63]. A position close to the Piperales and the Aristolochiales is also suggested by the molecular-based cladistic analyses of Qui et al. [9].

Lactoridaceae. The Lactoridaceae are a monotypic family with a single species, *Lactoris fernandeziana*, with shrubby habit and trimerous, bisexual or unisexual flowers. This species is restricted to the Masatierra and the Masatuera of the Juan Fernández Islands [64].

Early and mid-Cretaceous pollen reported from Minnesota and other North American localities as lactoridaceous [65] are shed single and do not provide convincing evidence of the family. A more convincing report includes the dispersed pollen grains of *Lactoripollenites africanus* from the Late Cretaceous (Turonian–Campanian) of southwest Africa [66]. These grains are shed in permanent tetrads closely comparable in size and morphological features to those of extant *Lactoris*.

6.2.5 Laurales

There is some uncertainty as to the circumscription of the Laurales. Both Takhtajan [35] and Cronquist [36] included the Amborellaceae, Calycanthaceae (including Idiospermaceae), Gomortegaceae, Hernandiaceae, Lauraceae, Monimiaceae, and Trimeniaceae in the order. Takhtajan [35] further included the Chloranthaceae, the Lactoridaceae, and the Austrobaileyaceae. These three families are placed in the Piperales (Chloranthaceae) and the Magnoliales (Lactoridaceae, Austrobaileyaceae) in the classification of Cronquist [36]. In cladistic analyses based on morphological and anatomical data the Chloranthaceae and Austrobaileyaceae are grouped with the Laurales [12,13]. In some phylogenetic analyses based on molecular data the Lactoridaceae associate with the Aristolochiaceae, the Chloranthaceae and Amborellaceae with the Nymphaeales or other herbaceous magnoliids, and the Austrobaileyaceae with the Illiciales [8,9].

Numerous fossil leaves from the mid-Cretaceous of North America and Europe have been allied with the Laurales, and among these, the most reliably determined are those that are known from both venation and cuticular anatomy, or that have been associated with lauraceous reproductive structures (see below). From the Rose Creek locality (Early to Middle Cenomanian) of southern Nebraska, USA, Upchurch and Dilcher [57] suggested that leaves assigned to the genera *Crassidenticulum*, *Densinervum*, *Ladonia*, *Pabiania*, and *Pademophyllum* are probably attributable to extant Laurales. Similarly, based on leaf venation, epidermal patterns, and the presence of secretory cells in the mesophyll, Kvaček [67] suggests that a group of mid-Cretaceous leaves assigned mainly to the fossil genera *Myrtophyllum*, *Daphnophyllum*, *Laurophyllum*, *Proteophyllum*, *Grevilleophyllum*, and *Magnoliaephyllum*, but also to extant *Magnolia*, and *Eucalyptus* [68] are probably of lauralean affinity. Among the material from the Cenomanian of Bohemia, Czech Republic, three kinds of reproductive structures appear to be associated with this leaf complex. Leafy branches assigned to "*Eucalyptus*" *geinitzii* have attached globular infructescences comprised of numerous individual units. Leafy branches assigned to "*Eucalyptus*" *angusta* have an associated globular infructescence assigned to *Anthocephale chuchlensis* [67]. The infructescence is about 8 mm in diameter and is composed of about 40 sessile fruits borne together in a head on a stalk about 1 cm long.

Specimens described as *Myricanthium amentaceum* are elongate, spicate inflorescences up to about 4–5 cm in length, which bear some resemblance to the infructescences of *Prisca reynoldsii* from the Middle Cretaceous of North America. All of these reproductive structures are currently under investigation by H. Eklund and J. Kvaček.

Lauraceae. The Lauraceae comprise about 50 genera with between 2500 and 3500 species of trees and shrubs with small, trimerous, bisexual flowers. The pattern of relationships within the family is uncertain and a recent conspectus of a preliminary classification recognizes two major groups: the Perseeae and the Laureae [69]. The Lauraceae are pantropical, with a few species extending north and south into the temperate zones.

The fossil record of Lauraceae is extensive and a number of new reproductive fossils have recently been identified from the Cretaceous. The oldest unequivocal floral remains of the Lauraceae are two fragmentary flowers from the Puddle-dock locality in the Early Cretaceous (Early or Middle Albian) of Virginia, USA [3]. Based on these fragments the reconstructed flower comprises a perianth with an outer whorl of three broad, five-veined tepals, and an inner whorl of three narrower three-veined tepals. The androecium is interpreted as comprising three stamens, each with an associated pair of staminal appendages. The gynoecium is unknown. A small dispersed stamen of probable lauraceous affinity is also known from the same locality [3]. The stamen consists of an elongated filament that grades into the basifixed anther without a joint. The anther is tetrasporangiate with two pairs of pollen sacs that differ slightly in size and the level at which they are positioned. The smaller sacs are positioned slightly more distally. Each pollen sac dehisces by a separate apically-hinged valve [3].

The most thoroughly documented Cretaceous member of the Lauraceae is *Mauldinia mirablis* from the Elk Neck Beds (Early Cenomanian) of northeastern Maryland, USA [26]. *Mauldinia* flowers have a perianth of three smaller outer tepals, three larger inner tepals, and an androecium of nine fertile stamens in three whorls. There is also an additional inner whorl of three sterile staminodes. Each of the three inner stamens has an associated pair of appendages with clavate-sagittate heads. Anthers dehisce by two valves that are hinged distally. No pollen grains have been observed in the anthers. The gynoecium consists of a superior, unilocular, and unicarpellate ovary containing a single anatropous pendant ovule. The flowers are sessile and borne on flattened, bilobed, cladode-like units with up to five flowers on each unit. These cladode-like units are arranged helically on elongated inflorescence axes. Investigations of charcoalified material from the type locality have also shown conclusively that the wood of the *Mauldinia* plant can be assigned to the genus *Paraphyllanthoxylon* [70]. *Paraphyllantho-xylon* is one of the most widespread Cretaceous angiosperm woods and if other species are also attributable to the family then the Lauraceae was widespread and abundant during the Cretaceous.

Inflorescences and flowers closely resembling those of *Mauldinia mirabilis* were described as *Prisca reynoldsii* based on material from two localities in the Dakota Formation (latest Albian–earliest Cenomanian) of central Kansas, USA [71]. Elliptical leaves associated with *Prisca reynoldsii* are assigned to *Magnoliaephyllum* sp. and are very similar to the *"Eucalyptus" geinitzii* leaves known from the Late Cretaceous (Cenomanian) of Bohemia [67]. *Mauldinia* flowers and inflorescences have also been found in these Bohemian floras (H. Eklund and J. Kvaček, unpublished). More fragmentary reproductive material that is similar to *Prisca reynoldsii* also occurs at several other Dakota Formation localities [72]. The probable lauraceous affinities of *Prisca* have not yet been confirmed.

Specimens resembling the elongated inflorescences of *Prisca reynoldsii* are also known from the early Cenomanian of Utah and Texas but these are too poorly preserved to confirm their lauraceous affinity (PR Crane, unpublished). *Mauldinia*-like inflorescence fragments are also known from the Cenomanian–Turonian of Kazakhstan (S Frumin and EM Friis, unpublished), and from several localities of Santonian–Campanian age in North America (unpublished). All of these fossils remain to be described in detail.

Perseanthus crossmanensis, from the South Amboy Fire Clay (Raritan Formation) of New Jersey, USA, is a trimerous lauraceous flower of Turonian age [30]. The flower consists of two series of three tepals, with the outer tepals smaller than the inner. The androecium consists of three cycles of three stamens (anthers unknown) with an inner fourth cycle of three staminodes. The center of the flower is occupied by a unilocular carpel containing a single ovule. Exceptionally preserved spinulose pollen grains occur among the trichomes on some of the flowers [30].

The fossil record of the Lauraceae during the Tertiary is extensive and includes numerous reports of lauraceous wood, leaves, and fruits; however, because of the extreme difficulties in delimiting extant genera, assignments to modern taxa are problematic and should be treated with caution [69]. The Tertiary record of the Lauraceae also includes several fossil flowers. *Androglandula tennessensis* from the Middle Eocene Claiborne Formation [73] and *Litseopsis rottensis* and *Lindera rottensis* from the Miocene of Rott, Germany [74], are preserved as compressions. *Cinnamomum prototypum*, *C. felixii*, and *Trianthera eusideroxyloides* are preserved as casts in the Early Tertiary Baltic Amber [75]. While macrofossils occur abundantly in Cretaceous and Tertiary floras the record of dispersed pollen is extremely sparse mainly due to the poorly developed exine of the grains and low pollen production in most taxa [39].

Monimiaceae. The Monimiaceae comprise 34 genera with about 440 species of trees, shrubs, and lianas distributed in tropical and subtropical regions, especially in the Southern Hemisphere [76]. The Monimiaceae exhibit enormous variation in floral structure with unisexual as well as bisexual flowers, and the number of floral parts varies from a few to almost 2000 in *Tambourissa ficus* [38,76].

The family has a poorly documented fossil record. Leaves assigned to the Monimiaceae are described from the Late Cretaceous (Santonian) of Germany

[77,78]. Early Tertiary records include leaves from the Paleocene of Antarctica [79] and *Laurelia*-type pollen from the Oligocene of New Zealand [39,80].

Hernandiaceae. The Hernandiaceae are a pantropical family comprising five genera and about 60 species: *Hazomalania* (1 sp.), *Hernandia* (about 22 spp.), *Illigera* (about 20 spp.), *Gyrocarpus* (3 spp.), and *Sparattanthelium* (about 13 spp.) of trees, shrubs, and vines with small unisexual or bisexual flowers [81].

The fossil record of the family is poor with only a single published record of fossil leaves from the Upper Miocene of Venezuela assigned to the genus *Gyrocarpus* [82]. A small epigynous flower from the Early Cretaceous of Portugal exhibits a combination of features comparable to that of some extant Hernandiaceae, but the systematic affinity of this fossil has not yet been fully documented [2].

Gomortegaceae. The Gomortegaceae are a monotypic family with a single species, *Gomortega nitida*. *Gomortega* is a tree with epigynous and bisexual flowers, distributed in southern Chile [83]. The family has no published fossil record.

Calycanthaceae. The family Calycanthaceae comprises four extant genera of trees and shrubs: *Idiospermum* (1 sp.), *Calycanthus* (2 spp.), *Sinocalycanthus* (1 sp.), and *Chimonanthus* (4 spp.) [84]. Flowers of the Calycanthaceae are bisexual and multipartite. *Sinocalycanthus* and *Chimonanthus* are both restricted to southern and eastern China, while *Idiospermum* is restricted to northeastern Australia. *Calycanthus* has two species, one in eastern North America and the other in western North America [84].

The most reliable fossil record of the family is based on a single well-preserved flower, *Virginianthus calycanthoides*, from the Puddledock locality in the Early Cretaceous (Early or Middle Albian) of Virginia, USA [23]. The flower is small, bisexual, about 2–3 mm long, and consists of about 12 tepals and 30–40 closely-spaced laminar and extrorse stamens borne around the rim of a deep hypanthial cup. The stamens are separated from the gynoecium by a zone of numerous sterile structures that may be interpreted as staminodes. Pollen grains observed in situ in the stamens are monocolpate and coarsely reticulate with low, widely spaced columellae, and a distinct footlayer; endexine is very thin or absent in non-apertural regions and well developed under the aperture. The gynoecium is composed of 18–26 carpels borne in three to four series on the inner surface of the hypanthium. Spherical secretory cells occur in most floral tissues including stamens and carpels. The fossil flower is closely comparable to flowers of extant Calycanthaceae (including *Idiospermum*), but is distinct from all extant members of the family, particularly in features of the stamens and pollen [23].

Fossil flowers with calycanthoid features were also reported from the Turonian of New Jersey, eastern North America [31]. One of these has a cupulate recepta-cle and spirally arranged bracts. The androecium consists of a whorl of extrorse and tetrasporangiate stamens. The gynoecium is multicarpellate with carpels in a spiral arrangement at the base of the floral cup. The flower shows some similarity to modern *Calycanthus* and *Chimonanthus*, but the systematic affinity of the fossil has not been established with certainty [31]. Another flower from the Turonian of New Jersey, "taxon B" shares features with modern Eupomatiaceae and Calycanthaceae [31] (see Eupomatiaceae). Other records of the family are scarce and include fruits from the Middle Miocene of Germany assigned to the genus *Calycanthus* [85].

Amborellaceae. The Amborellaceae are a monotypic family with the single species, *Amborella trichopoda*, endemic to New Caledonia. *Amborella* is a woody shrub with unisexual flowers with 5–8 tepals, up to 14 stamens in the staminate flowers and 5–8 carpels in the pistillate flowers [86]. The family has no published fossil record.

Trimeniaceae. According to Endress and Sampson [87] the Trimeniaceae consist of five to nine species in two genera: *Trimenia* (3–7 spp.) and *Piptocalyx* (2 spp.). Both genera are trees with small, simple, mostly bisexual flowers. The species are distributed from eastern Australia to Celebes and the southwest Pacific.

The fossil record of the family is scarce. The oldest potential record is dispersed *Trimenia*-like pollen from the Albian–Cenomanian of Brazil assigned to *Cretacaeisporites scabratus* [39]. This species was further reported from the Turonian of west Africa [88]. Dispersed trimeniaceous pollen grains have also been reported from the Campanian–Maastrichtian of Australia. These are assigned to the pollen species *Periporopollenites fragilis* and compared to the Trimeniaceae on the basis of similarities to pollen of extant *Trimenia* [89]. Pollen from the Early Tertiary of Australia assigned to *Periporopollenites demarcatus* was also referred to the Trimeniaceae [90], while pollen from the Oligocene–Miocene assigned to the same species was referred to the Caryophyllaceae [91].

Chloranthaceae. The Chloranthaceae consist of four genera and about 77 species distributed in tropical regions of the world. Three genera are restricted to the Old World: *Ascarina* (12 spp.), *Chloranthus* (18 spp.), and *Sarcandra* (2 spp.); the other genus occurs mainly in the New World: *Hedyosmum* (46 spp.) [92]. The extant species form two distinct groups: the *Ascarina–Hedyosmum* group including shrubs and larger trees with unisexual and anemophilous flowers, and the *Chloranthus–Sarcandra* group including herbs or small shrubs with bisexual and entomophilous flowers. All species have small, simple flowers lacking a well-developed perianth. Flowers have often been regarded as advanced and probably derived by reduction from larger, multipartite forms [34,36]. Alternatively they have been interpreted as primitively simple [93] and perhaps representing the basic floral condition in angiosperms as a whole.

Systematically the Chloranthaceae are normally placed in the Piperales [36] or in the Laurales [35], and Endress [94] suggested that it could perhaps be interpreted as a link between the Laurales and the Piperales. Connection to the Laurales is supported mainly by similarities to members of the Trimeniaceae and Amborellaceae [12,13,94].

The Chloranthaceae have an extensive fossil record, particularly in Early to Late Cretaceous sediments, and have been reported from most parts of the world [39]. Early Cretaceous chloranthoid fossils include several taxa closely related to *Ascarina* and *Hedyosmum*, while the earliest unequivocal fossils closely related to the *Chloranthus* and *Sarcandra* are from the early part of the Late Cretaceous.

Dispersed pollen grains of *Clavatipollenites* Couper [95] are among the oldest unequivocal angiosperms in the fossil record with the earliest record in the Late Hauterivian of Israel [96]. The grains are monocolpate, or more rarely trichotomocolpate, and reticulate with a distinct beaded or spiny suprareticulate ornamentation. Pollen wall ultrastructure shows a three-layered ektexine with a distinct footlayer and a thin endexine except under the apertures where the endexine is markedly expanded. In aperture configuration, pollen morphology, reticulum ornamentation, and pollen wall ultrastructure, *Clavatipollenites*-type pollen closely resembles that of extant *Ascarina* [34,56,80]. During the Early Cretaceous *Clavatipollenites* pollen was distributed virtually worldwide [2,28,34,39,89,95,97–102].

Several fossil floral structures have been linked to *Clavatipollenites* pollen. The genus *Couperites* was established to accommodate small fruiting units with pollen of the *Clavatipollenites* type adhering to the stigma and surface of fruit [28]. The fruiting units are monocarpellate and one-seeded with a single anatropous and pendant seed. The stigmatic area is sessile and poorly differentiated, and the fruit wall is characterized by bulging spherical resin bodies. The anatropous organization of the seed contrasts with the orthotropous organization that is characteristic for modern Chloranthaceae. Whether the anatropous condition in *Couperites* is ancestral or derived within the chloranthoids is as yet unsettled. The type species, *Couperites mauldinensis*, was described from Early Cenomanian strata of Maryland, USA [28]. Another species, *Couperites* sp., has been recovered in Early or Middle Albian sediments of Virginia, USA (Figs. 6.1a,b). This species co-occurs with anthers containing well-preserved pollen of the *Clavatipollenites*-type (Figs. 6.2a–d). Several other unicarpellate and one-seeded fruiting units associated with *Clavatipollenites* pollen have been recovered from Barremian–Aptian (and perhaps older) strata of Portugal [2]. Anthers containing *Clavatipollenites* have also been reported from Aptian strata of Argentina [101].

Another conspicuous chloranthoid element in the Early Cretaceous floras is dispersed pollen grains assigned to the genus *Asteropollis* [103]. The pollen grains are monoaperturate and reticulate with a beaded or spiny suprareticulate ornamentation. The aperture is star-shaped and variable, trichotomocolpate to hexachotomocolpate. Pollen wall ultrastructure shows a three-layered ektexine with a distinct footlayer, and a thin endexine except under the aperture where it is strongly thickened. Based on aperture configuration, wall ornamentation, pollen morphology, and pollen wall ultrastructure Walker and Walker [34] suggested a close relationship between dispersed *Asteropollis* pollen and modern *Hedyosmum*. The oldest record of *Asteropollis* is from the Barremian–Aptian. Cretaceous records include specimens from the Barremian–Aptian to Cenomanian of

Fɪɢ. 6.1a,b. *Couperites* sp. from the Early Cretaceous (Early to Middle Albian) Puddle-dock flora, Virginia, USA. PP44021. **a** Fruit with spiny fruit wall; apical part abraded. ×58. **b** Detail of fruit surface showing spines. ×240

Europe [104–106] and from the Barremian–Aptian of Egypt [107]. Further Cretaceous records are from the Albian to Turonian of North America [98,103,108–112], the Albian of South America, and the Albian? to Turonian of Australia [113–115].

Fossil floral structures have now been linked to *Asteropollis*-type chloranthoid pollen. The material includes small pistillate flowers, dispersed stamens, and perhaps also a staminate inflorescence axis. The pistillate flowers are minute, distinctly triangular in shape, and epigynous with three small tepals at the top of the ovary. The gynoecium is apparently unicarpellate and one-seeded and the ovary wall is characterized by having an irregular growth of the hypanthium tissue resulting in small openings (windows) on the faces of the ovary. Pollen grains of the *Asteropollis* type occur abundantly on the surface of the pistillate flowers. The organization and structure of the flowers is closely similar to that of flowers of extant *Hedyosmum*, and among extant plants a similar irregular growth of hypanthium has been observed only in flowers of *Hedyosmum*. The best-preserved pistillate flowers are from Barremian–Aptian localities in Portugal (Figs. 6.3a–d). Anthers containing *Asteropollis* pollen were found associated with the pistillate flowers at the Buarcos and Torres Vedras localities, Portugal. A small staminate inflorescence axis probably composed of numerous unistaminate and naked flowers comparable to that of extant *Hedyosmum* co-occurs with the pistillate flowers at the Torres Vedras locality, Portugal [2]. A pistillate *Hedyosmum*-like flower similar to the Portuguese material has also been discovered from the Early or Middle Albian of Virginia, USA (Fig. 6.4a). This flower also has numerous *Asteropollis* pollen grains adhering to the surface of the ovary (Figs. 6.4b–d) as observed in the Portuguese flowers.

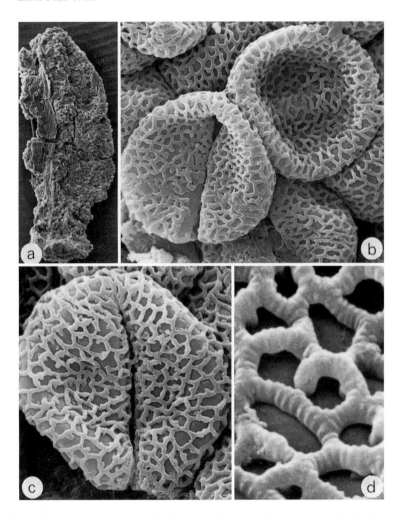

FIG. 6.2a–d. Dispersed stamen with *Clavatipollenites* pollen from the Early Cretaceous (Early or Middle Albian) Puddledock flora, Virginia, USA. PP44022. **a** Stamen. ×58. **b** Pollen grains in situ showing one grain in proximal view and one grain in distal view (with colpus). ×2145. **c** Distal view of pollen grain. ×3090. **d** Detail of pollen surface showing open reticulum with beaded muri. ×1415

Fossil leaves from the Cenomanian of Nebraska assigned to the genus *Crassidenticulum* are particular similar to extant *Ascarina* and *Hedyosmum*, and may represent an extinct genus of the Chloranthaceae [57]. Fossil leaves of *Densinervum kaulii* described from the same locality also share features with the Chloranthaceae and may be another extinct member of the family [57].

The oldest fossils related to the *Chloranthus/Sarcandra* species complex include stephanocolpate and reticulate pollen grains from the Middle Albian of

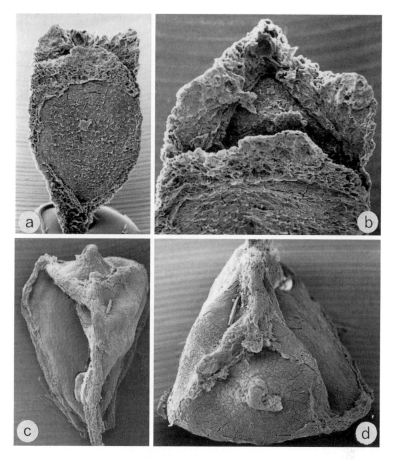

Fig. 6.3a–d. *Hedyosmum*-like pistillate flowers from the Early Cretaceous (Barremian-Aptian?) flora of Buarcos (a–b. S101600) and Torres Vedras (c–d. S101749), Portugal. **a** Flower in lateral view showing window in ovary wall. ×48. **b** Apical view of flower showing three tepals at top of the ovary. ×103. **c** Lateral view of flower. ×48. **d** Apical view of flower showing triangular shape and scars from tepals. ×68

eastern North America assigned to the dispersed pollen species *Stephanocolpites fredericksburgensis* [34,103] later transferred to *Hammenia* [111]. The oldest reproductive organs assignable to this complex include *Chloranthus*-like seeds from the Cenomanian–Turonian of Kazakhstan (S Frumin and EM Friis, unpublished) as well as floral structures of *Chloranthistemon* from the Turonian of North America [29] and the Santonian–Campanian of Sweden [25,116]. The genus *Chloranthistemon* was established to accommodate dispersed, tripartite androecia closely resembling those of extant *Chloranthus* [25]. The genus originally included a single species, *C. endressii* [25]. Subsequent studies of Late Cretaceous fossils from Sweden have provided more complete specimens of

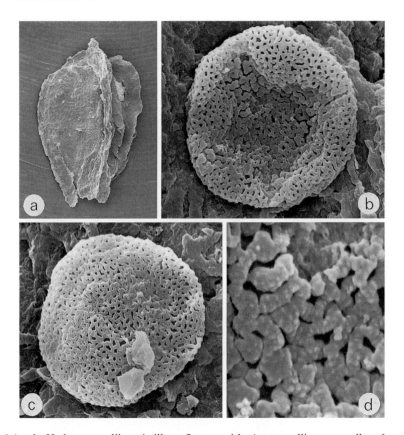

FIG. 6.4a–d. *Hedyosmum*-like pistillate flower with Asteropollis-type pollen from the Early Cretaceous (Early or Middle Albian) Puddledock flora, Virginia, USA. PP44116. **a** Flower in lateral view. ×34. **b** Distal view of pollen showing the starshaped aperture. ×2105. **c** Proximal view of pollen grain. ×2500. **d** Detail of pollen surface showing dense reticulum with beaded/spiny muri. ×9230

C. endressii as well as a new species, *C. alatus* [116]. The new material includes bisexual flowers borne in a decussate and opposite arrangement in the axils of bracts on elongated inflorescence axes. Flowers are minute and naked with a unilocular gynoecium formed from a single carpel. The tripartite androecium is borne abaxially on the gynoecium. Androecial lobes are free at the base in both species, and have distinctly expanded apical extension of the connective. In *C. endressii* the apical expansions are fused to form a massive shield. In *C. alatus* they are free, but in this species they are expanded apically as well as laterally and form characteristic wing-like projections at the top of the androecial lobes. Anther dehiscence is valvate, pollen sacs are very small, and sterile tissue is massive. These features indicate that entomophily was probably characteristic of these early *Chloranthus*-like plants. Pollen grains in the two Swedish species of *Chloranthistemon* have an aperture configuration unlike that of any modern Chloranthaceae. Apertures are spiraperturate in *C. endressii*, while *C. alatus* has a peculiar aperture arrangement with two opposite and decussately arranged colpi or furrows [116]. The organization of the androecium in the two Swedish *Chloranthistemon* species is also unlike that of any extant *Chloranthus* indicating a wider variation in both pollen morphology and floral morphology in *Chloranthus*-like plants than is known for extant forms. A third species of *Chloranthistemon*, *C. crossmanensis*, has been described from the Late Cretaceous of New Jersey, USA [29]. This species differs from the two Swedish species in having androecial lobes fused at the base. Pollen in this species is unknown [29].

While the Cretaceous record of the Chloranthaceae is extensive the Tertiary record is more restricted, and the decrease in distribution to the present relictual occurrence probably took place rather early in the history of the family.

Austrobaileyaceae. The Austrobaileyaceae are a monotypic family with a single species of liana, *Austrobaileya scandens*, distributed in northern Queensland, Australia [117]. Flowers of *Austrobaileya* are bisexual with spirally arranged parts [117]. The family has no published fossil record.

6.2.6 *Magnoliales*

The Magnoliales sensu Takhtajan [35] include the families: Annonaceae, Canellaceae, Degeneriaceae, Eupomatiaceae, Himantandraceae, Magnoliaceae, Myristicaceae, and Winteraceae. In the classification of Cronquist [36] the order further includes the Austrobaileyaceae and the Lactoridaceae. Cladistic analyses based on morphological and anatomical data by Donoghue and Doyle [12,13] generally supported the Takhtajan classification with the exception that the Winteraceae were placed together with Illiciaceae and the Schisandraceae in the winteroid clade [12,13]. Magnoliaceae, Annonaceae, and Degeneriaceae are also resolved as closely related in some phylogenetic analyses based on molecular data [e.g., 8].

Among the magnolialean families, the Magnoliaceae and Annonaceae have flowers with multicarpellate gynoecia and multistaminate androecia. Several fossils from the mid-Cretaceous, which appear to have multicarpellate gynoecia, may be closely related to these two families. Among these are several reproductive structures, described originally as bennettitalean "flowers" but now considered to be angiosperms based on reinterpretations of the structures at the center of the flowers as carpels rather than seeds [118]. The most completely understood of these fossils is *Lesqueria elocata* from the Dakota Formation of central Kansas, USA, and the Woodbine Formation of northern Texas, USA [118], which was originally assigned to the Bennettitales as *Williamsonia elocata* [119]. *Lesqueria* is a multifollicle up to about 9 cm long (including the receptacle below the gynoecium) and about 3–7 cm in diameter at the level of the gynoecium. The gynoecium consists of 175–250 follicles borne helically in a tight, more or less spherical head.

Each follicle is short-stalked, many-seeded, and bears two terminal prolongations (one on either side of the suture). The mass of follicles is borne at the apex of a long receptacle that is covered in numerous persistent, helically-arranged flaps. Comparisons with extant taxa show similarities to extant Magnoliaceae, Annonaceae, Austrobaileyaceae, and other Magnoliidae, but do not indicate a close relationship to a specific angiosperm family. In particular, there is no extant group which shows such a long receptacle with persistent flaps below the gynoecium [118].

Other fossils that were formerly included in the Bennettitales but seem to exhibit a general relationship to extant Magnoliales are listed by Crane and Dilcher [118, p. 395]. The best understood of these are: "*Williamsonia*" *recentior* from the "Upper Blairmore" flora (Middle to Late Albian) of southwestern Alberta, Canada [118,120–122], and *Palaeanthus problematicus* [118,123,124] from the Amboy Clays (Raritan Formation, Middle to Late Cenomanian) of New Jersey, USA. *Palaeanthus* in particular resembles fossil gynoecia from the Hoisington locality in central Kansas, which are associated with *Liriophyllum* leaves and are presumed to have been produced by a plant closely related to *Archaeanthus* [118, see below]. Another possible magnolialean fructification similar to that of *Archaeanthus problematicus* was described from the Late Cretaceous of Japan as *Protomonimia kasai-nakajhongii* [125]. The specimen is a permineralised fruiting structure with well-preserved anatomical details and has numerous carpels helically arranged on a slightly concave receptacle. Each carpel contains many seeds.

While *Lesqueria* and the plants discussed above all have relatively large fruiting structures (typically 3–7 cm long), there are also earlier records of much smaller, multicarpellate gynoecia from the Puddledock locality (Early or Middle Albian) in eastern Virginia, USA [3]. These gynoecia are apparently immature, up to about 3.5 mm long, 2 mm in diameter, and consist of an elongate, more or less conical receptacle bearing numerous helically arranged fruitlets. Some of these specimens show scars of probable stamens and other floral organs on the swollen receptacle immediately below the gynoecium [3]. Pollen grains found attached to the surface of these gynoecia and receptacles are monocolpate and reticulate [3].

Dispersed pollen of probable magnolialean affinity has been described as *Lethomasites fossulatus* from the Barremian–Early Aptian of the Potomac Group, eastern North America [126]. These pollen grains are large (>50 μm long), boat-shaped and monocolpate. The tectum is perforated by small foveolae or fossulae and is supported by a finely granular layer. Comparison with the pollen of extant magnoliids suggests that the affinities of these grains are with the core group of Magnoliales including the families Magnoliaceae, Annonaceae, Degeneriaceae, and Myristicaceae [126,127].

Magnoliaceae. The family Magnoliaceae comprises seven extant genera distributed predominantly in southern and eastern Asia (including Malaysia) and the Americas (from eastern North America southward to Brazil). The Magnoliaceae are divided into two subfamilies [128]. Subfamily Magnolioideae comprises six genera (*Magnolia, Manglietia, Pachylarnax, Kmeria, Elmerrillia, Michelia*) with about 160 species. About one third of the approximately 120 species of *Magnolia* occur in the New World from southeastern North America to southern Brazil. The remaining species in the subfamily are restricted to southeastern Asia and Malaysia [128]. Subfamily Liriodendroideae includes the single genus *Lirioden-dron* with two extant species: *L. chinense* in China and *L. tulipifera* in eastern North America.

Fossil leaves assigned to the Magnoliaceae are common during both the Cretaceous and Tertiary, but except for the distinctive foliage of *Liriodendron* [129], the comparative basis for these assignments is not clearly established. Fossil seeds attributable to the Magnoliaceae are common in Tertiary strata from Europe and Asia [51,53,62,130–136]. These Tertiary fossils are generally referred to modern genera, and seeds assigned to *Magnolia* and *Liriodendron* are prominent components of many Tertiary floras. The Magnoliaceae are also extensively represented during the Tertiary by a variety of fossil woods assigned to *Magnoliaceoxylon, Magnolioxylon*, and *Liriodendroxylon*, as well as to extant genera [137–139], and several records of pollen comparable to *Magnolia* and *Liriodendron* [39].

The Cretaceous record of the Magnoliaceae is much less extensive than in the Tertiary, but the presence of the family during the Late Cretaceous is well documented. The earliest probable magnoliaceous remains are fossil multifollic-ular fruits (*Archaeanthus linnenbergeri*) with associated leaves (*Liriophyllum kansense*), tepals (*Archaepetala beekeri, A. obscura*), and stipular bud scales (*Kalymmanthus walkeri*) from the Dakota Formation (latest Albian–earliest Cenomanian) of central Kansas, USA [140]. Each multifollicle is terminal on a vegetative branch and consists of 100–130 loosely packed, helically arranged follicles on an elongated receptacle up to 13 cm long. The width of the receptacle and follicle, at the base of the flower, is about 7 cm. Follicles are stalked with a short rounded tip and dehisced along the single adaxial suture. Ovules are nu-merous in each carpel, and mature follicles contain 10–18 seeds. At maturity both the follicles and seeds were shed. Numerous small elliptical scars below the base of the lowermost follicle indicate the presence of numerous stamens. Six to nine larger angular scars indicate the position of inner tepals, while three large ellip-tical scars, which delimit the base of the flower, indicate the position of three outer tepals. A prominent scar below the base of the flower is thought to mark the former attachment of the stipular bud scale.

The *Archaeanthus* plant differs most conspicuously from extant Magnoliaceae in the presence of numerous small seeds per follicle and the abscission of the follicle at maturity. Thus the dispersal biology of *Archaeanthus* was probably less specialized than that of any extant Magnoliaceae. In contrast, the arrangement of floral parts, as inferred from the scars at the base of the receptacle, is almost identical to that of certain extant species of *Magnolia* and this may reflect basic similarities in pollination biology. Based on these similarities, as well as the distinctive bilobed stipular bud scales which are probable a synapomorphy of the extant family, there is no doubt that *Archaeanthus* should be considered either within or very closely related to extant Magnoliaceae.

The structure identified as a fossil *Magnolia* petal from the Dakota Sandstone flora [141] may be from a plant similar to *Archaeanthus*, while leaves similar to those of the *Archaeanthus* plant are known from the Kassler Sandstone (latest Albian) of the South Platte Formation near Morrison, Colorado, USA. These leaves have been assigned to *Liriophyllum populoides*, and differ from those of *L. kansense* in having a prominent alate appendage attached to the basal part of the petiole, which may be a pair of modified adnate stipules. Leaves of both *L. kansense* and *L. populoides* differ from those of extant *Liriodendron* in having a strong midrib that runs to the leaf apex and divides into a pair of prominent veins that form the leaf margin in the lower portion of the leaf sinus [140]. While this pattern differs from that in extant *Liriodendron*, *Liriophyllum* leaves are nevertheless more similar to those of *Liriodendron* than to leaves of any other living genus. Fossil leaves with a pattern of leaf venation that very closely resembles that of extant *Liriodendron* are also present in the Dakota Formation (latest Albian–earliest Cenomanian), and continue through the Late Cretaceous [140].

In addition to *Archaeanthus*, two other multifollicular fruits have been described from the Cretaceous that may be attributable to the Magnoliaceae. *Magnoliaestrobus gilmouri* [142] is based on a single specimen from the Kardlok locality of western Greenland, which is of Late Cretaceous (Cenomanian–Santonian) age. The specimen is about 9 cm long and 3.5 cm wide with 30 open follicles that are filled with sediment: additional details have not been described. *Litocarpon beardii* [143], from the Haslam Formation (Late Santonian–Early Campanian) of Vancouver Island, British Columbia, Canada, is based on a single specimen showing about 19 helically arranged follicles borne on an elongated, slender receptacle. The multifollicle is about 4 cm in diameter but the length is unknown. Each follicle is laterally winged. Dehiscence was along the dorsal suture. Seeds are also winged and developed from anatropous ovules attached to parietal placentae along the ventral suture. A possible liriodendroid relationship for *Litocarpon* was suggested by Delevoryas and Mickle [143], but the seed coat is different from extant and extinct *Liriodendron* in lacking the characteristic palisade-shaped crystal cells of the endotesta. The features of *Litocarpon* do not match completely any other magnoliaceous taxon and the systematic affinity of the fossil remains to be established with certainty.

Winged magnoliaceous seeds have also been reported from Late Cretaceous strata of Asia and eastern North America [144]. These seeds are assigned to three species in the genus *Liriodendroidea* and, except for the presence of the wing, they are similar to those of extant *Liriodendron* in all details of seed coat structure and organization of the seeds. Critical seed characters includes anatropous organization, the presence of a distinct heteropyle, and endotestal seed coat structure with distinct palisade-shaped schlerenchyma cells containing fibrous lignifications and cubic crystals. The genus *Liriodendroidea* was established by Knobloch and Mai [145] based on dispersed seeds from the Maastrichtian of

Eisleben, Germany. The oldest reported *Liriodendroidea* is *L. alata* from the Cenomanian–Turonian of Kazakhstan [144]. Other species are *L. protogea* [146], from the Santonian of Quedlinburg, Germany, *L. latirapha* and *L. carolinensis* from the Campanian of the Neuse River, North Carolina, USA [144], undescribed seeds from the Santonian of Upatoi Creek, Georgia, USA (Crane and Herendeen, unpublished), and *L. germanica* [146] from the Maastrichtian of Eisleben, Germany. Wings have not been reported for the two European species (*L. germanica, L. protogea*). However, in the Kazakhstan and Neuse River material wings are present only in well-preserved seeds. The lack of a wing in the European species may reflect poor preservation. The presence of winged seeds in Cretaceous *Liriodendron*-like plants indicates a shift in dispersal strategy to winged follicles in modern liriodendroid Magnoliaceae. A further report of putative magnoliaceous seeds from the Cretaceous are specimens assigned to *Manglietia* from the Late Cretaceous of Walbeck, Germany [146].

Annonaceae. The Annonaceae comprise about 128 genera with about 2300 species of shrubs, trees, or climbers. Flowers are typically bisexual with a trimerous perianth and multipartite androecium and gynoecium [147]. The family is pantropical with two genera (*Asimina, Deeringothamnus*) extending into temperate regions. The pattern of relationships within the family is uncertain and Kessler [147] recognizes 14 informal groups.

Seeds of Annonaceae are distinct with a fibrous seed coat and a pronounced rumination of the endosperm that forms densely spaced transverse ridges on the endosperm surface. They are extensively represented in the Early Tertiary and have been reported from several floras in Europe, as well as in floras from Egypt and Pakistan, where they have been assigned to species of *Alphonsea, Annona, Anonaspermum, Asimina,* and *Uvaria* [130–131,148–150].

Despite the distinct seed characters and good fossilization potential seeds of Annonaceae are rare in the Cretaceous and appear rather late with records from the Maastrichtian of tropical Africa (Nigeria and Senegal) [151,152]. The earliest reliable palynological evidence of the family is also from the Maastrichtian [39] and includes *Malmea*-type pollen (*Foveomorphomonocolpites humbertoides*) from the Maastrichtian and Paleocene of Columbia [39,153].

Degeneriaceae. The Degeneriaceae consist of a single genus *Degeneria* with two species, *D. vitiensis* and *D. roseiflora*, both endemic to Fiji. Both species are trees with large, multipartite, bisexual flowers [154]. The family has no published fossil record.

Myristicaceae. The Myristicaceae comprise 17 genera with about 370 species of shrubs and trees with small, inconspicuous, unisexual flowers [155]. The family is pantropical with each genus restricted to an individual continent [155].

The fossil record of the Myristicaceae is poor and the Cretaceous record includes only a few putative myristicaceous taxa. Fossil wood assigned to *Myristicoxylon* was described from the Late Cretaceous of the Sahara [156], and

Walker and Walker [34] noted that several Early Cretaceous pollen grains referred to as aff. *Clavatipollenites* exhibit considerable similarity with pollen of extant Myristicaceae (e.g., *Virola*). Several records of myristicaceous pollen were reported from the Tertiary of Africa [39] and a single record of a myristicaceous seed (*Myristicarpum*) has been published from the Late Tertiary of Europe [157].

Himantandraceae. The family Himantandraceae include a single genus, *Galbulimima* (2 spp.), of large trees with multipartite bisexual flowers [158]. *Galbulimima* is distributed in the Papuan region and Queensland, Australia [158]. The family has no published fossil occurrence.

Eupomatiaceae. The Eupomatiaceae consist of a single genus *Eupomatia* with two species of shrubby plants: *E. laurina* in New Guinea and *E. bennettii* in eastern Australia [159]. Flowers are perianthless, bisexual, and with numerous stamens, staminodes, and carpels [159].

Two fossil flowers from the Late Cretaceous (Turonian) of eastern North America exhibit a combination of floral characters that indicates a close relationship with the Eupomatiaceae and perhaps other magnolialean or lauralean taxa, but their systematic affinity has not been established with certainty [31]. One flower (taxon "A") has a distinct receptacular cup with overlapping bracts and numerous laminar and incurved stamens along the rim of the cup, and numerous conduplicate carpels at the base of the floral cup. The other flower (taxon "B") shows features comparable to those of Eupomatiaceae and to the Calycanthaceae. The flower consists of a floral cup covered with imbricate bracts and numerous carpels at the base of the floral cup. Stamens are not preserved [31]. *Eupomatia*-type pollen has been reported from the Maastrichtian of California [56]. Muller [39] noted that the only objection to this record is the present distribution of the genus.

Canellaceae. The family Canellaceae consist of 5 genera and about 16 species: *Canella* (1 or 2 spp.), *Cinnamodendron* (about 5 spp.), *Cinnamosma* (3 spp.), *Pleodendron* (2 spp.), and *Warburgia* (3 spp.) of trees and shrubs with bisexual, hypogynous flowers. The species are distributed in tropical Africa, Madagascar, tropical South America, and the Caribbean [160].

A single fossil record of the Canellaceae is based on Early Tertiary (Oligocene) pollen assigned to *Pleodendron* from Puerto Rico [161], but according to Muller [39] this record is inadequately documented.

Winteraceae. The family includes about 4–9 genera and about 65 species of small shrubs and trees with a relictual distribution in South and Central America, Madagascar, New Zealand, Australia, and New Caledonia [38,162]. Flowers are typically bisexual and exhibit considerable plasticity with a variable number of floral parts even within the same species [38].

No reproductive structure assignable to the Winteraceae has been identified in the fossil record, but several distinctive pollen tetrads, leaves, and wood of winteraceous affinity have been described from Cretaceous and Tertiary floras.

The oldest records include dispersed tetrads of *Walkeripollis gabonensis* pollen from the Early Cretaceous (Late Barremian–Early Aptian) of Gabon [163,164]. A slightly younger record includes dispersed tetrads from the Aptian–Albian of Israel [165]. A possible winteraceous affinity was also suggested for the dispersed pollen of *Afropollis* from the Aptian–Albian of Northern Gondwana and dispersed pollen of *Schrankipollis* from the Aptian of eastern North America [164], but in both of these pollen types the grains are shed singly and the winteraceous affinity is more uncertain [164]. Fossil leaves reported from the Cenomanian of Nebraska as "New genus A" also exhibit great similarity with leaves of extant Winteraceae but lack several of the defining characters necessary to include them in the family [57]. Fossil wood of winteraceous affinity was reported from the Late Cretaceous (Maastrichtian) of California [166].

Several reports of dispersed pollen tetrads assigned to the extinct genus *Pseudowinterapollis* (comparable to *Pseudowintera*, *Belliolum*, and *Bubbia*) are known from the Late Cretaceous (Campanian–Maastrichtian) of Australia and the Tertiary of New Zealand [80,89,167]. Other winteraceous pollen records include species of the dispersed pollen genus *Harrisipollenites* (comparable to *Exospermum* and *Zygogynum*) from the Tertiary (Late Oligocene–Pleistocene) of New Zealand [168], as well as *Drimys*- and *Bubbia*-type pollen from the Late Tertiary (Early Miocene) of South Africa [169,170]. Among several leaf fossils reported from the Tertiary, Vink [162] accepts only the leaves of *Drimys patagonica* from the Early Miocene of Patagonia as a reliable record.

6.2.7 Nymphaeales

In the classification system of Takhtajan [35] and Cronquist [36] the Nymphaeales comprise four families: the Barclayaceae, Cabombaceae, Nymphaeaceae, and the Ceratophyllaceae. Recent molecular-based cladistic analyses generally support a nymphaealean clade that excludes the Ceratophyllaceae [8–10]. The Nymphaeales have typically been regarded as an advanced group within the Magnoliidae [35,36], but in several recent studies they have been placed in a basal or near basal position [10,34].

Numerous fossil leaves with a generalized nymphaealean form have been described from Cretaceous strata [171,172]. However, most of these are impression fossils and poorly documented. Recently, several different species of nymphaealean seeds have been discovered from the Early Cretaceous of Portugal and eastern North America (E.M. Friis, K.R. Pedersen, and P.R. Crane, unpublished). The seeds closely resemble seeds of extant *Brasenia* in shape and wall structure, but they apparently lack a distinct embryotega. The presence of an embryotega or operculum is a distinctive feature in the Nymphaeales, and the only exception is in *Barclaya*. The fossils seeds may thus represent a basal nymphaealean line with affinity to both Barclayaceae and the Cambombaceae.

Barclayaceae. The Barclayaceae include a single genus, *Barclaya* (4 spp.), of aquatic herbs distributed in tropical Indomalaysia, Thailand, and Burma [173].

The genus has bisexual entomophilous flowers with numerous parts. The genus is sometimes included in the Nymphaeaceae [173,174].

There is no unequivocal fossil record of the family (but see above). The systematic affinity of the *Barclayopsis urceolata* seeds described from the Maastrichtian and Paleocene of Europe by Knobloch and Mai [145] is not fully documented [52,53]. The only other fossil record of the Barclayaceae is a fruit from the Eocene of southern England assigned to the fossil genus *Protobarclaya* by Reid and Chandler [131]. According to Collinson et al. [149] seeds preserved in the fruit show clear nymphaealean characters with an apical micropylar cap, but the material is too poorly preserved to allow further examination and more detailed evaluation of the systematic affinity.

Cabombaceae. The Cabombaceae are a small family with six species in two genera, *Cabomba* (5 spp.) and *Brasenia* (1 sp.), distributed mainly in tropical and subtropical regions of the world. They are aquatic herbs with small solitary, bisexual, entomophilous flowers with a variable number of floral parts [175].

The Cabombaceae has an extensive fossil record from the Tertiary and Quaternary, based exclusively on the characteristic seeds. The fossil record indicates that the family was much more diverse during the Tertiary than it is in the Recent. In addition to species attributed to the extant genera *Cabomba* and *Brasenia*, the fossil material includes species assigned to several fossil genera including *Braseniella*, *Dusembaya*, *Palaeonymphaea*, and *Sabrenia* [53,174,176]. The genus *Brasenia*, which is now monotypic, is represented in the Tertiary floras of Europe by at least 20 species ranging from the Eocene [53,136,177], and several species are also recorded from Asia ranging in age from the Oligocene [177].

Numerous *Brasenia*-like seeds have recently been identified from the Early Cretaceous (see above), but these differ apparently from seeds of extant Cabombaceous in the lack of a distinct embryotega. It is interesting to note, however, that small epigynous flowers of possible lauralean (hernandiaceous?) affinity recovered from several Early Cretaceous localities in Portugal have monocolpate and finely striate pollen very similar to pollen of extant *Cabomba* [2]. In floral morphology, however, these flowers are markedly different from those of extant Cabombaceae.

Nymphaeaceae. The Nymphaeaceae contain five genera and 60–80 species: *Euryale* (1 sp.), *Nuphar* (7–25 spp.), *Nymphaea* (about 50 spp.), *Ondinea* (1 sp.), and *Victoria* (2 spp.) of aquatic herbs with large, bisexual, multipartite flowers. Some classifications [173] also include *Barclaya* in the family. *Nymphaea* has a worldwide distribution, and Nuphar is widely distributed in the Northern Hemisphere, while the three other genera have a much more restricted distribution, with *Ondinea* occurring in Australia, *Victoria*, in South America, and *Euryale* in India [173,174].

The Cretaceous record of the Nymphaeaceae is poor and based on only a few pollen records and a single seed record. The oldest record of dispersed pollen grains is *Nymphaea zanzibarensis*-type pollen (*Zonosulcites scollardensis* and *Z.*

parvus) from the Late Cretaceous (Maastrichtian) of Canada [39,178]. Putative nymphaeaceous seeds were reported from the Late Turonian of Middle Europe [146]. Late Cretaceous (Maastrichtian) *Nuphar*-like pollen reported from the Cretaceous–Tertiary boundary Teapot Dome locality [179] differs from pollen of extant Nymphaeaceae in lacking the characteristic monocolpate aperture and may be of pandanaceous affinity [180].

The Tertiary record of the Nymphaeaceae is extensive. Seeds are particularly well represented and include numerous species referable to the modern genera *Nymphaea*, *Nuphar*, and *Euryale* as well as many species assignable to the extinct genera *Eoeuryale*, *Irtyshenia*, *Nikitinella*, *Palaeoeuryale*, *Pseudoeuryale*, *Tavdenia*, and *Tomskiella* [52,74].

6.2.8 Piperales

The Piperales comprise the two families Piperaceae and Saururaceae, which are placed together in most classificatory schemes [35,36]. The two families have also been shown to be closely related based on phylogenetic analyses of both morphological and molecular data [8–10,12,13].

The Piperales were undoubtedly differentiated by the Early Tertiary (see below), but the group does not have an extensive Cretaceous fossil record. Dispersed fossil pollen grains of the *Tucanopollis/Transitoripollis* type are monocolpate, tectate with verrucate and echinate supratectal sculpture, and with a continuous tectum. The pollen wall exhibits a granular infratectal exine structure and a thin endexine in non-aperturate regions that expands toward the aperture. Pollen of the *Tucanopollis/Transitoripollis* complex was interpreted by Doyle and Hotton [127] as probably produced by early Piperales, based on similarities to the pollen of extant taxa, but they also noted considerable similarities with pollen of some Laurales. *Tucanopollis/Transitoripollis*-type pollen is particularly common in Early Cretaceous sediments and has been reported in Albian strata of Europe and North America [127,181], Barremian strata of Africa [127], and Barremian–Aptian of South America [182]. The only direct evidence of the likely producers of *Tucanopollis/Transitoripollis* pollen is provided by small fruits of *Appomattoxia ancistrophora* from the Early Cretaceous (Early or Middle Albian) of Virginia, eastern North America, which have *Transitoripollis*-like pollen on the stigmatic surfaces [183]. The fruits are small, unicarpellate, and unilocular with a single pendulous orthotropous seed. The fruit wall is thin and bears numerous, densely spaced, unicellular spines with distinctive curved tips. Pollen on the stigmatic surface is monocolpate, and tectate with a granular to columellate infrastructure. The pollen morphology and ultrastructure of these grains is most similar to that of Early Cretaceous pollen assigned to *Transitoripollis similis* [181], although other species of dispersed *Transitoripollis* and *Tucanopollis* pollen are also very similar in shape, colpus form, sculpture, and exine architecture. The systematic affinity of *Appomattoxia* has not been established with certainty and the fossil apparently exhibits a mosaic of features from several families. The orthotropous organization and the thin seed coat of the *Appomattoxia* seeds corresponds to that in extant Piperales, but seeds of Piperales are erect and fruits are typically formed from three carpels. The unicarpellate fruits as well as the pendant and orthotropous seed organization is shared with members of the Chloranthaceae and Circaeasteraceae. The spine-like hairs with curved tips on the fruit surface of *Appomattoxia* also resemble similar structures seen in extant *Circaeaster*. However, seeds in comparable Chloranthaceae have a thick seed coat, and pollen in Circaeasteraceae is tricolpate.

Piperaceae. The Piperaceae comprise five genera: *Zippelia* (1 sp.), *Piper* (about 1000 spp.), *Sarcorhachis* (4 spp.), *Macropiper* (9 spp.), and *Peperomia* (about 1000 spp.). Both *Piper* and *Peperomia* are pantropical. *Zippelia* is restricted to southeastern Asia and Malaysia, while *Macropiper* is restricted to the South Pacific. *Sarcorhachis* is confined to the Neotropics [184]. The Piperaceae are resolved as the sister group to the Saururaceae by recent phylogenetic studies [8–10,12,13].

Numerous fossil leaves from the Cretaceous and Tertiary have been assigned to the genus *Piper* but in the absence of detailed comparative studies these cannot be regarded as reliable records of the family. Small reticulate seeds assigned to *Peperomia* (*P. sibirica*) were described from Miocene strata of the Tambov and Omsk regions in Siberia [51]. According to Mai [53] *Peperomia* has also been recovered from the Tertiary of Europe, but no details of these fossils are provided.

Saururaceae. The Saururaceae comprise four genera: *Saururus* (1 sp.), *Houttuynia* (1 sp.), *Anemopsis* (1 sp.), and *Gymnotheca* (2 spp.). *Saururus* has one species in eastern North America and another in eastern Asia. *Anemopsis* is confined to western North America. *Houttuynia* is restricted to eastern and southeastern Asia, while *Gymnotheca* is restricted to southwestern China [185].

Fruits assigned to *Saururus* (*S. biloba*) are common in the European and Siberian Tertiary floras ranging in age from the Late Eocene to the Miocene [53,136]. The only possible Cretaceous record of the family is based on the stem *Saururopsis niponensis* Stopes and Fujii [186]. Re-examination is required to confirm the suggested systematic affinities of this material.

6.3 Discussion and Conclusions

6.3.1 Differentiation and Modernization of Magnoliid Lineages

As reviewed above, knowledge of Cretaceous magnoliids, and particularly their flowers, has expanded dramatically in the last decade. This new information has substantially changed our perspective on the pattern of early angiosperm diversification. Prior to 1980, evidence for lineages of extant angiosperms during the Cretaceous was sparse and almost entirely based on leaves and dispersed pollen grains [39]. However, the data now available document that many of the major lineages at the magnoliid grade had already differentiated by around the Early–Late Cretaceous boundary (about 95 million years BP). By this time angiosperm diversity included a range of magnoliids, as well as both monocots [187,188] and several lineages of eudicots [189]. Magnoliid families that are unequivocally present by the earliest Cenomanian include Calycanthaceae, Chloranthaceae, Lauraceae, Magnoliaceae, and Winteraceae. The existence of such taxonomic diversity implies that many other magnoliid lineages, leading, for example, to extant taxa that have no Cretaceous record, must also have diverged by the mid-Cretaceous.

Another striking result to emerge from recent paleobotanical studies is that among the families known to be present during the mid-Cretaceous, there are several taxa that seem to show almost no change in their floral structure over time. In the Chloranthaceae, for example, the occurrence of distinctive *Hedyosmum*-like fruits from the Barremian–Aptian–Albian and *Chloranthus*-like flowers from the Turonian to Santonian/Campanian provides two of the most remarkable examples of stasis in the angiosperm fossil record. Similarly, the closely comparable basic floral architecture of *Archaeanthus* and extant *Magnolia* (Magnoliaceae), *Virginianthus* and extant Calycanthaceae, and *Mauldinia* and extant Lauraceae, also document that many of the major aspects of floral form in extant Magnoliidae have persisted virtually unchanged for 100 million years or more.

6.3.2 Pollination and Dispersal in Early Magnoliids

Although mid-Cretaceous flowers are generally much smaller than those of their living counterparts, strong architectural similarities to the floral structures of their extant relatives imply basically similar pollination mechanisms. In dispersal, however, differences are more pronounced, and in general, dispersal in mid-Cretaceous magnoliids appears to be less specialized than their extant counterparts. For example, in *Archaeanthus*, dispersal seems to have occurred both by shedding of the follicles from the elongated receptacle axis, and also by shedding of the numerous small seeds through the open adaxial (ventral) suture. The winged *Liriodendron*-like seeds of *Liriodendroidea* [144] and *Litocarpon* [143] also appear to have been similar to those of *Archaeanthus* in having been shed from the follicles at maturity. These apparently generalized dispersal syndromes contrast markedly with the specialized, often bird-dispersed seeds characteristic of extant Magnolioideae, and the indehiscent, samaroid fruits characteristic of *Liriodendron*. In extant *Magnolia* and its allies there are usually few (generally two to eight) relatively large seeds with a well-developed sclerotic layer. At maturity these seeds are often red or pink, hang on a thread derived from the funicle, and are exposed to potential dispersers by dehiscence along the dorsal (abaxial) suture. In *Liriodendron* the indehiscent, one- or two-seeded mature carpels have a long, flat, distal wing and are shed from the receptacle at maturity.

A further indication of generalized dispersal in Cretaceous magnoliids is provided by the very thin fruit wall of many mid-Cretaceous chloranthoid fruits. In the related extant taxa the mechanical protective layer is generally located in the fruit wall (endocarp). However, in the fossils there appears to be little woody tissue in the fruit wall, and the mechanical layer is confined to the seed coat. Similarly, there is little mechanical tissue in the lauraceous fruits of *Mauldinia* or putative piperalean fruits of *Appomattoxia*. The most obviously specialized of the fruits currently known from the mid-Cretaceous are those of *Appomattoxia*, which have numerous spine-like hairs on their surface. Among extant taxa such fruits are often associated with animal dispersal.

The recognition of greater similarities in pollination than in dispersal between mid-Cretaceous Magnoliidae and their living relatives is consistent with the observation by Endress [190] that among extant magnoliids pollination interactions generally involve animal groups with relatively ancient, and often significantly pre-angiosperm, fossil records. In contrast, dispersal interactions for the same plants generally involve animal groups that underwent their major diversification during the Late Cretaceous and Tertiary (e.g., birds, mammals). These observations suggest that the evolution of pollination and dispersal systems among magnoliid angiosperms has been to some extent decoupled, with more pronounced changes occurring in dispersal rather than pollination over the past 80–100 million years.

6.3.3 Patterns of Floral Evolution

Neobotanical studies have emphasized the great diversity of floral form and biology among extant magnoliids, which ranges from large multipartite, bisexual, insect-pollinated flowers to small, few-parted, unisexual wind-pollinated forms. The available paleobotanical data show that similar variability in floral form was established early in angiosperm evolution. By the mid-Cretaceous angiosperm flowers included small, bisexual, few-parted, regular trimerous flowers (*Mauldinia*); larger, bisexual, multipartite flowers with an elongated receptacle (*Archaeanthus*); small, bisexual, multipartite flowers with a cup-shaped hypanthium (*Virginianthus*); small, bisexual, few-parted, bilaterally symmetrical flowers (*Chloranthistemon*); and small, unisexual, few-parted flowers with an inferior ovary (*Hedyosmum*-like).

Given this extreme variation among extant and fossil taxa, identifying the likely basic floral condition in angiosperms as a whole is intimately connected with clarifying phylogenetic patterns at the magnoliid grade. While the fossil record alone cannot conclusively determine the kinds of flowers that are likely to be basic within angiosperms, taken together current paleobotanical data and phylogenetic analyses challenge the view that the earliest angiosperms had large *Magnolia*-like flowers. In particular, the emerging molecular evidence, and the very early occurrence of a variety of small simple flowers, is not compatible with that interpretation. These data suggest instead that the flowers of the earliest angiosperms were small, simple (perhaps unisexual), and lacking or possessing an undifferentiated perianth. Stamens and carpels would have been few and perhaps irregular in number. Stamens would have had a poorly differentiated filament and a well-developed anther with valvate dehiscence. Carpels would have been unilocular with one or two ovules. This revised paradigm for the early evolution of angiosperm flowers provides a new basis for examining the evolution of floral form. Rather than placing the emphasis on pervasive patterns of reduction, it suggests instead that floral evolution among basal angiosperms has involved the repeated duplication of tepals, stamens, and carpels, and numerous independent derivations of polymerous flowers.

References

1. Friis EM, Endress PK (1990) Origin and evolution of angiosperm flowers. Adv in Bot Res 17:99–162
2. Friis EM, Pedersen KR, Crane PR (1994) Angiosperm floral structures from the Early Cretaceous of Portugal. Pl Syst Evol [Suppl] 8:31–49
3. Crane PR, Friis EM, Pedersen KR (1994) Palaeobotanical evidence on the early radiation of magnoliid angiosperms. Pl Syst Evol [Suppl] 8:51–72
4. Crane PR, Friis EM, Pedersen KR (1995) The origin and early diversification of angiosperms. Nature 374:27–33
5. Les DH (1988) The origin and affinities of the Ceratophyllaceae. Taxon 37:326–345
6. Taylor DW, LJ Hickey (1992) Phylogenetic evidence for the herbaceous origin of angiosperms. Pl Syst Evol 180:137–156
7. Nixon KC, Crepet WL, Stevenson D, Friis EM (1994) A reevaluation of seed plant phylogeny. Ann Mo Bot Gard 81:484–533
8. Chase MW, Soltis DE, Olmstead RG, Morgan D, Les DH, Mishler BD, Duvall MR, Price RA, Hills HG, Qui Y-L, Kron KA, Rettig JH, Conti E, Palmer JD, Manhart JR, Sytsma KJ, Michaels HJ, Kress WJ, Karol KG, Clark WD, Hedrén M, Gaut BS, Jansen RK, Kim K-J, Wimpee CF, Smith JF, Furnier GR, Strauss SH, Xiang Q-Y, Plunkett GM, Soltis PS, Swensen SM, Williams SE, Gadek PA, Quinn CJ, Eguiarte LE, Golenberg E, Learns GH, Graham SW, Barrett SCH, Dayanandan S, Albert V (1993) Phylogenetics of seed plants: an analysis of nucleotide sequences from the plastid gene rbcL. Ann Mo Bot Gard 80:528–580
9. Qui Y-L, Chase MW, Les DH, Parks CR (1993) Molecular phylogenetics of the Magnoliidae: Cladistic analyses of nucleotide sequences of the plastid gene rbcL. Ann Mo Bot Gard 80:587–606
10. Doyle JA, Donoghue MJ, Zimmer EA (1994) Integration of morphological and ribosomal RNA data on the origin of angiosperms. Ann Mo Bot Gard 81:419–450
11. Taylor DW, Hickey LJ (1996) Evidence for and implication of an herbaceous origin for angiosperms. In: Taylor DW, Hickey LJ (eds) Flowering plant origin, evolution and phylogeny. Chapman & Hall, New York, pp 232–266
12. Donoghue MJ, Doyle JA (1989) Phylogenetic analysis of angiosperms and the relationships of Hamamelidae. In: Crane PR, Blackmore S (eds) Evolution, systematics and fossil history of the Hamamelidae. I, Introduction and "lower" Hamamelidae. Clarendon, Oxford, pp 17–45
13. Donoghue MJ, Doyle JA (1989) Phylogenetic studies of seed plants and angiosperms based on morphological characters. In: Fernholm B, Bremer K, Jörnwall H (eds) The hierarchy of life. Excerpta Medica, Amsterdam, pp 181–193

14. Taylor DW, Hickey LJ (1990) An Aptian plant with attached leaves and flowers: Implications for angiosperms origin. Science 247:702–704
15. Hill CR (1996) A plant with flower-like organs from the Wealden of the Weald (Lower Cretaceous), southern England. Cretaceous Res 17:27–38
16. Friis EM, Skarby A (1981) Structurally preserved angiosperm flowers from the Upper Cretaceous of Southern Sweden. Nature 291:485–486
17. Friis EM, Skarby A (1982) *Scandianthus* gen. nov., angiosperm flowers of saxifragalean affinity from the Upper Cretaceous of southern Sweden. Ann Bot 50:569–583
18. Friis EM (1983) Upper Cretaceous (Senonian) floral structures of Juglandalean affinity containing Normapolles pollen. Rev Palaeobot Palynol 39:161–188
19. Friis EM (1984) Preliminary report on Upper Cretaceous angiosperm reproductive organs from Sweden and their level of organization. Ann Mo Bot Gard 71:403–418
20. Friis EM (1985) Structure and function in Late Cretaceous angiosperm flowers. Niels Bohr Volume Biol Skr Dan Vid Selsk 25:1–37
21. Friis EM, Crane PR, Pedersen KR (1986) Floral evidence for Cretaceous chloranthoid angiosperms. Nature 320:163–164
22. Friis EM, Crane PR, Pedersen KR (1988) Reproductive structure of Cretaceous Platanaceae. Biol Skr Dan Vid Selsk 31:1–56
23. Friis EM, Eklund H, Pedersen KR, Crane PR (1994) *Virginianthus calycanthoides* gen. et sp. nov. A calycanthaceous flower from the Potomac Groups (Early Cretaceous) of eastern North America. Int J Plant Sci 155:772–785
24. Crane PR, Friis EM, Pedersen KR (1986) Angiosperm flowers from the Lower Cretaceous: Fossil evidence on the early radiation of the dicotyledons. Science 232:852–854
25. Crane PR, Friis EM, Pedersen KR (1989) Reproductive structure and function in Cretaceous Chloranthaceae. Pl Syst Evol 165:211–226
26. Drinnan AN, Crane PR, Friis EM, Pedersen KR (1990) Lauraceous flowers from the Potomac Group (mid-Cretaceous) of eastern North America. Bot Gaz 151:370–384
27. Drinnan AN, Crane PR, Pedersen KR, Friis EM (1991) Angiosperm flowers and tricolpate pollen of buxaceous affinity from the Potomac Group (mid-Cretaceous) of eastern North America. Am J Bot 78:153–176
28. Pedersen KR, Crane PR, Drinnan AN, Friis EM (1991) Fruits from the mid-Cretaceous of North America with pollen grains of the *Clavatipollenites* type. Grana 30:577–590
29. Herendeen PS, Crepet WL, Nixon KC (1993) *Chloranthus*-like stamens from the Upper Cretaceous of New Jersey. Am J Bot 80:865–871
30. Herendeen PS, Crepet WL, Nixon KC (1994) Fossil flowers and pollen of Lauraceae from the Upper Cretaceous of New Jersey. Pl Syst Evol 189:29–40
31. Crepet WL, Nixon KC (1994) Flowers of Turonian Magnoliidae and their implications. Pl Syst Evol [Suppl] 8:73–91
32. Friis EM, Crepet WL (1987) Time of appearance of floral features. In: Friis EM, Chaloner WG, Crane PR (eds) The origins of angiosperms and their biological consequences. Cambridge University Press, Cambridge, pp 145–179
33. Crepet WL (1996) Timing in the evolution of derived floral characters: Upper Cretaceous (Turonian) taxa with tricolpate and tricolpate derived pollen. Rev Palaeobot Palynol 90:339–359
34. Walker JW, Walker AG (1984) Ultrastructure of Lower Cretaceous angiosperm pollen and the origin and early evolution of flowering plants. Ann Mo Bot Gard 71:464–521

35. Takhtajan A (1980) Outline of the classification of flowering plants (Magnoliophyta). Bot Rev 46:225–359
36. Cronquist A (1981) An intergrated system of classification of flowering plants. Columbia University Press, New York
37. Huber H (1993) Aristolochiaceae. In: Kubitzki K, Rohwer JG, Bittrich V (eds) The families and genera of vascular plants. II Flowering plants—dicotyledons. Magnoliid, Hamamelid and Caryophyllid families. Springer, Berlin Heidelberg New York, pp 129–137
38. Endress PK (1994) Diversity and evolutionary biology of tropical flowers. Cambridge University Press, Cambridge
39. Muller J (1981) Fossil pollen records of extant angiosperms. Bot Rev 47:1–142
40. Kulkarni AR, Patil KS (1977) *Aristolochioxylon prakashii* from the Deccan Inter-trappean Beds of wardha district, Maharashtra. Geophytol 7:44–49
41. Kolakovsky AA (1957) Pervoe dopolnenie k kodorskoj flore (Meore-Atara) (in Russian). Tr Such Bot Cada 10
42. Kolakovsky AA (1964) Pliotsenovaja flora Pitsundy (in Russian). Tr Such Bot Cada 14
43. Pimenova HV (1954) Sarmatskaja flora Amvrosievki (in Russian). Tr Inst geol Nauk AN USSR, ser Stratigr i Paleontol 8
44. Czeczott H (1951) Srodkowo-miocenska flora Zalesiec kolo Wisniowca (in Polish). 1. Acta Geol Pol 2(3):349–445
45. MacGinitie HD (1953) Fossil plants of the Florrissant Beds, Colorado. Publ Carnegie Inst Wash 599:1–197
46. MacGinitie HD (1969) The Eocene Green River flora of northwestern Colorado and northeastern Utah. Univ Calif Publ Geo Sci 83:1–140
47. MacGinitie HD (1974) An early middle Eocene flora from the Yellowstone-Absaroka Volcanic province, northwestern Wind River Basin, Wyoming. Univ Calif Publ Geo Sci 108:1–103
48. Endress PK (1994) Evolutionary aspects of the floral structure in *Ceratophyllum*. Pl Syst Evol [Suppl] 8:175–183
49. Dilcher DL (1989) The occurrence of fruits with affinity to Ceratophyllaceae in Lower and mid-Cretaceous sediments. Am J Bot 6 [Suppl]:162
50. Dorofeev PI (1963) Treticnye flory Zapadnoj Sibiri (in Russian). Izd Akad nauk SSSR, Moskva
51. Dorofeev PI (1988) Miocenovye flory Tambovskoi oblasti (in Russian). Nauka, Leningrad
52. Mai DH (1985) Entwicklung der Wasser- und Sumpfpflanzen-Gesellschaften Europas von der Kreide bis ins Quartär. Flora 176:449–511
53. Mai DH (1995) Tertiäre Vegetationsgeschichte Europas. Gustav Fischer, Jena
54. Herendeen PS, Les DH, Dilcher DL (1990) Fossil *Ceratophyllum* (Ceratophyllaceae) from the Tertiary of North America. Am J Bot 77:7–16
55. Keng H (1993) Illiciaceae. In: Kubitzki K, Rohwer JG, Bittrich V (eds) The families and genera of vascular plants. II Flowering plants—dicotyledons. Magnoliid, Hamamelid and Caryophyllid families. Springer, Berlin Heidelberg New York, pp 344–347
56. Chmura CA (1973) Upper Cretaceous (Campanian-Maastrichtian) angiosperm pollen from the Western San Joaquin Valley, California, U.S.A. Palaeontographica 141B:89–171
57. Upchurch GR, Dilcher DL (1990) Cenomanian angiosperm leaf megafossils, Dakota Formation, Rose Creek Locality, Jefferson County, southeastern Nebraska. US Geol Surv Bull 1915:1–55

58. Tiffney BH, Barghoorn ES (1979) Fruits and seeds of the Brandon Lignite IV. Illiciaceae. Am J Bot 66:321–329
59. Keng H (1993) Schizandraceae. In: Kubitzki K, Rohwer JG, Bittrich V (eds) The families and genera of vascular plants. II Flowering plants—dicotyledons. Magnoliid, Hamamelid and Caryophyllid families. Springer, Berlin Heidelberg New York, pp 589–592
60. Gregor H-J (1981) *Schisandra geissertii* nova spec.—ein exotisches Element in Elsässer Pliozän (Sessenheim, Brunssumien). Mitt. badischen Landesvereins Naturkunde und Naturschutz 12:241–247
61. Mai DH, Walther H (1985) Die obereozänen Floren des Weisselster-Beckens (Bezirk Leipzig). Abh Staatl Mus Mineral Geol Dresden 33:1–220
62. Manchester SR (1994) Fruits and seeds of the Middle Eocene Nut Beds Flora, Clarno Formation, Oregon. Palaeontogr Am 58:1–205
63. Carlquist S (1990) Wood anatomy and the relationships of Lactoridaceae. Am J Bot 77:1498–1504
64. Kubitzki K (1993) Lactoridaceae. In: Kubitzki K, Rohwer JG, Bittrich V (eds) The families and genera of vascular plants. II Flowering plants—dicotyledons. Magnoliid, Hamamelid and Caryophyllid families. Springer, Berlin Heidelberg New York, pp 359–361
65. Zavada MZ, Taylor TN (1986) Pollen morphology of Lactoridaceae. Pl Syst Evol 154:31–39
66. Zavada MZ, Benzon JM (1987) First fossil evidence for the primitive angiosperm family Lactoridaceae. Am J Bot 74:1590–1594
67. Kvaček Z (1992) Lauralean angiosperms in the Cretaceous. Cour Forch-Inst Senckenberg, 147:345–367
68. Vélenovský J (1885) Die Flora der Böhmischen Kreideformation. 4. Beitr Paläont Österreich-Ungarns, Orient 5(1):62–75
69. Rohwer JG (1993) Lauraceae. In: Kubitzki K, Rohwer JG, Bittrich V (eds) The families and genera of vascular plants. II Flowering plants—dicotyledons. Magnoliid, Hamamelid and Caryophyllid families. Springer, Berlin Heidelberg New York, pp 366–391
70. Herendeen PS (1991) Charcoalified angiosperm wood from the Cretaceous of eastern North America and Europe. Rev Palaeobot Palynol 70:225–239
71. Retallack G, Dilcher DL (1981) Early angiosperm reproduction: *Prisca reynoldsii* gen. et sp. nov. from mid-Cretaceous coastal deposits, Kansas, USA. Palaeonotographica 179B:103–137
72. Kovach WL, Dilcher DL (1988) Megaspores and other dispersed plant remains from the Dakota Formation (Cenomanian) of Kansas, U.S.A. Palynology 12:89–119
73. Taylor DW (1988) Eocene floral evidence of Lauraceae: Corroboration of the North American megafossil record. Am J Bot 75:948–957
74. Weyland H (1938) Beiträge zur kenntnis der rheinischen Tertiär flora. III. Palaeontographica 83B:123–171
75. Conwentz H (1886) Die Flora des Bernsteins, 2. Die Angiospermen des Bernsteins. Wilhelm Engelmann, Leipzig. pp 140
76. Philipson WR (1993) Monimiaceae. In: Kubitzki K, Rohwer JG, Bittrich V (eds) The families and genera of vascular plants. II Flowering plants—dicotyledons. Magnoliid, Hamamelid and Caryophyllid families. Springer, Berlin Heidelberg New York, pp 426–437

77. Rüffle L (1965) Monimiaceen-Blätter im älteren Senon von Mitteleuropa. Geologie 14:78–89
78. Rüffle L, Knappe H (1988) Ökologische und paläogeographische Bedeutung der Oberkreideflora von Quedlinburg, besonders einiger Loranthaceae und Monimiaceae. Hall Jb f Geowiss 13:49–65
79. Dusén P (1908) Über die tertiäre Flora der Seymour-Insel. Wiss Erg Schwed Südpol Exped 1901–1903 3(3):1–27
80. Couper RA (1960) New Zealand Mesozoic and Cainozoic plant microfossils. New Zeal Geol Surv Palaeontol Bull 32:1–87
81. Kubitzki K (1993) Hernandiaceae. In: Kubitzki K, Rohwer JG, Bittrich V (eds) The families and genera of vascular plants. II Flowering plants—dicotyledons. Magnoliid, Hamamelid and Caryophyllid families. Springer, Berlin Heidelberg New York, pp 457–467
82. Berry EW (1937) *Gyrocarpus* and other fossil plants from the Cumarebo field in Venezuela. J Washington Acad Sci 27:501–550
83. Kubitzki K (1993) Gomortegaceae. In: Kubitzki K, Rohwer JG, Bittrich V (eds) The families and genera of vascular plants. II Flowering plants—dicotyledons. Magnoliid, Hamamelid and Caryophyllid families. Springer, Berlin Heidelberg New York, pp 318–320
84. Kubitzki K (1993) Calycanthaceae. In: Kubitzki K, Rohwer JG, Bittrich V (eds) The families and genera of vascular plants. II Flowering plants—dicotyledons. Magnoliid, Hamamelid and Caryophyllid families. Springer, Berlin Heidelberg New York, pp 197–200
85. Mai DH (1987) Neue Arten nach Früchten und Samen aus dem Tertiär von Nordwestsachsen und der Lauzitz. Fedd Repert 98:105–126
86. Philipson WR (1993) Amborellaceae. In: Kubitzki K, Rohwer JG, Bittrich V (eds) The families and genera of vascular plants. II Flowering plants—dicotyledons. Magnoliid, Hamamelid and Caryophyllid families. Springer, Berlin Heidelberg New York, pp 92–93
87. Endress PK, Sampson FB (1983) Floral structure and relationships of the Trimeniaceae (Laurales). J Arnold Arb 64:447–473
88. Herngreen GFW (1973) Palynology of Albian-Cenomanian strata of borehole 1-QS-1-MA, State of Maranhao, Brazil. Pollen Spores 15:515–555
89. Dettmann ME (1994) Cretaceous vegetation: the microfossil record. In: Hill RS (ed) History of the Australian vegetation: Cretaceous to Recent. Cambridge University Press, Cambridge, pp 143–170
90. Macphail MK, Alley NF, Truswell EM, Sluiter IRK (1994) Early Tertiary vegetation: evidence from spores and pollen. In: Hill RS (ed) History of the Australian vegetation: Cretaceous to Recent. Cambridge University Press, Cambridge, pp 328–367
91. Blackburn DT, Sluiter IRK (1994) The Oligo-Miocene coal floras of southeastern Australia. In: Hill RS (ed) History of the Australian vegetation: Cretaceous to Recent. Cambridge University Press, Cambridge, pp 328–367
92. Todzia CA (1993) Chloranthaceae. In: Kubitzki K, Rohwer JG, Bittrich V (eds) The families and genera of vascular plants. II Flowering plants—dicotyledons. Magnoliid, Hamamelid and Caryophyllid families. Springer, Berlin Heidelberg New York, pp 281–289
93. Wettstein RR von (1907) Handbuch der Systematischen Botanik. (2 ed) Franz Deuticke, Leipzig

94. Endress PK (1987) The Chloranthaceae: reproductive structures and phylogenetic position. Bot Jb Syst 109:153–226
95. Couper RA (1958) British Mesozoic microspores and pollen grains. Palaeontographica 103B:75–179
96. Brenner GJ (1996) Evidence for the earliest stage of angiosperm pollen evolution: A paleoequatorial section from Israel. In: Taylor DW, Hickey LJ (eds) Flowering plant origin, evolution and phylogeny. Chapman & Hall, New York, pp 91–115
97. Herngreen GFW (1975) Palynology of Middle and Upper Cretaceous strata in Brazil. Med R Geol D n s 26:39–116
98. Doyle JA, Robbins EI (1977) Angiosperm pollen zonation of the continental Cretaceous of the Atlantic Coastal Plain and its application to deep wells in the Salisbury embayment. Palynology 1:43–78
99. Raine JI, Speden IG, Strong CP (1981) New Zealand. In: Reyment RA, Bengtson P (eds) Aspects of mid-Cretaceous regional geology. Academic, London, pp 221–267
100. Hughes NF, McDougall AB (1987) Records of angiospermid pollen entry into the English Early Cretaceous succession. Rev Palaeobot Palynol 50:255–272
101. Archangelsky S, Taylor TN (1993) The ultrastructure of in situ Clavatipollenites pollen from the Early Cretaceous of Patagonia. Am J Bot 80:879–885
102. Dettmann ME (1989) Antarctica: Cretaceous cradle of austral temperate rainforests? In: Crame JA (ed) Origins and Evolution of the Antarctic Biota. Geol Soc Spec Publ 47:89–105
103. Hedlund RW, Norris G (1968) Spores and pollen grains from Fredericksburgian (Albian) strata, Marshall County, Oklahoma. Pollen Spores 10:129–159
104. Groot JJ, Groot CR (1962) Plant microfossils from Aptian, Albian and Cenomanian deposits of Portugal. Comun Serv Geol Portugal 46:133–176
105. Pais J, Reyre Y (1981) Problèmes posés par la population sporo-pollinique d'un niveau à plantes de la série de Buarcos (Portugal). Soc Geol Portugal Boletim 22:35–40
106. Singh C (1983) Cenomanian microfloras of the Peace River area, northwestern Alberta. Res Coun Alta Bull 44:1–322
107. Schrank E (1987) Paleozoic and Mesozoic palynomorphs from Northeast Africa (Egypt and Sudan) with special reference to Late Cretaceous pollen and dinoflagellates. Berliner geowiss Abh (A) 75.1:249–310
108. Phillips PP, Felix CJ (1971) A study of Lower and Middle Cretaceous spores and pollen from the southeastern United States. II. Pollen. Pollen Spores 13:447–473
109. Davies EH, Norris G (1976) Ultrastructural analysis of exine and apertures in angiospermous colpoid pollen (Albian, Oklahoma). Pollen Spores 28:129–144
110. Doyle JA (1969) Cretaceous angiosperm pollen of the Atlantic Coastal Plain and its evolutionary significance. J Arnold Arbor 50:1–35
111. Ward JV (1986) Early Cretaceous angiosperm pollen from the Cheyenne and Kiowa Formations (Albian) of Kansas, USA. Palaeontographica 202B:1–81
112. Srivastava SK (1977) Microspores from the Fredericksburg Group (Albian) of the southern United States. Paléobiologie Continentale 6(2):1–119
113. Dettmann ME (1973) Angiospermous pollen from Albian to Turonian sediments of eastern Australia. Spec Publ Geol Soc Aust 4:3–34
114. Burger D (1980) Palynological studies in the Lower Cretaceous of the Surat Basin, Australia. Spec Publ Geol Soc Aust 4:3–34
115. Burger D (1980) Palynological studies in the Lower Cretaceous of the Surat Basin, Australia. Bur Miner Resour Geol Geophys (Canberra) Bull 189:1–106

116. Eklund H, Friis EM, Pedersen KR (1996) Late Cretaceous reproductive organs of chloranthaceous affinity from Scania, southern Sweden. Pl Syst Evol (in press)

117. Endress PK (1993) Austrobaileyaceae. In: Kubitzki K, Rohwer JG, Bittrich V (eds) The families and genera of vascular plants. II Flowering plants—dicotyledons. Magnoliid, Hamamelid and Caryophyllid families. Springer, Berlin Heidelberg New York, pp 138–140

118. Crane PR, Dilcher DL (1984) *Lesqueria*: an early angiosperm fruiting axis from the mid-Cretaceous. Ann Mo Bot Gard 71:384–402

119. Lesquereux L (1892) Flora of the Dakota Group. Monogr US Geol Surv 17:1–400

120. Dawson JW (1886) On the Mesozoic floras of the Rocky Mountain Region of Canada. Proc Trans R Soc Canada 3:1–22

121. Seward AC (1917) Fossil plants. Vol 3, Pteridospermae, Cycadofilices, Cordaites, Cycadophyta. Cambridge University Press, Cambridge

122. Bell WA (1956) Lower Cretaceous floras of western Canada. Mem Geol Surv Branch Dept Mines Canada 285:1–331

123. Newberry JS (1886) Notes on the later extinct floras of North America. Am J Sci 46:401–407

124. Newberry JS (1895) The flora of the Amboy Clays. Hollick A (ed) Monogr US Geol Surv 26:1–260

125. Nishida H, Nishida M (1988) *Protomonimia kasai-nakajhongii* gen. et sp. nov.: a permineralized magnolialean fructification from the mid-Cretaceous of Japan. Bot Mag Tokyo 101:397–437

126. Ward JV, Doyle JA, Hotton CL (1989) Probable granular magnoliid angiosperm pollen from the Early Cretaceous. Pollen Spores 33:113–132

127. Doyle JA, Hotton CL (1991) Diversification of early angiosperm pollen. In: Blackmore S, Barnes SH (eds) Pollen and spores, patterns of diversity. Systematics Association Special Volume 44. Clarendon, Oxford, pp 169–195

128. Nooteboom HP (1993) Magnoliaceae. In: Kubitzki K, Rohwer JG, Bittrich V (eds) The families and genera of vascular plants. II Flowering plants—dicotyledons. Magnoliid, Hamamelid and Caryophyllid families. Springer, Berlin Heidelberg New York, pp 391–401

129. Baghai NL (1988) Liriodendron (Magnoliaceae) from the Miocene Clarkia flora of Idaho. Am J Bot 75:451–464

130. Chandler MEJ (1964) The Lower Tertiary floras of southern England. IV. A summary and survey of findings in light of recent botanical observations. Br Mus (Nat Hist), London

131. Reid EM, Chandler MEJ (1933) The flora of the London Clay. Br Mus (Nat Hist), London

132. Dorofeev PI (1970) Treticnye flory Urala (in Russian). Izd Nauka, Leningrad

133. Dorofeev PI (1974a) Liriodendron. In: Takhtajan A (ed) Iskopaemye cvetkovye rastenija SSSR. I (in Russian). Izd Nauka, Leningrad, pp 18–20

134. Dorofeev PI (1983) Dva novych vida Liriodendron iz treticnych otlozenii SSSR (in Russian). Bot Zurn 68:1401–1408

135. Tiffney BH (1977) Fruits and seeds of the Brandon Lignite: Magnoliaceae. Bot J Linn Soc 75:299–323

136. Friis EM (1985) Angiosperm fruits and seeds from the Middle Miocene of Jutland (Denmark). Biol Skr Dan Vid Selsk 24(3):1–165

137. Wheeler E, Scott RA, Barghoorn ES (1977) Fossil dicotyledonous wood from Yellowstone National Park. J Arnold Arbor 58:280–306

138. Scott RA, Wheeler E (1982) Fossil wood from the Eocene Clarno Formation of Oregon. Int Assoc Wood Anat Bull ns 3:135–154

139. Cevallos-Ferriz SRS, Stockey RA (1990) Vegetative remains of the Magnoliaceae from the Princeton chert (Middle Eocene) of British Columbia. Can J Bot 68:1327–1339

140. Dilcher DL, Crane PR (1984) *Archaeanthus*: An early angiosperm from the Cenomanian of the western interior of North America. Ann Mo Bot Gard 71:351–383

141. Hollick A (1903) A fossil petal and a fossil fruit from the Cretaceous (Dakota Group) of Kansas. Bull Torrey Bot Club 30:102–105

142. Seward AC, Conway VM (1935) Fossil plants from Kingigtok and Kagdlunguak, West Greenland. Meddel Grønland 93(5):1–41

143. Delevoryas T, Mickle JE (1995) Upper Cretaceous magnoliaceous fruit from British Columbia. Am J Bot 82:763–768

144. Frumin S, Friis EM (1996) Liriodendroid seeds from the Late Cretaceous of Kazakhstan and North Carolina, USA. Rev Palaeobot Palynol 94:39–55

145. Knobloch E, Mai DH (1984) Neue Gattungen nach Früchten und Samen aus dem Cenoman bis Maastricht (Kreide) von Mitteleuropa. Fedd Repert 95:3–41

146. Knobloch E, Mai DH (1986) Monographie der Früchte und Samen in der Kreide von Mitteleuropa. Rozpravy ústředního ústavu geologického, Praha 47:1–219

147. Kessler PJA (1993) Annonaceae. In: Kubitzki K, Rohwer JG, Bittrich V (eds) The families and genera of vascular plants. II Flowering plants—dicotyledons. Magnoliid, Hamamelid and Caryophyllid families. Springer, Berlin Heidelberg New York, pp 93–129

148. Chandler MEJ (1954) Some Upper Cretaceous and Eocene fruits from Egypt. Bull Br Mus (Nat Hist) Geol 2:147–187

149. Collinson ME, Boulter MC, Holmes PL (1993) Magnoliophyta ("Angiospermae"). In: Benton MJ (ed) The fossil record 2. Chapman & Hall, London, pp 809–841

150. Tiffney BH, McClammer JU (1988) A seed of the Anonaceae from the Palaeocene of Pakistan. Tert Res 9:13–20

151. Chester KIM (1955) Some plant remains from the Upper Cretaceous and Tertiary of West Africa. Ann Mag Nat Hist (12)8:498–503

152. Monteillet J, Lappartient J-R (1981) Fruits et graines du Crétacé supérior des carrières de Paki. Rev Palaeobot Palynol 34:331–344

153. Sole de Porta N (1971) Algunos generos nuevos de polen procedentes de la Formación Guaduas (Maastrichtiense-Paleoceno) de Colombia. Studia Geol, Salamanca 2:133–143

154. Kubitzki K (1993) Degeneriaceae. In: Kubitzki K, Rohwer JG, Bittrich V (eds) The families and genera of vascular plants. II Flowering plants—dicotyledons. Magnoliid, Hamamelid and Caryophyllid families. Springer, Berlin Heidelberg New York, pp 290–291

155. Kühn U, Kubitzki K (1993) Myristicaceae. In: Kubitzki K, Rohwer JG, Bittrich V (eds) The families and genera of vascular plants. II Flowering plants—dicotyledons. Magnoliid, Hamamelid and Caryophyllid families. Springer, Berlin Heidelberg New York, pp 457–467

156. Boureau E (1950) Étude paléoxylologique du Sahara (IX). Sur un *Myristicoxylon princeps* n. gen., nov. sp., du Danien d'Asselar. Bull Mus Hist Nat Paris II 22:523–528

157. Gregor H-J (1978) Die miozänen Frucht- und Samen-Floren der Oberpfälzer Braunkohle. I. Funde aus den sandigen Zwischenmitteln. Palaeontographica 167B:1–103

158. Endress PK (1993) Himantandraceae. In: Kubitzki K, Rohwer JG, Bittrich V (eds) The families and genera of vascular plants. II Flowering plants—dicotyledons. Magnoliid, Hamamelid and Caryophyllid families. Springer, Berlin Heidelberg New York, pp 338–341

159. Endress PK (1993) Eupomatiaceae. In: Kubitzki K, Rohwer JG, Bittrich V (eds) The families and genera of vascular plants. II Flowering plants—dicotyledons. Magnoliid, Hamamelid and Caryophyllid families. Springer, Berlin Heidelberg New York, pp 296–298

160. Kubitzki K (1993) Canellaceae. In: Kubitzki K, Rohwer JG, Bittrich V (eds) The families and genera of vascular plants. II Flowering plants—dicotyledons. Magnoliid, Hamamelid and Caryophyllid families. Springer, Berlin Heidelberg New York, pp 200–203

161. Graham A, Jarzen DM (1969) Studies in neotropical paleobotany. I. The Oligocene communities of Puerto Rico. Ann Mo Bot Gard 56:308–357

162. Vink W (1993) Winteraceae. In: Kubitzki K, Rohwer JG, Bittrich V (eds) The families and genera of vascular plants. II Flowering plants—Dicotyledons. Magnoliid, Hamamelid and Caryophyllid families. Springer, Berlin Heidelberg New York, pp 630–638

163. Doyle JA, Hotton CL, Ward JV (1990) Early Cretaceous tetrads, zonosulcate pollen, and Winteraceae. I. Taxonomy, morphology, and ultrastructure. Am J Bot 77:1544–1557

164. Doyle JA, Hotton CL, Ward JV (1990) Early Cretaceous tetrads, zonosulcate pollen, and Winteraceae. II. Cladistic analysis and implications. Am J Bot 77:1558–1568

165. Walker JW, Brenner GJ, Walker AG (1983) Winteraceous pollen in the Lower Cretaceous of Israel: early evidence of a magnolialean angiosperm family. Science 220:1273–1275

166. Page VM (1979) Dicotyledonous wood from the Upper Cretaceous of Central California. J Arnold Arbor 60:323–349

167. Specht RL, Dettmann ME, Jarzen DM (1992) Community associations and structure in the Late Cretaceous vegetation of southeast Australasia and Antarctica. Palaeogeogr Palaeoclimatol Palaeoecol 94:283–309

168. Mildenhall DC, Crosbie YM (1979) Some porate pollen from the Upper Tertiary of New Zealand. N Z J Geol Geophys 22:499–508

169. Coetzee JA, Muller J (1984) The phytogeographic significance of some extinct Gondwana pollen types from the Tertiary of the southwestern Cape (South Africa). Ann Mo Bot Gard 71:1088–1099

170. Coetzee JA, Praglowski J (1987) Winteraceae pollen from the Miocene of the southwesten Cape (South Africa). Grana 27:27–37

171. Teixeira C (1945) Nymphéacées fossiles du Portugal. Serviços Geológicos de Portugal, Lisbon

172. Saporta G de (1894) Flore fossile du Portugal. Nouvelles contributions à la flore Mezoique. Accompagnées d'une notice stratigraphique par Paul Choffat. Imprimerie de l'Academie Royale des Sciences, Lisbon

173. Schneider El, Williamson PS (1993) Nymphaeaceae. In: Kubitzki K, Rohwer JG, Bittrich V (eds) The families and genera of vascular plants. II Flowering plants—dicotyledons. Magnoliid, Hamamelid and Caryophyllid families. Springer, Berlin Heidelberg New York, pp 486–493

174. Collinson ME (1980) Recent and Tertiary seeds of the Nymphaeaceae *sensu lato* with a revision of *Brasenia ovula* (Brong.) Reid and Chandler. Ann Bot 46:603–632

175. Williamson PS, Schneider EI (1993) Cabombaceae. In: Kubitzki K, Rohwer JG, Bittrich V (eds) The families and genera of vascular plants. II Flowering plants— dicotyledons. Magnoliid, Hamamelid and Caryophyllid families. Springer, Berlin Heidelberg New York, pp 157–161

176. Dorofeev PI (1973) Systematics of ancestral forms of *Brasenia*. Paleontol J 7:219–227

177. Dorofeev PI (1974) Nymphaeales. In: Takhtajan A (ed) Iskopaemye cvetkovye rastenija SSSR. I (in Russian). Izd Nauka, Leningrad, pp 52–85

178. Srivastava SK (1969) Assorted angiosperm pollen from the Edmonton Formation (Maastrichtian), Alberta, Canada. Can J Bot 47:975–989

179. Wolfe JA (1991) Palaeobotanical evidence for a June "impact" at the Cretaceous/ Tertiary boundary. Nature 352:420–423

180. Nichols DJ (1992) Plants at the K/T boundary. Nature 356:295

181. Góczán F, Juhász M (1984) Monosulcate pollen grains of angiosperms from Hungarian Albian sediments. I. Acta Bot Hung 30:289–319

182. Regali MSP (1989) *Tucanopollis*, um género novo das angiospermas primitivas. Bol Geoscié Petrobras 3:395–402

183. Friis EM, Pedersen KR, Crane PR (1995) *Appomattoxia ancistrophora* gen. et sp. nov., a new Early Cretaceous plant with similarities to *Circaeaster* and extant Magnoliidae. Am J Bot 82:933–943

184. Tebbs MC (1993) Piperaceae. In: Kubitzki K, Rohwer JG, Bittrich V (eds) The families and genera of vascular plants. II Flowering plants—dicotyledons. Magnoliid, Hamamelid and Caryophyllid families. Springer, Berlin Heidelberg New York, pp 516–520

185. Cheng-Yih W, Kubitzki K (1993) Saururaceae. In: Kubitzki K, Rohwer JG, Bittrich V (eds) The families and genera of vascular plants. II Flowering plants— dicotyledons. Magnoliid, Hamamelid and Caryophyllid families. Springer, Berlin Heidelberg New York, pp 586–588

186. Stopes MC, Fujii K (1910) Studies on the structure and affinities of Cretaceous plants. Phil Trans 201B:1–90

187. Doyle JA (1973) The monocotyledons: their evolution and comparative biology. V. Fossil evidence on early evolution of the Monocotyledons. Quart Rev Biol 48:399–413

188. Herendeen PS, Crane PR (1995) The fossil history of the monocotyledons. In: Rudall PJ, Cribb PJ, Cuttler DF, Humphries CJ (eds) Monocotyledons: systematics and evolution. R Bot Gard Kew, pp 1–21

189. Drinnan AN, Crane PR, Hoot SB (1994) Patterns of floral evolution in the early diversification of non-magnoliid dicotyledons (eudicots). Pl Syst Evol [Suppl] 8:93–122

190. Endress PK (1990) Evolution of reproductive structures and functions in primitive angiosperms (Magnoliidae). Mem New York Bot Gard 55:5–34

Molecular Phylogenetic Relationships Among Angiosperms: An Overview Based on *rbcL* and 18S rDNA Sequences

Douglas E. Soltis[1], Carola Hibsch-Jetter[1], Pamela S. Soltis[1], Mark W. Chase[2], and James S. Farris[3]

7.1 Introduction

During only the past 4 years, two large data sets of DNA sequences have greatly clarified the broad picture of angiosperm relationships and evolution. By far the more extensive of these two data sets is that based on the chloroplast gene *rbcL*, with sequences representing 499 species of seed plants [1]. More recently, a smaller data set representing 223 angiosperms has been compiled for the nuclear 18S ribosomal RNA gene [2]. The general structural features, rate of evolution, and phylogenetic utility of both genes have been previously reviewed [3,4] and are not discussed here. For the first time, species representing the diversity of angiosperms have been sequenced for both a chloroplast and a nuclear gene. Visual inspection of the *rbcL* and 18S rDNA topologies suggests a high degree of overall concordance. Particularly noteworthy is the support provided by the 18S rDNA topology for those nontraditional relationships suggested by the *rbcL* trees. For example, 18S rDNA sequence analysis confirms the *rbcL*-based inferences that Droseraceae and Nepenthaceae are closely related to the Caryophyllidae and that a number of members of Rosidae and Dilleniidae have as their closest relatives members of Asteridae.

We first provide an overview of the topologies resulting from broad phylogenetic analyses of *rbcL* [1] and 18S rDNA sequences [2]. We not only emphasize the striking similarities between these chloroplast- and nuclear-based trees but also point to areas of discrepancy in the hope that these differences will serve as a stimulus for additional research. Second, although *rbcL* and 18S rDNA trees appear highly concordant, we consider more thoroughly the congruence of *rbcL* and 18S rDNA data sets for angiosperms. To test the congruence of the two data sets, we assembled *rbcL* and 18S rDNA sequences for a nearly identical suite of

[1] Department of Botany, Washington State University, Pullman, WA 99164-4238, USA
[2] Laboratory of Molecular Systematics, Royal Botanic Gardens, Kew, Richmond, Surrey TW9 3AB, UK
[3] Naturhistoriska Riksmusect, Molekylärsystematiska Laboratoriet, P.O. Box 50087 S-104 05 Stockholm, Sweden

226 angiosperms, plus six outgroups, and conducted a homogeneity test. Using these separate data matrices, we also examined the effects of combining large data sets by conducting parsimony jackknife analyses [5] on the separate and combined data sets.

7.2 Visual Comparison of the *rbcL* and 18S rDNA Topologies

7.2.1 Overview

The *rbcL* and 18S rDNA trees of Chase et al. [1] and Soltis et al. [2] are summarized in Figs. 7.1 and 7.2. Both suggest that a division exists in the angiosperms that agrees with palynological and chemical data [6–8]: taxa having uniaperturate pollen and ethereal oils (the monosulcates) and taxa having triaperturate pollen and tannins/alkaloids (the eudicots). The monosulcates include the woody magnoliids, monocots, and paleoherbs; the eudicots are strongly supported and comprise all other dicot groups. In the *rbcL* trees, the monosulcates form a weakly supported but distinct clade that is sister to the eudicot clade. In contrast, the shortest 18S rDNA trees also suggest a monosulcate "grade" at the base of the angiosperms that is terminated by a eudicot clade. In both trees, Schisandraceae and Illiciaceae, taxa having triaperturate pollen, appear among the monosulcates. As noted by Doyle et al. [9], however, the tricolpate condition in Illiciaceae and Schisandraceae differs from that in the eudicots. Thus, both *rbcL* and 18S rDNA sequence data confirm the view [9] that the tricolpate condition in these two families is not homologous to that in the eudicots. Several of the 18S rDNA searches place one group of paleoherbs (Aristolochiaceae–Lactoridaceae–Chloranthaceae) at the base of the eudicots. This is true, for example, in the tree summarized in Fig. 7.2 (see Paleoherb B). Other searches, however, have placed this second group of paleoherbs with the monosulcates, as would be expected [2].

In the large eudicot clade, both the *rbcL* and 18S rDNA topologies place many of the same families at the base: Nelumbonaceae, Platanaceae, Proteaceae, Sabiaceae, ranunculids (Ranunculaceae and segregate families, Menispermaceae, Papaveraceae, Berberidaceae, Fumariaceae), Trochodendraceae, and Tetracentraceae. Although the placement of some of these taxa at the base of the eudicots by *rbcL* sequence data (e.g., Sabiaceae, Proteaceae) was initially considered somewhat surprising, the similar placement of these taxa by 18S rDNA sequences reinforces evidence of their antiquity as inferred from the fossil record (e.g., P. Crane, personal communication).

Considering the higher eudicots, analyses of *rbcL* and 18S rDNA sequences again are in agreement in suggesting the presence of two major clades: Rosidae and Asteridae *s. l.* These two large clades correspond to a basic division in embryology and corolla condition: nontenuinucellate ovule/polypetalous corolla versus tenuinucellate ovule/sympetalous corolla, although exceptions to these

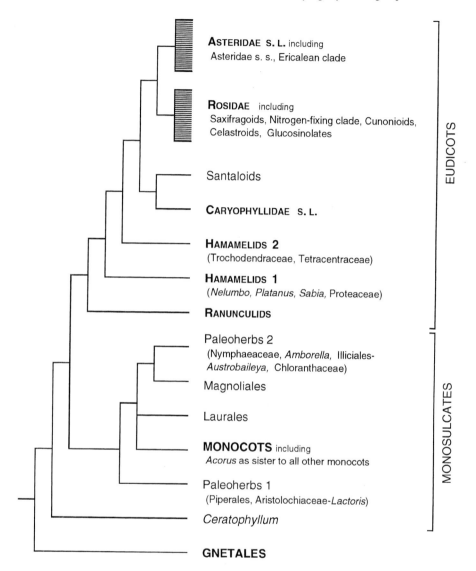

FIG. 7.1. Summary of the major clades identified in the strict consensus of 3900 shortest trees for 499 taxa based on *rbcL* sequence data [1]. (Modified from Figure 1B of [1])

character states exist (e.g., some Asteridae s. l. have free petals). The presence of a Rosidae clade distinct from Asteridae s. l. is also noteworthy in that modern classification schemes (e.g., [10]) have suggested that Asteridae may be derived from within the rosids. Both molecular topologies suggest instead the possibility of a common ancestor of these two large clades. Furthermore, both the *rbcL* and 18S rDNA trees strengthen the notion of an expanded Asteridae (Asteridae *s. l.*)

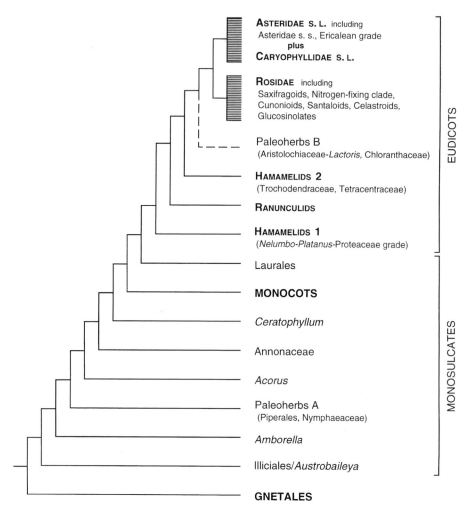

ASTERIDAE S. L. including
Asteridae s. s., Ericalean grade
plus
CARYOPHYLLIDAE S. L.

ROSIDAE including
Saxifragoids, Nitrogen-fixing clade,
Cunonioids, Santaloids, Celastroids,
Glucosinolates

Paleoherbs B
(Aristolochiaceae-*Lactoris,* Chloranthaceae)

HAMAMELIDS 2
(Trochodendraceae, Tetracentraceae)

RANUNCULIDS

HAMAMELIDS 1
(*Nelumbo-Platanus*-Proteaceae grade)

Laurales

MONOCOTS

Ceratophyllum

Annonaceae

Acorus

Paleoherbs A
(Piperales, Nymphaeaceae)

Amborella

Illiciales/*Austrobaileya*

GNETALES

EUDICOTS

MONOSULCATES

FIG. 7.2. Summary of the major clades identified in the strict consensus of 5294 shortest trees for 228 taxa based on 18S rDNA sequence data [2]. *Dashed line* to paleoherbs B indicates that the position of these taxa varies among searches; in some searches they appear with the monosulcates. (Modified from Figure 1 of [2])

clade that in addition to Asteridae s. s. includes traditional rosids (e.g., Apiaceae, Araliaceae, Garryaceae, Pittosporaceae), dilleniids (e.g., Ericaceae, Primulaceae, Actinidiaceae, Theaceae, Diapensiaceae, Sarraceniaceae), and the hamamelid *Eucommia.* This concept was suggested late in the nineteenth century [11], subsequently abandoned by some [10], and then recently supported by cladistic analysis of morphological and chemical data [12] as well as *rbcL* sequences [13]. Within Asteridae *s. l.,* both data sets identify an ericalean assemblage (as either a distinct clade or grade) that includes many traditional Dilleniidae. Both data sets also suggest the monophyly of the following: Apiales, Dipsacales, Boraginales, and Lamiales.

Analysis of 18S rDNA and *rbcL* sequences also recovered an expanded Caryophyllidae (Caryophyllidae *s. l.*). In addition to the core families of Caryophyllales plus Plumbaginaceae and Polygonaceae, Caryophyllidae *s. l.* also contains two families of carnivorous plants, Droseraceae and Nepenthaceae. Although both *rbcL* and 18S rDNA sequences support the monophyly of Caryophyllidae *s. l.*, the placement of Caryophyllidae *s. l.* differs between the two topologies. In the former [1], Caryophyllidae *s. l.* are embedded within the Rosidae, although this position varies from search to search (compare the A and B topologies of Chase et al. [1]). In contrast, the analyses of 18S rDNA sequences [2] place Caryophyllidae *s. l.* within Asteridae *s. l.* However, analyses of 18S rDNA sequences that focused primarily on eudicots also placed Caryophyllidae *s. l.* within the Rosidae [14].

In both the *rbcL* and 18S rDNA trees, Hamamelidae are grossly polyphyletic and comprise several phylogenetically distant lineages. For example, Trochodendraceae, Tetracentraceae, and Platanaceae consistently appear near the base of the eudicots. In contrast, several other families of Hamamelidae appear in several different portions of the Rosidae clade. Hamamelidaceae, Cercidiphyllaceae, and Daphniphyllaceae are part of a saxifragoid or Saxifragales clade [14], whereas Betulaceae, Urticaceae, Moraceae, and Ulmaceae are components of a nitrogen-fixing clade. Eupteleaceae appear with the ranunculids in both analyses (see also [15]), a placement apparently not suggested based on morphology; as noted, Eucommiaceae are nested within Asteridae *s. l.*

Dilleniidae are also clearly polyphyletic in both the *rbcL* and 18S rDNA trees, with families of this subclass occurring in several distinct clades within Rosidae, Asteridae *s. l.*, and Caryophyllidae *s. l.* For example, Nepenthaceae are part of Caryophyllidae *s. l.* while several dilleniid families occur within Asteridae *s. l.* (e.g., Sarraceniaceae, Diapensiaceae, Ericaceae, Theaceae, Primulaceae, Clethraceae, Styracaceae, Actinidiaceae, Symplocaceae, Ebenaceae, and Balsaminaceae). Even considering only the Rosidae clade, traditional Dilleniidae are not monophyletic. For example, Capparaceae and Bataceae are found within the glucosinolate clade, and members of Malvales (e.g., Malvaceae, Bombacaceae, Tiliaceae) form a distinct clade.

Although the separate *rbcL* and 18S rDNA analyses suggest similar overall patterns of angiosperm relationship and many of the same clades, these major clades do not have strong internal support. Although it is impractical to conduct decay analyses on such large data sets, the parsimony jackknife approach recently developed by Farris et al. [5] has been applied to both the *rbcL* data set [5] and to the 18S rDNA data set [2]. In these studies, none of the large major clades received strong jackknife support. With this overview of angiosperm phylogeny as inferred from *rbcL* and 18S rDNA analyses, we now summarize the smaller clades recovered by these analyses.

7.2.2 First-Branching Angiosperms

Although both *rbcL* and 18S rDNA sequences agree in placing monosulcate taxa as sister to the eudicots either as a distinct clade or as a series of successive sister-groups (a monosulcate grade), the two topologies differ regarding the first-branching angiosperms. Analyses of *rbcL* sequences place the paleoherb *Ceratophyllum* as sister to all other angiosperms, but 18S rDNA sequences identify several woody magnoliids (*Amborella* and a clade of *Austrobaileya*, *Illicium*, and *Schisandra*) as sister to all other flowering plants. Evidence from morphology and the fossil record can be used to support either hypothesis [1,2]).

Although we can neither distinguish between these possibilities nor rule out the possibility of some other family as sister to all other angiosperms, it is noteworthy that both topologies support the general view of Endress [16] that the first angiosperms may have had relatively small flowers with few or a moderate number of parts and varied phyllotaxy. This contrasts with the more traditional view [17] of the earliest angiosperms as having large, strobiloid, *Magnolia*-like flowers, an elongate floral axis, and numerous spirally arranged parts.

Within the monosulcates, several monophyletic groups are strongly supported by both the *rbcL* and 18S rDNA analyses: Austrobaileyaceae–Illiciaceae–Schisandraceae; Piperales (Piperaceae–Saururaceae); and Lactoridaceae–Aristolochiaceae. The close relationship of Lactoridaceae to Aristolochiaceae agrees with recent morphological studies [16]. A close relationship between Lauraceae and Calycanthaceae is also suggested by analyses of both data sets. The close relationship of Austrobaileyaceae to Schisandraceae–Illiciaceae is particularly noteworthy in that the former family has often been placed in Magnoliales while the latter two families form Illiciales.

The monocots are only sparsely represented in 18S rDNA analyses; thus, it is premature to discuss relationships within this group. Nonetheless, several clades are strongly supported by both data sets (e.g., Zingiberales, Liliales, and higher commelinids). Both *rbcL* and 18S rDNA sequences also concur in suggesting that *Acorus* may be phylogenetically isolated from other monocots, certainly from other Araceae.

7.2.3 Ranunculids

Both *rbcL* and 18S rDNA sequence data suggest a ranunculids clade containing Lardizabalaceae, Berberidaceae, Ranunculaceae, Menispermaceae, Eupteleaceae, Fumariaceae, Sargentodoxaceae, and Papaveraceae [15,18]. This clade represents the core of Ranunculales [10] and corresponds closely to the Berberidales of Thorne [19] and Ranunculiflorae of Dalhgren [20]. Analyses of *rbcL* and 18S rDNA, as well as *atpB* sequences, place Eupteleaceae (Hamamelidae) within this clade [15]. Also part of this clade in the 18S rDNA analyses is Sargentodoxaceae, a family typically placed in Ranunculales and allied with Lardizabalaceae [10]. The broad analyses of *rbcL* sequences placed Sargentodoxaceae with Fabaceae

[1], but this result was determined to be an error based on misidentified leaf material [18; Qiu and Chase, unpublished data]. A reanalysis of *rbcL* sequences using plant material known to be from *Sargentodoxa* places the genus within this ranunculid clade [15,18].

7.2.4 Saxifragoids

Within the large Rosidae clade, several monophyletic groups are apparent in the shortest trees revealed by both analyses. One of the larger clades is the saxifragoids or Saxifragales [14], which is unusual in that it contains members of three of Cronquist's [10] subclasses: Crassulaceae, Haloragaceae, Saxifragaceae *s. s.*, and the genera *Tetracarpaea*, *Penthorum*, *Itea*, *Ribes*, and *Pterostemon* (Rosidae); Daphniphyllaceae, Hamamelidaceae, Cercidiphyllaceae (Hamamelidae); and Paeoniaceae (Dilleniidae). The composition of this clade and the morphology, embryology, and chemistry of its members are discussed in more detail by Soltis and Soltis [14].

7.2.5 Glucosinolates

Both *rbcL* and 18S rDNA analyses also support the monophyly of the glucosinolate-producing plants, with the exception of *Drypetes* (Euphorbiaceae). This result is further demonstrated in more focused analyses of both *rbcL* [21] and 18S rDNA sequences (Rodman et al., unpublished data) and by morphological studies [22,23]. Thus, the glucosinolate families Tropaeolaceae, Capparaceae, Brassicaceae, Bataceae, Caricaceae, Limnanthaceae, Akaniaceae, Moringaceae, Resedaceae, and Bretschneideraceae should be considered a monophyletic group, distinct from *Drypetes* (Euphorbiaceae), which represents a second independent origin of the glucosinolate pathway.

7.2.6 Celastroids

A celastroid clade is also suggested by analyses of both *rbcL* and 18S rDNA data sets. This clade includes Celastraceae, *Brexia*, *Parnassia*, and *Lepuropetalon*. The latter three genera represent members of Saxifragaceae *s. l.* The morphological and embryological data that support the monophyly of this eclectic group, which includes the smallest terrestrial angiosperm (*Lepuropetalon*), herbaceous perennials (*Parnassia*), and shrubs and small trees (Celastraceae, *Brexia*), have been reviewed by Morgan and Soltis [24].

7.2.7 Nitrogen-Fixing Clade

Species of only ten families of angiosperms are known to form symbiotic associations with nitrogen-fixing bacteria in root nodules (all three subfamilies of Fabaceae, Betulaceae, Casuarinaceae, Coriariaceae, Datiscaceae, Elaeagnaceae, Myricaceae, Rhamnaceae, Rosaceae, and Ulmaceae). These families are distributed among four of Cronquist's [10] six subclasses of dicotyledons, implying that many are only distantly related. Recent phylogenetic analyses of *rbcL* sequences reveal, however, that representatives of all ten of these families occur together in a single clade, a nitrogen-fixing clade [25]. In addition to these ten families, this clade also contains several families not known to form associations with nitrogen-fixing bacteria, including Moraceae, Cannabaceae, Urticaceae, Polygalaceae, Fagaceae, Begoniaceae, and Cucurbitaceae.

Analyses of 18S rDNA sequences suggest a similar, but not identical, alliance of these nitrogen-fixing families. The 18S rDNA-based nitrogen-fixing clade represents a subset of the taxa present in the *rbcL*-based nitrogen-fixing clade and includes Betulaceae, Casuarinaceae, Datiscaceae, Elaeagnaceae, Rhamnaceae, and Ulmaceae, all families that contain taxa forming symbiotic associations with nitrogen-fixing bacteria. Other families known to form such associations (e.g., Coriariaceae and Myricaceae) and that appeared in the *rbcL*-based nitrogen-fixing clade were not included in the 18S rDNA analysis. Also part of the 18S rDNA nitrogen-fixing clade are Begoniaceae, Moraceae, Urticaceae, and Cucurbitaceae, families found to be part of this clade in *rbcL* analyses. However, neither Rosaceae nor Fabaceae, two families possessing members forming nitrogen-fixing symbioses, are included in the 18S rDNA nitrogen-fixing clade although both are part of the *rbcL*-based clade.

7.2.8 Santaloids

Analyses of both *rbcL* and 18S rDNA data sets consistently reveal a monophyletic santaloid clade corresponding closely to circumscriptions of Santalales [10]. Santalales comprise seven families (Eremolepidaceae, Loranthaceae, Misodendraceae, Olacaceae, Opiliaceae, Santalaceae, and Viscaceae) that are widely considered to form a natural group based on morphology [10] and have formed a clade in previous, smaller analyses of 18S rDNA sequences [4,25]. Although the monophyly of the santaloids appears well supported, the relationship of this clade to other eudicots remains enigmatic, differing between the two *rbcL* trees [1] as well as among the four 18S rDNA topologies [2].

7.2.9 Malvoids

Bombacaceae, Tiliaceae, and Malvaceae form a malvoid or Malvales clade in the broad analyses of *rbcL* and 18S rDNA sequences, in agreement with taxonomic treatments based on morphology [10]. All these families are core members of Malvales [10,18,26] and share a suite of morphological features.

7.2.10 Cunonioids

Bauera and *Ceratopetalum* (Cunoniaceae), *Eucryphia* (Eucryphiaceae), and the carnivorous plant family Cephalotaceae consistently form a clade in both *rbcL* and 18S rDNA analyses. A close relationship between Cunoniaceae and Eucryphiaceae was also revealed by a cladistic analysis of morphological features [27,28]. The relationships of Cephalotaceae have long been enigmatic. The *rbcL*-based inference that this family is closely related to Cunoniaceae and Eucryphiaceae was considered somewhat surprising but is bolstered by 18S rDNA data. *Sloanea* (Elaeocarpaceae) is also part of this clade in the 18S rDNA trees but was not represented in the *rbcL* analyses [1]. *Sloanea* also appears in this clade, in our separate and combined *rbcL* and 18S rDNA analyses using data sets of 232 taxa (see following). Other taxa that appear closely allied with Cunoniaceae, Eucryphiaceae, and Cephalotaceae in *rbcL* analyses include Tremandraceae and Oxalidaceae; however, neither family was included in the 18S rDNA analyses.

7.3 Separate and Combined Analyses of *rbcL* and 18S rDNA Data Sets: Congruence and Phylogenetic Relationships

The foregoing overview based on visual comparison of the *rbcL* [1] and 18S rDNA trees [2] suggests that the two topologies are largely congruent. In addition to the numerous larger clades shared by both *rbcL* and 18S rDNA topologies, many of the same terminal sister-group relationships are also revealed. However, several areas obviously disagree, the most noteworthy being the discrepancy in the first-branching angiosperms and the placement of Caryophyllidae *s. l.* Considerable discussion and debate in recent years have focused on the congruence of multiple data sets and whether different data sets should be analyzed separately or combined and analyzed together [29–38]. Combining data sets in phylogenetic analyses can facilitate the retrieval of "true" clades, but this can also yield phylogenetic estimates that are spurious if there are major discrepancies between the data sets. Several factors can result in discrepancy, including sampling error, different phylogenetic histories between the organisms and the molecule, and distinct stochastic processes acting on the characters in each data set [30,38].

A number of approaches have been proposed to test for congruence of multiple data sets. Although visual inspection of resultant trees certainly is one method of crudely ascertaining congruence between data sets, much more objective and rigorous approaches have been proposed [32,34–36,39]. However, the application of some of these methods to large data sets such as the large *rbcL* and 18S rDNA data sets for angiosperms is problematic. Even though the 18S and *rbcL* trees for angiosperms appear largely congruent, we tested the congruence of these data sets using the partition randomness test provided in PAUP* 4.0 [39].

To evaluate the effects of combining large data sets, we compared levels of support for clades in the separate *rbcL* and 18S topologies with those in the tree resulting from analysis of the combined *rbcL*/18S data matrix. We wanted to determine whether increased resolution and support are achieved when *rbcL* and 18S rDNA sequences are combined compared to the analysis of separate *rbcL* and 18S rDNA data sets. Although implementing decay analyses [40,41] with such large data sets is impractical, the parsimony jackknife approach [5] provides an estimate of the internal support for these topologies and can be accomplished with large data sets in a matter of minutes. Basically, the jackknife is a resampling approach, similar to the bootstrap, in which a data set is resampled to generate replicate data sets. Each replicate is analyzed, and the proportion of replicates supporting a given conclusion (a clade) is considered a measure of support. Jackknife percentages can therefore be interpreted in a manner similar to bootstrap percentages. Parsimony jackknifing [5] involves only addition of taxa and no branch swapping.

To assess congruence and to implement the parsimony jackknife comparison, we first constructed *rbcL* and 18S rDNA data sets of 232 taxa for which both *rbcL* and 18S rDNA sequences were available for either the same species or the same genus. These *rbcL* and 18S rDNA data sets were then combined into a single data matrix. Represented in this matrix are 226 angiosperm genera (plus six outgroups from Gnetales) from 160 families that represent well the diversity of angiosperms. We used *rbcL* and 18S rDNA sequences for the same species, if possible; 164 of the 18S rDNA sequences represent the same species sequenced for *rbcL*, including 97 species of eudicots for which the same DNA samples were used for both *rbcL* and 18S rDNA sequencing. In the remaining 68 cases, different species of the same genus were used. The separate *rbcL* and 18S rDNA data sets, as well as the combined data matrix, are available upon request.

We also conducted cursory parsimony searches with all three data sets using PAUP* 4.0 simply to determine whether the general topologies obtained with the data sets constructed here concur with those of Chase et al. [1] for *rbcL* and Soltis et al. [2] for 18S rDNA. All parsimony searches were conducted as follows. First, 500 replicate heuristic searches using NNI branch swapping were conducted, saving five trees per replicate. Using the shortest trees obtained from these initial searches as starting trees, we then searched using TBR branch swapping and saving all most parsimonious trees. After 2 weeks of analysis on Power Macintosh or Quadra 950 computers, the searches were stopped. It is almost certain that we did not find the shortest trees via this approach. The trees from these parsimony searches are not presented but generally agree with those summarized previously for *rbcL* [1] and 18S rDNA [2].

7.3.1 Congruence

Issues of congruence and whether or not multiple data sets should be combined to form a single data set have been the focus of considerable debate (reviewed earlier in this chapter). Visual comparison of the *rbcL* trees of Chase et al. [1] and 18S rDNA trees of Soltis et al. [2] certainly suggests that these two molecular data sets are congruent. Using the 232-taxon data matrices constructed for *rbcL*, 18S rDNA, and *rbcL* plus 18S rDNA, we assessed congruence using the partition randomness test option implemented on PAUP* 4.0; tree lengths obtained from partitioning the data by gene were compared to those obtained from random partitions of the data following Farris et al. [5]. This combinability test found the 232-taxon *rbcL* and 18S rDNA data sets to be incongruent ($P < 0.05$).

This statistical incongruence suggests that the *rbcL* and 18S rDNA data sets should not be combined. However, significant incongruence between matrices does not mean incombinability. We have found, for example, that multiple sequence data sets derived from the chloroplast genome are statistically incongruent although the topologies retrieved from each are highly similar (D. Soltis and L. Johnson, unpublished data). Furthermore, the sensitivity of this homogeneity test has not been explored rigorously, particularly for large data sets. Perhaps this test is too conservative for large data sets, with a different placement of only one or a few clades sufficient to indicate statistically significant incongruence when the overall picture is one of general congruence.

Several examples of clear incongruence between the *rbcL* and 18S rDNA data sets are obvious, based on the shortest trees and the parsimony jackknife results. For example, Caryophyllidae *s. l.* appear as part of Asteridae *s. l.* in the shortest 18S rDNA trees but not in the *rbcL* trees, but this placement is not supported by the jackknife analysis. Sequencing errors may also contribute to incongruence between data sets. For example, the placement of *Sargentodoxa* is a source of incongruence because of the misidentification of leaf material used for the *rbcL* sequence (see following). In addition, several of the 18S rDNA sequences (particularly some of the older sequences) appear to contain errors and should be resequenced [2]. Thus, several factors other than true differences in evolutionary history could explain the incongruence of these 18S rDNA and *rbcL* data sets. For the reasons noted earlier, we pursued the analysis of the combined *rbcL*/18S rDNA data set as an exercise in handling large data sets.

7.3.2 *Parsimony Jackknife Analyses*

Using the 232-taxon data matrices constructed for *rbcL*, 18S rDNA, and *rbcL* plus 18S rDNA, we conducted parsimony jackknife analyses [5] to assess internal support for the topologies. In the three analyses discussed here, 1000 replicates were conducted, and only clades supported by jackknife values of 50% or greater were retained. Values less than 50% are here referred to as "no support," although some nodes with low jackknife values may be present in all shortest trees and even in all trees that are several to many steps longer [2]. Thus, the parsimony jackknife may represent a conservative measure of internal support.

Most of the larger clades receiving jackknife support greater than 50% are given in Table 7.1. The complete parsimony jackknife results for all three data sets are available on request. Comparison of the jackknife values for the separate *rbcL* and 18S rDNA data sets reveals much higher overall internal support from the *rbcL* data. Many more clades have jackknife values of 50% or greater with *rbcL* than with 18S rDNA sequences. This result is not surprising given the slower rate of evolution and lower signal of 18S rDNA compared to *rbcL* [4]. Most of the clades having jackknife values of 50% or more with 18S rDNA are a subset of those observed for *rbcL*. However, some clades receiving jackknife support are unique to the 18S rDNA data set (e.g., saxifragoids or Saxifragales, Aristolochiaceae–Lactoridaceae, but are nonetheless present and supported in some *rbcL* analyses with different sampling). When the *rbcL* and 18S rDNA data sets are combined, the number of supported clades increases dramatically compared to the separate *rbcL* and 18S rDNA data sets, with fewer taxa part of a large basal polytomy in the combined analysis. For example, the number of genera in polytomies in the parsimony jackknife strict consensus tree is 90, 46, and 30 in the 18S, *rbcL*, and *rbcL*/18S rDNA analyses, respectively.

In many instances, clades supported by a jackknife value of 50% or more in the separate analyses are more strongly supported in the combined analysis (Table 7.1). Of the 106 clades compared in the parsimony jackknife analyses, 87 received increased support in the analysis of the combined *rbcL*/18S rDNA data set, and of the 35 large clades noted in Table 7.1, 32 received increased support. In some instances, the increase in the jackknife value is small (1%–2%), but in many cases the increase is greater than 5% and certainly reflects the positive interaction of the two data sets. For example, the clade of Zingiberales has a jackknife value of 94% in the *rbcL* analysis, 68% with 18S rDNA, and 99% in the analysis of the combined data set. On a smaller scale, the clade of Magnoliaceae–Annonaceae has a jackknife value of 81% with *rbcL*, 76% with 18S rDNA, and 94% with the combined data set. In other cases, a clade has jackknife support of 50% or more with one data set, and receives no support (i.e., less than 50%) with the other data set (although the relationship is seen in the shortest trees obtained with the second data set), yet the combined data set reveals increased support. For example, Caryophyllidae *s. l.* receives jackknife support (65%) only with *rbcL*, although this clade is also present in the shortest 18S rDNA trees; with the combined data set, the jackknife value increases to 89%. Similarly, although the glucosinolate clade is present in the shortest trees resulting from analysis of the separate *rbcL* and 18S rDNA data sets, it receives jackknife support only in the *rbcL* analysis (72%); in the analysis of the combined data set, the jackknife value increases to 82%.

In several instances increased jackknife support is observed for a clade in the combined analysis although it appears as a grade in the shortest trees from one of the separate data sets. The cornoid clade of *Camptotheca, Nyssa, Cornus, Philadelphus,* and *Hydrangea* is supported by a jackknife value of 58% in the *rbcL* analysis. Although *Philadelphus* and *Hydrangea* are sisters in the 18S rDNA analysis, they form a grade with the remaining genera in the shortest 18S rDNA trees. In the combined analysis, the jackknife support for the cornoid clade is substantially higher than with *rbcL* alone, 70%. Similarly, *Nelumbo* and *Platanus* form a clade in the *rbcL* trees (jackknife value of 51%) and are part of a grade or clade in the 18S rDNA trees (but do not form sister taxa); in the analysis of

TABLE 7.1. Subset of the clades supported by jackknife values of 50% or more in at least one of the three sequence data sets (*rbcL*, 18S, *rbcL*/18S)

Clade	*rbcL* (%)	18*S* (%)	*rbcL*/18*S* (%)
*Glucosinolate	72	—	82
*Glucosinolate-1	95	—	98
*Glucosinolate-2	66	—	91
*Glucosinolate-3	97	55	99
*Caryophyllidae s. l.	65	—	89
*Caryophyllidae s. l.-1	58	—	73
*Caryophyllidae s. l.-2	99	62	100
*Santaloid	79	—	93
*Santaloid-1	—	76	85
*Santaloid-2	91	—	93
Austrobaileyaceae/Amborellaceae/Illiciales/ Nymphaeaceae	77	—	72
*Amborellaceae/Nymphaeaceae	88	—	93
*Austrobaileyaceae/Illiciales	92	92	95
*Cunonioid	97	82	100
Saxifragoid	—	63	—
*Saxifragaceae s. s.	71	—	87
†Saxifragoid-1	—	—	55
**Tetracarpaea–Penthorum*–Haloragaceae	93	51	96
*Lamiales	60	—	83
†Ericales	—	—	64
†Asterales	—	—	53
*Cornoid	58	—	70
Urticales	55	—	55
*Piperales	94	95	97
*Ranunculids	67	—	75
*Ranunculids (-Pap)	80	—	88
*Ranunculids-1	55	—	67
*Ranunculids-2	99	93	100
†Malvoid	—	—	71
†Sapindales/Malvales	—	—	55
*Celastroid	99	51	100
Aristolochiaceae/Lactoridaceae	—	84	84
*Magnoliaceae/Annonaceae	81	76	94
Asparagales	83	—	80
*Zingiberales	94	68	99

Run times for the three data sets (*rbcL*, 18S, *rbcL*/18S) were 1568.3, 578.9, and 2392.0 s, respectively. This table contains most of the larger clades detected, but very few of the three-taxon clades and none of the numerous two-taxon clades receiving jackknife support are given. An asterisk indicates a clade that receives a higher jackknife value in the analysis of the combined *rbcL*/18S data set than in either individual data set; †indicates a "novel" clade that does not have jackknife support (i.e.,<50%) in either individual data set, but does in the analysis of the combined *rbcL*/18S data set. The names given for clades should be considered informal, given simply for convenience. Taxon composition of clades is given in Appendix I (see ps. 177,178).

the combined data set, the jackknife support for this sister relationship increases to 65%.

Although the parsimony jackknife analysis of the combined *rbcL*/18S rDNA matrix generally identifies those clades supported by jackknife values of 50% or more in the separate *rbcL* and 18S rDNA analyses, this is not always the case. The saxifragoid clade is moderately to strongly supported in the 18S rDNA analysis (63%), but it does not obtain jackknife support in the combined analysis, despite the fact that this clade is present in the shortest *rbcL* trees of Chase et al. [1] and in the shortest trees obtained in our parsimony searches of the combined *rbcL*/18S rDNA data set. In contrast, some clades are supported by jackknife values of 50% or more in the combined analysis although they receive no support (i.e., <50%) from either separate data set. An example of a novel or uniquely supported clade is the ericalean clade (see Table 7.1). This clade has a jackknife value of 64% in the combined *rbcL*/18S rDNA analysis but does not appear with a jackknife value greater than 50% in either of the separate *rbcL* or 18S rDNA analyses. The shortest *rbcL* trees of Chase et al. ([1]; see also [13]) and 18S rDNA trees of Soltis et al. [2], as well as the shortest trees obtained here in parsimony searches of the separate 232-taxon data sets, recover an ericalean clade and grade, respectively.

Another example of a novel or uniquely supported clade that receives jack-knife support of 50% or greater only in the analysis of the combined *rbcL*/18S rDNA matrix is Asterales, consisting of *Tagetes, Tragopogon, Corokia, Roussea, Lobelia*, and *Campanula* (see Table 7.1). Analyses of the separate data sets revealed high jackknife support for two component subclades of Asterales, *Tag-etes–Tragopogon–Corokia* and *Lobelia–Campanula*, but not the entire clade. *Roussea* (an enigmatic Escalloniaceae) was not allied with a high jackknife value with any of these genera in the separate analyses. This Asterales clade is present in the shortest trees resulting from the analyses of the separate data sets, but it is only by combining the two data sets that sufficient characters (base substitutions) are present to provide strong jackknife support. Other novel clades that receive jackknife support only when the *rbcL* and 18S rDNA data sets are combined include saxifragoid-1 and malvoids (Table 7.1). Eleven such novel or uniquely supported clades were detected in the jackknife analysis of the combined data sets; five of these are given in Table 7.1.

The *rbcL* and 18S rDNA data sets often complement each other, strongly supporting different portions of a given clade. This complementarity is well illustrated by the santaloids (Table 7.1), a clade observed in the shortest trees resulting from the separate analyses of the *rbcL* and 18S rDNA data sets that has high jackknife support (79%) only in the *rbcL* analysis. In the jackknife analysis of the combined data set, this support increases to 93%. Conversely, within santaloids only the 18S rDNA analysis reveals the santaloid-1 subclade (jackknife value of 76%). This subclade is present in the shortest *rbcL* trees but does not receive jackknife support from the *rbcL* data set. In the combined analysis, this clade is present with a higher jackknife value (85%) than observed in the 18S rDNA analysis alone.

Another example that illustrates the interplay of the 18S rDNA and *rbcL* data sets is provided by the ranunculids clade, although this example is complicated somewhat by *Sargentodoxa* (see following). The shortest trees based on separate analyses of *rbcL* and 18S rDNA sequences contain a ranunculids clade; this clade receives jackknife support (67%) from the *rbcL* data set but not the 18S rDNA data set. In the jackknife analysis of the combined data matrix, the support for the monophyly of this ranunculids clade increases to 75%. The jackknife support for some of the subclades within the ranunculids, such as the ranunculids-1 clade, also increases in the combined analysis (*rbcL*/18S rDNA, jackknife value of 67%; *rbcL* only, jackknife value of 55%; 18S rDNA only, no jackknife support). Because of some conflict between the *rbcL* and 18S rDNA data sets, however, the support for some clades actually decreases in the analysis of the combined data set. For example, although the analysis of *rbcL* sequences shows *Akebia* and *Sinofranchetia* to be sister taxa (jackknife of 96%) with *Decaisnea* their sister (jackknife of 99%), with 18S rDNA these relationships are reversed: *Sinofranchetia–Decaisnea*, 69%, and *Sinofranchetia–Akebia*, 64%. The analysis of the combined data set reveals a sister-group relationship between *Akebia* and *Sinofranchetia*, but with a lower jackknife value (75%). The relationships among members of the ranunculids based on multiple molecular dats sets are discussed in more detail by Hoot and Crane [15].

We also attempted to ascertain the impact of erroneous sequences in a combined analysis, with the idea that possible sequencing errors (particularly for 18S rDNA) might be the cause for some of the differences between the *rbcL* and 18S rDNA trees. The 18S rDNA sequences of *Potamogeton* and *Lilium* were included here, although both sequences seem to be dubious and should be resequenced. In the 18S rDNA analysis, these genera appear as strongly supported sisters (jackknife value of 98%) and in turn are part of a well-supported clade (jackknife value of 76%) with *Illicium*, an unlikely relationship for these taxa. The *rbcL* analysis, in contrast, reveals the expected close relationship of *Lilium* to other Liliaceae (*Clintonia–Lilium*, jackknife value of 100%). In the analysis of the combined *rbcL* and 18S rDNA data sets, a more weakly supported (jackknife value of 73%) sister relationship between *Lilium* and *Clintonia* is preserved. A strongly supported relationship is not seen between *Potamogeton* and any other taxon in the jackknife analysis of the combined data set, although in the shortest trees obtained in the parsimony analysis of the combined data set, *Potamogeton* is associated with other monocots. Thus, the signal from the *rbcL* sequences is apparently strong enough in this instance to override the influence of the 18S rDNA sequences that likely contain errors.

Similarly, the *rbcL* sequence of *Sargentodoxa* from Chase et al. [1] has been shown to be based on misidentified leaf material; in actuality it is a sequence from a legume whose identity has not been determined [14,17]. In our analysis of the *rbcL* data set, the mislabeled "*Sargentodoxa*" appears as the strongly supported (jackknife of 99%) sister of *Pisum*, as would be expected based on the material used. For the 18S rDNA data matrix, we used a sequence actually obtained from *Sargentodoxa* [17]; as expected, it appears in a clade of ranunculids with moderate support (jackknife value of 64%). In the jackknife analysis of the combined *rbcL*/18S rDNA data matrix, the position of *Sargentodoxa* as sister to *Pisum* is retained with high jackknife support (99%). Thus, while an analysis of a combined data set may in some instances override the signal when one sequence is in error (e.g., *Lilium*, *Potamogeton*), this is not always the case (e.g., *Sargentodoxa*). We have also seen, in all these instances in which problematic sequences were employed (the erroneous sequences of *Lilium* and *Potamogeton*, the mislabeled *Sargentodoxa*), that the *rbcL*-based inference is preserved in the analysis of the combined data set, indicating that *rbcL* is supplying the stronger phylogenetic signal. This is not surprising given the greater number of clades with high internal support that this gene provides compared to 18S rDNA [4].

7.4 Prospectus

Congruence of trees derived from multiple data sets is the single most significant measure of phylogenetic relationship [31,39,42]. The similarity between the *rbcL* [1] and 18S rDNA topologies [2] suggests that major features of angiosperm phylogeny have indeed been correctly identified. Other recent studies of higher-level relationships within the angiosperms similarly demonstrate highly concordant 18S rDNA and *rbcL* topologies ([14,43]; L. Johnson et al., unpublished data). Additional comparisons will be possible when taxon sampling for 18S rDNA approaches that of *rbcL*; 18S rDNA sequences are currently being obtained for Magnoliidae (D. and P. Soltis et al., unpublished data) and monocots (W. Hahn et al., unpublished data), two groups that to date have been poorly sampled.

Combining *rbcL* and 18S rDNA data sets for a similar suite of 232 taxa increases the support (as measured by jackknife values) for many clades, often dramatically. Analyses of the combined *rbcL*/18S rDNA data set generally combine the well-supported clades present in the separate *rbcL* and 18S rDNA trees. Combining data sets also reveals novel clades (with jackknife values greater than 50%) that did not receive jackknife support in the analyses of the separate data sets. This initial attempt at analyzing a combined *rbcL*/18S rDNA data matrix also reveals greater overall resolution of relationships and a much higher number of jackknife-supported clades than did analysis of either data set separately. Complementarity and increased support for clades in the combined *rbcL*/18S rDNA analysis is comparable to that observed with combined data sets in more focused studies [44,45].

In addition to producing trees with increased resolution and higher internal support, the combined data set also showed improvements in computer run times. The 500 replicate searches with NNI branch swapping were completed most quickly with the combined *rbcL*/18S rDNA data set, with the largest number of parsimony-informative characters, and most slowly with the 18S rDNA data set, with the fewest. Furthermore, with the combined *rbcL*/18S rDNA data set, subsequent swapping on the starting trees with tree bisection–reconnection (TBR) was completed in a few days. Swapping was not completed with either the 18S rDNA or *rbcL* data sets used here; our previous analyses of 228 18S rDNA sequences never swapped to completion despite use of more than 2 years of computer time. Thus, although perhaps counterintuitive, this result illustrates another advantage of combining sequences in studies of this magnitude: the time required for data analysis actually decreases with the addition of more characters. The ultimate advantage of decreased run times is the opportunity for more thorough searches.

Simulation studies [46] have recently been used to test the ability of current methods to recover complex phylogenies. Using a model tree based on the estimated phylogeny of 228 angiosperms inferred from 18S rDNA sequences [2], Hillis [46] accurately reconstructed the model phylogeny with only 5000 nucleotides of sequence data, considerably fewer nucleotides than are needed to reconstruct some four-taxon trees [47]. The unexpected ease of phylogeny reconstruction for this large data set appears to involve the dispersal of homoplasy across many branches of the tree, allowing the phylogenetic signal to be detected [46]. Large data sets may therefore be more tractable than previously considered [48,49]. The exploratory analysis of the combined *rbcL*/18S rDNA data set reported here further suggests that a well-resolved, strongly supported phylogenetic framework for angiosperms not only is possible but may be achieved perhaps within 2 years.

By combining large data sets of comparable taxa for several gene sequences, the overall picture of angiosperm relationships will be further refined. Given that accurate reconstructions may be possible with as few as 5000 bp [46], improved estimates of angiosperm phylogeny are certainly achievable, with *rbcL* and 18S rDNA sequences together providing more than 3000 bp of sequence data. Sequence data are rapidly emerging for a second chloroplast gene, *atpB* (V. Savolainen, S. Hoot, M. Chase et al., unpublished data), which is more than 1400 bp in length and similar in rate of evolution to *rbcL* [15,18]. Efforts are also under way to increase the number of phylogenetically informative characters available from the nuclear genome by sequencing 26S rDNA (more than 3000 bp) for the same taxa sequenced for 18S rDNA and *rbcL*. Analyses of such large data sets constructed from comparable suites of taxa (and DNAs) for these and other chloroplast and nuclear genes will ultimately result in an improved estimate of angiosperm phylogeny.

Acknowledgments. This research was supported in part by a grant from the National Science Foundation (DEB 9307000). We thank D. Swofford for access to PAUP*4.0. We also thank S. Hoot, S.-M. Chaw, and W. Hahn for the use of several unpublished 18S rDNA sequences and M. Fay, S. Swensen, K. Cameron, J. Hartwell, K. Kron, and C. Morton for several unpublished *rbcL* sequences.

References

1. Chase MW, Soltis DE, Olmstead RG, Morgan D, Les DH, Mishler BD, Duvall MR, Price RA, Hills HG, Qiu Y-L, Kron KA, Rettig JH, Conti E, Palmer JD, Manhart JR, Sytsma KJ, Michaels HJ, Kress WJ, Karol KG, Clark WD, Hedrén M, Gaut BS, Jansen RK, Kim K-J, Wimpee CF, Smith JF, Furnier GR, Strauss SH, Xiang Q-Y, Plunkett GM, Soltis PS, Swensen S, Williams SE, Gadek PA, Quinn CJ, Eguiarte LE, Golenberg E, Learn GH Jr, Graham SW, Barrett SCH, Dayanandan S, Albert VA (1993) Phylogenetics of seed plants: an analysis of nucleotide sequences from the plastid gene *rbcL*. Ann MO Bot Gard 80:528–580
2. Soltis DE, Soltis PS, Nickrent DL, Johnson LA, Hahn WJ, Hoot SB, Sweere JA, Kuzoff RK, Kron KA, Chase MW, Swensen SM, Zimmer EA, Chaw S-M, Gillespie LJ, Kress WJ, Sytsma KJ (1997) Angiosperm phylogeny inferred from 18S ribosomal DNA sequences. Ann MO Bot Gard 84:1–49
3. Palmer JD, Jansen RK, Michaels HJ, Chase MW, Manhart JR (1988) Chloroplast DNA variation and plant phylogeny. Ann MO Bot Gard 75:1180–1206
4. Nickrent DL, Soltis DE (1995) A comparison of angiosperm phylogenies from nuclear 18S rDNA and *rbcL* sequences. Ann MO Bot Gard 82:208–234
5. Farris JS, Albert VA, Källersjö M, Lipscomb D, Kluge AG (1997) Parsimony jack-knifing outperforms neighbor-joining. Cladistics 12:99–124
6. Donoghue MJ, Doyle JA (1989) Phylogenetic analysis of angiosperms and the relationships of Hamamelidae. In: Crane PR, Blackmore S (eds) Evolution, systematics and fossil history of the Hamamelidae. Clarendon, Oxford, pp 17–45
7. Donoghue MJ, Doyle JA (1989) Phylogenetic studies of seed plants and angiosperms based on morphological characters. In: Fernholm B, Bremer K, Jornvall H (eds) The hierarchy of life: molecules and morphology in phylogenetic analysis. Elsevier, Amsterdam, pp 181–193
8. Doyle JA, Hotton CL (1991) Diversification of early angiosperm pollen in a cladistic context. In: Blackmore S, Barnes SH (eds) Pollen and spores. Clarendon, Oxford, pp 169–195
9. Doyle JA, Hotton CL, Ward JV (1990) Early Cretaceous tetrads, zonasulculate pollen, and Winteraceae. II. Cladistic analysis and implications. Am J Bot 77:1558–1568
10. Cronquist A (1981) An integrated system of classification of flowering plants. Columbia University Press, New York
11. De Candolle A (1873) Prodromus systematis naturalis regni vegetabilis, vol 17, Paris
12. Hufford L (1992) Rosidae and their relationships to other nonmagnoliid dicotyledons: a phylogenetic analysis using morphological and chemical data. Ann MO Bot Gard 79:218–248
13. Olmstead RG, Bremer B, Scott KM, Palmer JD (1993) A parsimony analysis of the Asteridae sensu lato based on *rbcL* sequences. Ann MO Bot Gard 80:700–722
14. Soltis DE, Soltis PS (1997) Phylogenetic relationships among Saxifragaceae sensu lato: a comparison of topologies based on 18S rDNA and *rbcL* sequences. Am J Bot 84:504–522
15. Hoot SB, Crane PR (1995) Inter-familial relationships in the Ranunculidae based on molecular systematics. Plant Syst Evol 9:119–131
16. Endress PK (1994) Floral structure and evolution of primitive angiosperms: recent advances. Plant Syst Evol 192:79–97

17. Cronquist A (1968) The evolution and classification of flowering plants. Houghton Mifflin, Boston
18. Hoot SB, Culham A, Crane PR (1995) Phylogenetic relationships of the Lardizabalaceae and Sargentodoxaceae: chloroplast and nuclear DNA sequence evidence. Plant Syst Evol 9:195–199
19. Thorne RF (1992) An updated classification of the flowering plants. Aliso 13:365–389
20. Dahlgren RT (1980) A revised system of classification of the angiosperms. Bot J Linn Soc 80: 91–124
21. Rodman JE (1991) A taxonomic analysis of glucosinolate-producing plants. Part 1. Phenetics. Syst Bot 16:598–618
22. Rodman JE (1991) A taxonomic analysis of glucosinolate-producing plants. Part 2. Cladistics. Syst Bot 16:619–629
23. Rodman JE, Price RA, Karol K, Conti E, Sytsma KJ, Palmer JD (1993) Nucleotide sequences of the *rbcL* gene indicate monophyly of mustard oil plants. Ann MO Bot Gard 80:686–699
24. Morgan DR, Soltis DE (1993) Phylogenetic relationships among members of Saxifragaceae sensu lato based on *rbcL* sequence data. Ann MO Bot Gard 80:631–660
25. Soltis DE, Soltis PS, Morgan DR, Swensen SM, Mullin BC, Dowd JM, Martin PG (1995) Chloroplast gene sequence data suggest a single origin of the predisposition for symbiotic nitrogen fixation in angiosperms. Proc Natl Acad Sci USA 92:2647–2651
26. Nickrent DL, Franchina CR (1990) Phylogenetic relationships of the Santalales and relatives. J Mol Evol 31:294–301
27. Takhtajan A (1987) System of Magnoliophyta. Academy of Sciences of the USSR, Leningrad
28. Hufford L, Dickison WC (1992) A phylogenetic analysis of Cunoniaceae. Syst Bot 17:181–200
29. Kluge AG (1989) A concern for evidence and a phylogenetic hypothesis of relationships among *Epicrates* (Boidae, Serpentes). Syst Zool 38:7–25
30. De Queiroz A, Donoghue MJ, Kim J (1995) Separate versus combined analysis of phylogenetic evidence. Annu Rev Ecol Syst 26:657–681
31. Miyamoto MM, Fitch WM (1995) Testing species phylogenies and phylogenetic methods with congruence. Syst Biol 44:64–76
32. Farris JS, Källersjö M, Kluge AG, Bult C (1995) Testing significance of incongruence. Cladistics 10:315–319
33. Lanyon SM (1993) Phylogenetic frameworks: towards a firmer foundation for the comparative approach. Biol J Linn Soc 49:45–61
34. Rodrigo AG, Kelly-Borges M, Berquist PR, Berquist PL (1993) A randomization test of the null hypothesis that two cladograms are sample estimates of a parametric phylogenetic tree. N Z J Bot 31: 257–268
35. Bull JJ, Huelsenbeck JP, Cunningham CW, Swofford DL, Waddell PJ (1993) Partitioning and combining data in phylogenetic analysis. Syst Biol 42:384–397
36. Mickevich MF, Farris JS (1981) The implications of congruence in *Menidia*. Syst Zool 30:351–370
37. Lutzoni F, Vilgalys R (1995) Integration of morphological and molecular data sets in estimating fungal phylogenies. Can J Bot 73:S649–S659
38. Huelsenbeck JP, Bull JJ, Cunningham CW (1996) Combining data in phylogenetic analysis. Trends Ecol Evol 11:152–158

39. Swofford DL (1991) When are phylogeny estimates incongruent? In: Miyamoto MM, Cracraft J (eds) Phylogenetic analysis of DNA sequences. Oxford University Press, New York

40. Bremer K (1988) The limits of amino acid sequence data in angiosperm phylogenetic reconstruction. Evolution 42:795–803

41. Donoghue MJ, Olmstead RG, Smith JF, Palmer JD (1992) Phylogenetic relationships of Dipsacales based on *rbcL* sequences. Ann MO Bot Gard 79:333–345

42. Penny D, Hendy MD (1986) Estimating the reliability of evolutionary trees. Mol Biol Evol 3:403–417

43. Kron KA (1996) Phylogenetic relationships of Empetraceae, Epacridaceae, and Ericaceae: evidence from nuclear ribosomal 18S sequence data. Ann Bot (Lond) 77:293–303

44. Olmstead RG, Sweere JA (1994) Combining data in phylogenetic systematics: an empirical approach using three molecular data sets in the Solanaceae. Syst Biol 43:467–481

45. Soltis DE, Kuzoff RK, Conti E, Gornall R, Ferguson K (1996) *mat*K and *rbcL* gene sequence data indicate that *Saxifraga* (Saxifragaceae) is polyphyletic. Am J Bot 83:371–382

46. Hillis DM (1996) Inferring complex phylogenies. Nature 383:130

46. Hillis DM, Huelsenbeck JP, Swofford DL (1994) Hobgoblin of phylogenetics? Nature 369:363–364

47. Hillis DM (1995) Approaches for assessing phylogenetic accuracy. Syst Biol 44:3–16

48. Patterson C, Williams DM, Humphries DJ (1993) Congruence between molecular and morphological phylogenies. Annu Rev Ecol Syst 24:153–188

Appendix I. Taxon composition of clades given in Table 7.1.

Clade	Genera
Glucosinolate	*Brassica, Cleome, Capparis, Tovaria, Koeberlinia, Batis, Limnanthes, Floerkia, Carica, Moringa, Tropaeolum, Bretschneidera, Akania*
Glucosinolate-1	*Brassica, Cleome, Capparis, Tovaria, Koeberlinia, Batis, Limnanthes, Floerkea*
Glucosinolate-2	*Carica, Moringa*
Glucosinolate-3	*Tropaeolum, Bretschneidera, Akania*
Caryophyllidae s. l.	*Plumbago, Polygonum, Coccoloba, Nepenthes, Drosera, Spinacia, Mollugo, Phytolacca, Mirabilis*
Caryophyllidae s. l.-1	*Plumbago, Polygonum, Coccoloba, Nepenthes, Drosera*
Caryophyllidae s. l.-2	*Spinacia, Mollugo, Phytolacca, Mirabilis*
Santaloid	*Misodendron, Schoepfia, Gaiadendron, Eubrachion, Santalum, Osyris, Opilia, Viscum, Ginalloa, Dendrophthora*
Santaloid-1	*Dendrophthora, Ginalloa, Viscum, Osyris, Santalum, Eubrachion*
Santaloid-2	*Gaiadendron, Schoepfia, Misodendron*
Austrobaileyaceae/Amborellaceae/ Illiciales/Nymphaeaceae	*Austrobaileya, Amborella, Illicium, Schisandra, Nymphaea, Nuphar*
Amborellaceae/Nymphaeaceae	*Amborella, Nymphaea, Nuphar*
Austrobaileyaceae/Illiciales	*Austrobaileya, Illicium, Schisandra*
Cunonioid	*Bauera, Ceratopetalum, Eucryphia, Sloanea, Cephalotus*
Saxifragoid	*Haloragis, Myriophyllum, Penthorum, Tetracarpaea, Itea, Pterostemon, Saxifragaceae s. s., Hamamelis, Disanthus, Exbucklandia, Liquidambar, Altingia, Paeonia, Daphniphyllum, Cercidiphyllum, Ribes, Sedum, Dudleya, Kalanchoe, Crassula*
Saxifragaceae s. s.	*Boykinia, Sullivantia, Heuchera, Saxifraga mertensiana, Saxifraga integrifolia, Chrysosplenium*
Saxifragoid-1	*Sedum, Dudleya, Kalanchoe, Crassula, Tetracarpaea, Penthorum, Haloragis, Myriophyllum*
Tetracarpaea-Penthorum-Haloragaceae	*Tetracarpaea, Penthorum, Haloragis, Myriophyllum*
Lamiales	*Buddleja, Lamium, Byblis, Vahlia, Montinia, Ipomoea*
Ericales	*Arctostaphylos, Vaccinium, Pyrola, Halesia, Styrax, Cobaea, Gilia, Diapensia, Galax, Ardisia, Camellia, Clethra, Cyrilla, Diospyros, Fouquieria, Impatiens, Manilkara, Symplocos, Sarracenia, Actinidia*
Asterales	*Tagetes, Tragopogon, Corokia, Roussea, Lobelia, Campanula*
Cornoids	*Hydrangea, Philadelphus, Cornus, Nyssa, Camptotheca*
Urticales	*Morus, Pilea, Ceanothus, Zelkova, Elaeagnus*
Piperales	*Houttuynia, Saururus, Piper, Peperomia*
Ranunculids	*Dicentra, Euptelea, Akebia, Sinofranchetia, Decaisnea, Caulophyllum, Nandina, Tinospora, Glaucidium, Hydrastis, Xanthorhiza, Ranunculus*
Ranunculids (minus Papaveraceae)	*Euptelea, Akebia, Sinofranchetia, Decaisnea, Caulophyllum, Nandina, Tinospora, Glaucidium, Hydrastis, Xanthorhiza, Ranunculus*

Clade	Genega
Ranunculids-1	*Caulophyllum, Nandina, Tinospora, Hydrastis, Xanthorhiza, Glaucidium, Ranunculus*
Ranunculids-2	*Akebia, Sinofranchetia, Decaisnea*
Malvoids	*Bombax, Gossypium, Muntingia*
Sapindales/Malvales	*Bombax, Gossypium, Muntingia, Koelreuteria, Acer*
Celastroids	*Brexia, Euonymus, Parnassia, Lepuropetalon*
Aristolochiaceae/Lactoridaceae	*Asarum, Saruma, Aristolochia, Lactoris*
Magnoliaceae/Annonaceae	*Magnolia, Isolona, Mkilua*
Asparagales	*Aristea, Gladiolus, Isophysis, Cyanella, Asparagus, Xanthorrhoea, Eucharis, Chlorophytum, Bowiea, Allium*
Zingiberales	*Canna, Heliconia, Maranta, Musa, Zingiber, Costus*

Evolution of MADS Gene Family in Plants

MITSUYASU HASEBE[1] and JO ANN BANKS[2]

8.1 Introduction

One of the biggest problems in systematics is the occasional discrepancy between molecular-based and morphology-based phylogenies. For both molecular and morphological data, it is difficult to distinguish homology from analogy, especially among distantly related taxa. Therefore, there are no a priori criteria to evaluate the different phylogenetic trees inferred from the different kinds of data. By understanding the molecular and genetic basis of morphological characters, these discrepancies may be reconciled for two reasons. First, a morphological character is generally a result of the interactions of many gene products. As these genes are identified, the number of characters from which a phylogenetic inference is made is greatly increased from one morphological character (e.g., petal) to many genes (e.g., all the genes that specify petal identity and morphogenesis). Second, by comparing the expression patterns of genes that are involved in the development of morphological characters among different taxa, it is possible to distinguish between relatedness by homology and relatedness by analogy. The greater resolution of such molecular developmental studies will greatly enhance our understanding of the genetic changes that have accompanied the evolution and divergence of organisms.

Genes that encode transcription factors, especially homeotic genes, have important roles in the regulation of animal [1,2] and plant morphogenesis [3]. In plants, flower development has been extensively studied at both the genetic and molecular levels. Several floral homeotic genes have been identified by mutation, cloned, and shown to be transcription factors, most of which belong to the MADS gene family ([4,5] and references therein). In this chapter, we discuss the evolution of the MADS gene family based on a newly compiled gene tree and examine how gene duplications within this family might have played a

[1] National Institute for Basic Biology, 38 Nishigonaka, Myoudaiji-cho, Okazaki 444, Japan
[2] Department of Botany and Plant Pathology, Purdue University, Lilly Hall of Life Sciences, West Lafayette, IN 47907-1155, USA

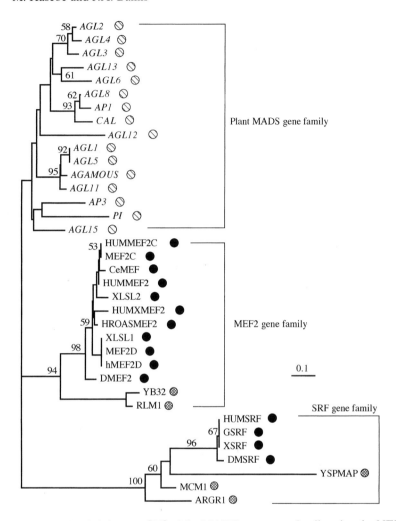

F𝐼G. 8.1. The neighbor joining tree [79] of the MADS supergene family using the NEIGH-BOR program [80]. The evolutionary distances were calculated with the PROTDIST program [80] using the 56 conservative amino acid residues between positions 2 and 57 (from the initial methionine codon of *AP1* [31]). The amino acid sequences could be aligned without any insertions or deletions. The branch length is proportional to the distances. *Scale bar* = 0.1 amino acids per residue. Bootstrap values calculated with the SEQBOOT program [80] are indicated for nodes supported in more than 50% of 100 bootstrap replicates. This is an unrooted tree. Symbols after the gene names indicate the origin of the genes: animals (*black circles*), fungi (*dark-hatched circles*), or plant (*lightly hatched circles*). All sequences were cited from the GenBank DNA database. *Brackets on right* indicate different groups of the MADS gene family

significant role in the evolution of morphological characters presently observed in plants.

8.2 MADS Supergene Family

The term MADS is an acronym for four genes that contain a conserved 56 amino acid sequence which is referred to as the MADS box [6]. These genes include the minichromosome maintenance gene (MCM1) from fungi, *AGAMOUS* (*AG*) and *DEFICIENS* (*DEF*) genes from plants, and the serum response factor gene (SRF) from animals. Molecular and biochemical studies of SRF and MCM1 have shown that these MADS genes code transcription factors and that the MADS box is necessary for DNA binding, dimerization, and accessory–factor interactions [7]. The X-ray crystal structure of SRF protein supports these results [8]. We have constructed a MADS gene tree using the 56 conserved amino acid sequences within the MADS box; the plant MADS genes are represented only by *Arabidopsis* genes. As shown in Fig. 8.1, this tree indicates that the MADS supergene family is composed of three families of genes: the SRF family; the myocyte enhancer factor 2 gene (MEF2) family; and the plant MADS family.

8.3 The SRF and MEF2 MADS Genes

The SRF family of MADS genes includes closely related animal SRF and three fungal genes, including YSPMAP, MCM1, and ARGR1. The SRF genes are involved in a signal transduction cascade that mediates the stimulation of cell division by growth factors [9]. The *Drosophila* SRF gene is related to tracheary development [10]. The MCM1 product is part of a protein complex that has three distinct functions: to regulate α- and a-cell-type-specific gene expression [11]; to promote morphological changes necessary for mating in response to secreted pheromones [11]; and to regulate arginine metabolism [12]. In the latter case, MCM1 protein works together with ARGRI, another MADS gene of the SRF family [12,13]. The YSPMAP gene is involved in cell type specification according to the brief description of this gene in the DNA databases.

The MEF2 family of MADS genes is composed of several animal MEF2 and two fungal genes. Many MEF2-related genes have been cloned from human, mouse, frog, and nematode, and all of them have high sequence similarities to form a monophyletic group (Fig. 8.1). All those genes characterized have been shown to be transcriptional activators involved in muscle development [7]. RLM1 from yeast is involved in the mitogen-activated protein kinase pathway [14] while the function of YB32 is unknown. None of the animal or fungal SRF and MEF2 genes contain the K (keratin coiled-coil domain-like) box [15] that is a partially conserved, approximately 70 amino acid sequence found in all plant

MADS genes. The K box can potentially form amphiphatic helices and thus may be involved in protein–protein interactions [5].

Figure 8.1 shows that the SRF and MEF2 families diverged from each other before the divergence of animals and fungi. We cannot speculate whether the original MADS gene was with or without the K box because the tree illustrated in Fig. 8.1 is unrooted. A characterization of MADS genes from protists, which diverged before the diversification of the plant, fungi, and animal kingdoms, should provide an answer to this question.

8.4 Plant MADS Genes

Based on the genetic analyses of several homeotic mutants that affect floral organ identity, the ABC model of flower morphogenesis was proposed [16–18]. This model explains how the four whorls of floral organs (i.e., sepals, petals, stamens, and gynoecium) are determined by the action of three classes of genes, referred to simply as the A, B, and C class genes. The A gene (e.g., APETALA1 [AP1]) and C gene (e.g., AG) specify sepal and gynoecium identity, respectively. Petal and stamen organ identities are mediated by combinations of A and B genes (e.g., AP1 plus PISTILLATA [PI] and AP3) or B and C genes (e.g., PI and AP3 plus AG), respectively. This model has been slightly modified with the recent discovery of new MADS genes that regulate ovule development and are thought to interact with other ABC class genes [19].

The cloning and sequencing of each of the ABC genes in Arabidopsis and Antirrhinum revealed that most of them are MADS genes [4,5,18,20]. The limited expression patterns of the ABC genes to specific parts of the developing flower is brought about by negative interactions between some of the ABC genes as well as by floral meristem identity genes and cadastral genes, most of which are not MADS genes [4,18]. The floral meristem identity genes are involved in the transition of the meristem from an inflorescence to a floral phase, which is a prerequisite for the induction of genes involved in flower development [21]. The cadastral genes limit the expression of organ identity genes to specific organs [18]. The combined action of the ABC floral organ identity genes, the floral meristem identity genes, and the cadastral genes can account for the patterning of the four whorls of floral organs in almost all angiosperms.

Previous molecular phylogenetic studies of the MADS gene family [22,23] have shown that this gene family can be divided into several groups, each group having functional similarity among its members. Several MADS genes from gymnosperms were subsequently reported [24] and shown to be phylogenetically related to some groups within the angiosperm MADS gene family. The inclusion of gymnosperm sequence data in the phylogenetic analysis indicated that the plant MADS groups diversified before the split of angiosperms and gymnosperms. Since this report [24], the number of MADS genes characterized from plants has almost doubled.

We have used the sequences of 75 plant MADS genes, representing 10 families, to construct a plant MADS gene family tree based on the amino acid sequences of MADS and K boxes. This tree (Fig. 8.2) indicates that the plant MADS genes can be divided into 12 monophyletic groups. Each group diversified before the divergence of monocots and dicots, and monophyly of each group was supported with more than 50% bootstrap probability. Each group is represented by the name of an *Arabidopsis* gene member. If a group does not have a member from *Arabidopsis*, the group was named for a randomly selected gene member. The phylogenetic relationships among the families of taxa used in Fig. 8.2 are shown in Fig. 8.3, which is based on *rbcL* DNA sequence data [25]. Next we discuss how these genes evolved based on our interpretation of the data presented in Figs. 8.2 and 8.3. The patterns of mRNA expression of representative plant MADS genes are shown in Fig. 8.4.

8.5 The *AGL2* Group

Monophyly of the AGL2 group is supported with 100% bootstrap values, but internal relationships in this group are not well resolved. This lack may reflect amino acid sequence diversity that correlates with functional differentiation within this group. The expression patterns of members within this group are quite different. For example, the *AGAMOUS*-like gene 3 (*AGL3*) is expressed throughout the whole plant except for the root [26], while the expression of *AGL2* and *AGL4* is confined to floral organs [27,28].

AGL2 expression is first observed in the flower primordium and continues to be expressed in all floral organs, including the ovules [27]. The expression expands to the seed coat and embryo. Its sister gene *AGL4* has a pattern of expression similar to *AGL2* [28]. The functions of *AGL2* and *AGL4* are unknown, although Flanagan and Ma [27] have speculated that their functions are similar to that of the yeast MCM1. MCM1 is expressed in three types of cells (a, α haploid cells, and diploid cell) but is necessary for specifying a cell type by associating with other cell-type-specific regulatory proteins [29]. Another possible function of *AGL2* is as an intermediate in the expression of floral meristem identity genes (e.g., *AP1* and *LEAFY* [*LFY*]) and the MADS floral organ identity genes (e.g., *PI, AP3,* and *AG*), because *AGL2* expression is initiated after the floral meristem identity genes and before the floral organ identity genes are expressed [28].

The tomato MADS box gene no. 5 (TM5), another *AGL2* member, is expressed in developing and mature petals, stamens, and gynoecium [30]. The function of the TM5 gene has been assessed by overexpressing TM5 antisense RNA in transgenic plants, which results in a variety of phenotypic alterations in the petals, stamens, and gynoecium, as expected on the basis of its expression pattern [30]. Some of the transformants are similar in phenotype to the *ap1* mutant of *Arabidopsis* [30], which has a flower within which a second flower with an elongated pedicel replaces the whorl of stamens [31]. The similarity in pheno-

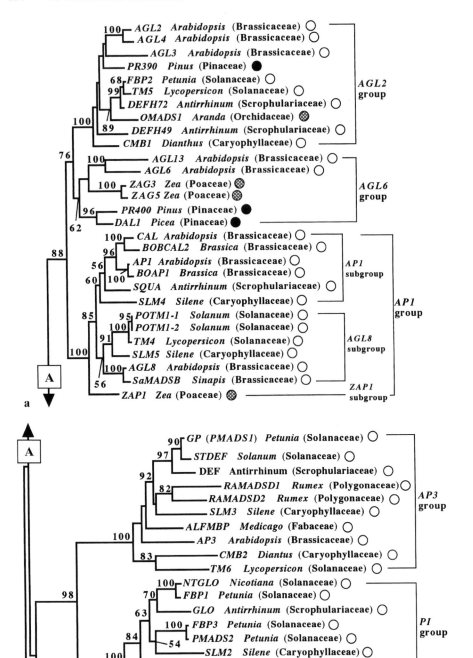

a

b

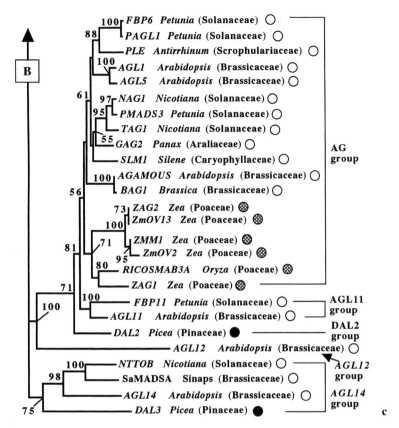

FIG. 8.2a–c. The neighbor joining tree [79] of plant MADS genes. Amino acid sequences of plant MADS genes cited from the DNA database were aligned using the CLUSTAL W program [81] and then were revised manually. (The alignment is available from M. Hasebe on request.) Sequences that have long deletions or insertions (*RAMADSP1*, *TDR3*, and *BAG15-2*) compared to other sequences and that could not be well aligned (*TDR8*) were omitted. The 146 amino acid residues corresponding to positions 8–61, 63–77, 85–88, 92, 94–110, 116–143, and 148–174 from the initial methionine codon of *API* [31] were used to calculate the evolutionary distances with the PROTDIS program [80]. The tree was obtained with the neighbor joining method [79] using the NEIGHBOR program [80]. The branch length is proportional to the distances. *Scale bar* = 0.1 amino acids per residue. Bootstrap values calculated with the SEQBOOT program [80] are indicated for nodes supported in more than 50% of 100 bootstrap replicates. This is an unrooted tree. Genera and families (in *brackets*) are indicated after gene names. Symbols after the gene names indicate the origin of the genes: dicots (*open circles*), monocots (*hatched circles*), or gymnosperms (*black circles*). All sequences were cited from the GenBank DNA database. *Brackets on right* indicate the different groups and subgroups of the plant MADS gene family. The basal (**c**), middle (**b**), and upper portions (**a**) of the tree are connected along the branch labeled *A* and *B*

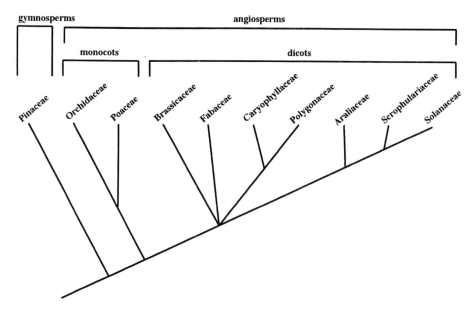

FIG. 8.3. Phylogenetic relationships among taxa used in Fig. 8.2. This tree is a consensus of two trees shown by Chase et al. [25] using *rbcL* DNA sequence data

type between *ap1*, a floral meristem identity gene, and some lines of tomato in which TM5 mRNA presumably is not functional, indicates that TM5 may be downstream of floral meristem identity genes, including the *AP1* homologue in tomato.

8.6 The *AGL6* Group

What we know about this interesting group (the AGL6 group) of MADS genes is based only on sequence information and a limited analysis of gene expression patterns; no mutations of any of these genes have been reported. The *AGL6* and *AGL13* genes of *Arabidopsis* are expressed only in ovules [32]. Because the expression of *AGL6* mRNA in flowers is not detected in an *ap1* or *ag* mutant background, *AGL6* is likely to be upregulated by *AP1* and *AG* [32]. The *Zea AG* genes no. 3 and 5 (*ZAG3* and *ZAG5*) are expressed in both female and male reproductive organs (ear and tassel) [33]. The *deficiens–agamous*-like gene no. 1 (*DAL1*) from the gymnosperm *Picea abies* is expressed in vegetative shoots and female and male inflorescences [24]. *PR400*, another gymnosperm (*Pinus radiata*) MADS gene registered in the DNA database, is also a member of this family; its pattern of expression has not been reported. A *DAL1* homologue has been recently cloned from the female inflorescence of *Gnetum parvifolium* (Shindou et al., unpublished results).

Group	gene	root	stem	leaf	i.p.	f.p.	se.p.	p.p.	st.p.	g.p.	ovule	seed	embryo	references
AGL2group	AGL2	?												[27]
	AGL4											?	?	[28]
	AGL3				?	?	?	?	?	?	?	?	?	[26]
AGL6group	TM5	?	?		?							?	?	[30]
	AGL13	?				?	?	?	?			?		[32]
	AGL6	?		?	?	?	?	?	?			?	?	[32]
AP1subgroup	AP1											?	?	[31]
	CAL										?	?	?	[45]
	SQUA											?	?	[46]
SLM4	SLM4											?	?	[47]
AGL8subgroup	AGL8				?							?	?	[48]
	SLM5				?							?	?	[47]
AP3group	AP3											?	?	[52]
	DEF	?	?								?	?	?	[49]
	SLM3	?	?									?	?	[47]
PIgroup	PMADS1	?	?									?	?	[56]
	PI	?	?									?	?	[51]
	GLO	?	?	?								?	?	[50]
	SLM2	?	?	?							?	?	?	[47]
	FBP1													[56]
AGL15group	AGL15	?	?											[32]
AGL17group	AGL17													[32]
AGgroup	FBP6	?			?						?	?	?	[57]
	PLE	?									?	?	?	[62]
	SLM1	?										?	?	[47]
	AGL1	?										?	?	[28]
	AGL5	?										?	?	[28]
	AG	?										?	?	[61]
AGL11group	FBP11	?	?									?	?	[59]
AGL12group	AGL12													[32]
AGL14group	AGL14													[32]

FIG. 8.4. Patterns of mRNA expression in representative plant MADS genes. Strong expression was designated by *heavy shading*, weak expression by *medium shading*, and undetectable expression by *unshaded (white) bars*. Question marks mean the parts have not been examined. *i.p.*, inflorescence primordia; *f.p.*, floral primordia; *se.p.*, sepal primordia; *p.p.*, petal primordia; *st.p.*, stamen primordia; *g.p.*, gynoecium primordia

Gnetum is often treated as a sister group of angiosperms based on morpholog-ical characters [34–37], although this relationship has not been supported by rRNA [38,39], *rbcL* [40], and noncoding region from chloroplast inverted repeat [41] sequence data. The evolutionary relationships of *Gnetum*, gymnosperms, and angiosperms, the evolution of angiosperm ovules from gymnosperm ovules, and the origin of the double integuments in angiosperms are all enigmatic ques-tions. Given that the *AGL6* genes investigated thus far are expressed in ovules and are present in angiosperms, gymnosperms, and *Gnetum*, these genes are excellent candidates for study in resolving these questions.

8.7 The *AP1* Group

As illustrated in Fig. 8.2, the *AP1* group is composed of three distinct monophyl-etic subgroups: the *AP1* subgroup, the *AGL8* subgroup, and the *Zea AP1* gene (*ZAP1*) subgroup (Fig. 8.2). Each subgroup is discussed individually next.

8.7.1 The AP1 Subgroup

Genetic and molecular analyses of the *AP1* gene of *Arabidopsis* indicate that *AP1* is necessary for the induction of the floral meristem from the inflorescence meristem [31,42,43,44]. The *CAULIFLOWER* gene (*CAL*) of *Arabidopsis* [43,45] has a similar function, and the *SQUAMOSA* gene (*SQUA*) of *Antirrhi-num* [46] has a function similar to that of *AP1*. Based on Figs. 8.2 and 8.3, the function of the *AP1* subgroup was present before the diversification of the Scrophulariaceae and Brassicaceae. Figure 8.2 clearly shows that *AP1* and *CAL* are sister genes that split after the diversification of the Scrophulariaceae and Brassicaceae. A complete defect in both genes (the *ap1 cal* double mutant) is severe, resulting in a cauliflower-like compound inflorescence; *cal* alone does not show any defects [43], while the phenotype of *ap1* alone is not as severe as the double mutant. One interpretation is that, following gene duplication, the func-tion of each gene was altered such that any function that was lost in one gene was retained in the other.

A null mutant of *SQUA* does not display the complete defects in flower meristem formation as are observed in the *ap1 cal* double mutant [46], although *SQUA* is expressed in all organs where *AP1* and *CAL* are expressed (Fig. 8.4). This indicates that other duplicated genes that have not yet been identified may be involved in flower induction in *Antirrhinum*. The *Silene latifolia* MADS gene no. 4 (*SLM4*) shows an expression pattern similar to that of *AP1* (Fig. 8.4), although its function is not known [47].

8.7.2 The AGL8 Subgroup

The *AGL8* subgroup diverged from the *AP1* subgroup after the divergence of monocots and dicots (Fig. 8.2). The *AGL8* gene is expressed in the inflorescence meristem and gynoecium and is negatively regulated by *AP1* [48]. Because mutations of *AGL8* do not exist, regulation of *AP1* by *AGL8* is not possible to assess. The expression pattern of *AGL8* is different from the genes in the *AP1* subgroup (Fig. 8.4), indicating that the functions of the *AP1* subgroup and *AGL8* subgroup diversified after the duplication of these genes. A comparison of the expression patterns of genes in the *AP1* and *AGL8* subgroups of *Arabidopsis* and *Silene* (Fig. 8.4) provides an interesting example of how genes diverge after their duplication. *SLM4*, a member of the *AP1* subgroup, and *SLM5*, a member of the *AGL8* subgroup, have been cloned from *Silene latifolia* and shown to have very similar patterns of expression. Both genes are expressed in the inflorescence meristem, floral meristem, sepal primordia, petal primordia, gynoecium primordia, and ovule [47], which is the sum of the fields of *AP1* plus *AGL8* expression (the former is expressed in the floral meristem, sepal primordia, petal primordia, and ovule, while the latter is expressed in the inflorescence meristem, gynoecium, and ovule) [31,48].

This observation leads us to hypothesize that the common ancestral gene of the *AP1* and *AGL8* subgroups had an expression pattern similar to that of *SLM4* and *SLM5*. After the gene duplication event, which resulted in formation of the *AP1* and *AGL8* subgroups, the expression patterns of the duplicated genes in *Arabidopsis* (*AGL8* and ancestral gene of *AP1* and *CAL*) were changed as is observed now, while the *Silene latifolia SLM4* and *SLM5* genes retained their original expression pattern. Although more sequence and expression information from additional plants is necessary to test this hypothesis, this example illustrates that diversification of gene expression patterns may lead to functional differentiation differently in different taxa.

8.7.3 The ZAP1 Subgroup

The *Zea AP1* gene (*ZAP1*) subgroup is represented by only one member, *ZAP1*. Its expression pattern has not been well studied, but the results of Northern analyses have shown that it is expressed in young flowers [33].

The three groups of plant MADS genes discussed thus far (*AGL2*, *AGL6*, and *AP1*) are monophyletic and diversified before the divergence of extant gymnosperms and angiosperms (see Fig. 8.2). The present data sets are insufficient to infer the function of their ancestral gene. The study of these genes in lower plants that diversified before the divergence of seed plants may, perhaps, identify the function of the ancestral gene. Further studies of these genes in seed plants also should be very interesting because the *AP1* subgroup of genes are among those involved in specifying sepal and petal organ identity [42,43], which are newly derived organs unique to the angiosperm lineage.

8.8 *AP3* and *PI* Groups

The *AP3* and *PI* groups of MADS genes comprise the B function genes in the ABC model, which are involved in petal and stamen organ identity and morphogenesis. In *Arabidopsis* and *Antirrhinum*, genes from each of these groups form a heterodimer and are autoregulating [49–51]. Based on Fig. 8.2, the *AP3* and *PI* groups are sister groups that diversified before the split of monocots and dicots. As petals are organs unique to angiosperms, the evolution of these two groups may have played an important role in the morphological evolution from gymnosperms to angiosperms. To date, no *AP3* or *PI* genes from gymnosperms have been reported, and indeed none may exist given that gymnosperm flowers lack petals. The search for *AP3* and *PI* genes in young inflorescences of *Gnetum*, which also lack petals, has also been unsuccessful (Shindou et al., unpublished data). An unexpected but interesting outcome from this analysis is that the *AP3* and *PI* groups diverged from a common ancestor early in plant MADS gene evolution, before the diversification of seed plants (see Fig. 8.1). The function of the ancestral gene may be assessed by studying the *AP3* and *PI* genes from the lower plants, which also diverged before the diversification of seed plants.

Only a single gene from each of the *AP3* and *PI* groups has been reported from *Arabidopsis* and *Antirrhinum* [49–52], as illustrated in Fig. 8.2. However, several genes have been isolated from *Petunia*; three fall within the *PI* group (floral binding protein 1 and 3, or *FBP1* and *FBP3*, and *PMADS2*) and one (petunia MADS gene no. 1 [*PMADS1*]) in the *AP3* group [53–56]. The *FBP3* and *PMADS2* were obtained from closely related *Petunia* strains and are likely to be alleles of each other. Interestingly, loss of *PMADS2* mRNA expression does not have an effect on the phenotype of the *Petunia* flower [56]. Transformed plants having no *FBP1* expression display an altered phenotype typical of a "B" type mutant, i.e., the petals and stamens are transformed to sepals and carpels, respectively [57]. The functional analysis of *PMADS2* and *FBP1* indicate that *FBP1* is necessary for B function and that the *PMADS2* has either lost its original function and has since become a nonfunctional pseudogene or has other unknown functions.

In the *gp* mutant of *Petunia*, which has a defective *PMADS1* gene [54], the petals are transformed to sepals but there is no effect on the stamen [58]. This phenotype is unusual for a class B mutant and may be the result of gene duplication. Figure 8.2 shows that there are two groups of *AP3* genes (one represented by *TM6* and the carnation MADS box gene no. 2 [*CMB2*] and the other by all other members of the *AP3* group), indicating that the ancestral *AP3* gene duplicated before the divergence of the Solanaceae and Caryophyllaceae (see Figs. 8.2 and 8.3). Therefore, there should be at least one unknown *Petunia* gene (Solanaceae) in the *TM6-CMB2* lineage. The supposed duplicated gene may have the function regulating stamen morphogenesis, and the combined action of the gene and *PMADS1* is necessary for complete B function in *Petunia*.

8.9 The *AG*, *AGL11*, and *DAL2* Groups

These three groups of plant MADS genes form a monophyletic group. The *DAL2* group is represented by a single gene from the gymnosperm *Picea abies* that is expressed in both male and female inflorescences, although its detailed expression pattern is unknown [24]. Based on Fig. 8.2, the *AG* and *AGL11* groups were formed by a gene duplication event that occurred after the diversification of conifers and angiosperms but before the branching of monocots from dicots. This period corresponds to the approximate time when angiosperms originated, indicating that this gene duplication might have had an important role during the course of evolution from gymnosperm flower to angiosperm flower. The functions of the genes within these groups also point to the importance of this duplication event in the evolution of the angiosperm ovule.

The *AGL11* and *FBP11* genes belong to the *AGL11* group and are specifically expressed in the ovule [32,59]. The effect of constitutively expressing *FBP11* in the transgenic petunia results in ectopic ovule formation on the dorsal side of the petal and ventral side of the sepal, indicating that *FBP11* regulates the position or initiates the developmental program of the ovule [60]. Within the *AG* group of MADS genes, *AG* from *Arabidopsis* and *PLENA* (*PLE*) from *Antirrhinum* have been studied at both genetic and molecular levels. Both *ag* and *ple* mutants lack stamens and gynoecium but form repeated whorls of sepals and petals [16,61,62], indicating that these genes are necessary for gynoecium and stamen development. However, normal ovules develop in the first whorl of double and triple mutant flowers including *ag-1* (*ap2-2 ag-1* and *ap2-2 ag-1 pi-1*), indicating that *AG* is not required for ovule development [63,64]. Based on these observations, it is possible that the AG group of genes is required for stamen and carpel development, while the *AGL11* group is necessary for ovule formation.

Based on Fig. 8.2, there are two possible scenarios that could explain the evolution of the *AG* and *AGL11* genes. The first is that the ancestral gene of the *AG* and *AGL11* groups had a positive role in the formation of stamens and gynoecium, including ovules. After gene duplication, these functions were divided among the *AG* group (stamen and gynoecium formation except for ovules) and *AGL11* group (ovule formation). A second possibility is that the ancestral gene was involved only in ovule development. After gene duplication, the *AG* group of genes obtained new functions involved in the elaboration of stamens and gynoecium. Further studies of gymnosperms should be useful in distinguishing between these possibilities as an elaborated gynoecium, which is lacking in gymnosperms, evolved during the course of evolution from gymnosperm to angiosperm. Other genes that have been shown to be involved in ovule development, including *AP2* [65,66], *SUPERMAN* [67,68], *BELL1* [66,69–71], and other genes [66, 72–74], are also likely to have been important in the evolution of the angiosperm flower, especially in the formation of an ovule with double integuments. A comparison of the functions of these genes in gymnosperms and angiosperms should give further insights into the origin of angiosperm type flowers.

Gene duplication is also evident within the *AG* group of MADS genes. The *AGL1* and *AGL5* sister genes diverged from the *AG* gene after the divergence of monocots and dicots (see Fig. 8.2). The expression of *AGL5* is different from the pattern observed for *AG*, indicating that the function(s) of *AGL5* have also diverged [28]. The *AGL5* gene is expressed only in the gynoecium, and the expression in stamens is not observed differently from *AG* expressed both of gynoecium and stamens. Both *AGL1* and *AGL5* are induced by *AG* as they are not expressed in an *ag* mutant background [28]. Furthermore, the *AGL5* gene promoter has an *AG* binding site, and the ectopic overexpression of *AG* results in the ectopic induction of *AGL5* in the leaves of the transgenic plants [28]. These observations raise several intriguing questions. How did the *AGL5* promoter obtain an *AG* binding site? Was the *AG* binding site inherited from the ancestral gene of *AG* and *AGL5*, or was the binding site obtained in the *AGL5* lineage after the duplication of these genes? Is *AG* itself autoregulated? A comparison of the regulatory regions of *AG*, *AGL1*, and *AGL5* should be informative in understanding the evolution of *cis*-acting sequences following gene duplication and their role in flower development and evolution.

8.10 Other Groups

The functions of genes belonging to other groups of plant MADS genes have not been studied extensively. The genes of the *AGL15* group have been cloned from *Arabidopsis* and *Brassica* and shown to be expressed only in the embryo [32,75]. The *AGL12* and *AGL17* groups are monotypic (see Fig. 8.2) and have been reported only from *Arabidopsis* [32]. Both genes are expressed only in roots, and their function is unknown [32]. The *AGL12* gene clusters with the *AG-AGL11-DAL2* monophyletic groups, a relation supported with 100% bootstrap probability (Fig. 8.2). Because *AGL12* is not expressed in reproductive organs [32], its function is speculated to be different from the *AG-AGL11-DAL2* groups. Based on Fig. 8.2, *AGL2* diverged from the *AG-AGL11-DAL2* groups before the diversification of seed plants. The final plant MADS group, *AGL14*, includes several genes. All those assayed are expressed in vegetative organs of the plant [24,32,76], and their function in development is unknown.

8.11 Concluding Remarks

Our discussion of the evolution of plant MADS genes has repeatedly stressed the importance of gene duplications. Our analysis indicates that there are at least two different evolutionary fates a pair of MADS genes might have following gene duplication. First, both the duplicated genes can diverge somewhat in function, yet work as a pair such that each is indispensable for their original or new function/s. The relationship between the *AP3* and *PI* sister groups is an example of this. Genes from each group form a heterodimer, and both are necessary for the development of petals and stamens. Unfortunately, the available data are not sufficient to know the function of the ancestral gene before duplication. The second fate of duplicated genes is for one sister gene to retain and the other to lose the original function of the ancestral gene; once the original function is lost the gene may either become a pseudogene or gain a new function(s). Examples of this have been indicated within the *AP1* and *AG* groups of plant MADS genes although the function of the ancestral genes is also unknown. More detailed studies of these groups are expected to not only show differences in expression pattern but also reveal functional diversity. The changes in the regulatory interactions among MADS genes have also played important roles in the evolution of plant morphology. The interactions between MADS genes and other regulatory genes, although poorly understood at the present time, also need to be considered.

This study shows that most of the plant MADS gene groups diversified before the divergence of gymnosperms and angiosperms. Studies of plant MADS genes in ferns indicate that most of the plant MADS groups diversified before the divergence between ferns and seed plants (Hasebe and Banks, unpublished results). The ancestor of ferns and seed plants is thought to be a *Psilophyton*-type plant that did not have floral organs (including ovules) and propagated via spores. What was the original function of plant MADS genes in *Psilophyton*-type plants or even older ancestors? To address this question, the function of MADS genes in lower plants, including pteridophytes, bryophytes, and green algae, must be assessed. The study of MADS genes in the fern *Ceratopteris* [77], the bryophyte *Physcomitrella* [78], and the green algae *Chlamydomonas* should provide information that is critical in understanding how MADS genes evolved and how their evolution has led to the present diversity in plants we observe today.

Acknowledgments. This research was supported in part by grants from the Ministry of Education, Science and Culture, Japan (08874123) and the National Science Foundation, USA. This is paper 15201 of the Purdue University Agricultural Experimental Station.

References

1. Bürglin TR (1995) The evolution of homeobox genes. In: Arai R, Kato M, Doi Y (eds) Biodiversity and evolution. National Science Museum Foundation, Tokyo, pp 291–336
2. Carroll SB (1995) Homeotic genes and the evolution of arthropods and chordates. Nature 376:479–485
3. Ramachandran S, Hiratsuka K, Chua N-H (1994) Transcription factors in plant growth and development. Curr Opin Genet Dev 4:642–646
4. Ma H (1994) The unfolding drama of flower development: recent results from genetic and molecular analyses. Genes Dev 8:745–756
5. Davies B, Schwartz-Sommer Z (1994) Control of floral organ identity by homeotic MADS-box transcription factors. In: Nover L (ed) Results and problems in cell differentiation, vol 20. Springer, Berlin Heidelberg New York, pp 235–258
6. Schwarz-Sommer Z, Huijser P, Nacken W, Saedler H, Sommer H (1990) Genetic control of flower development by homeotic genes in *Antirrhinum majus*. Science 250:931–936

7. Shore P, Sharrocks AD (1995) The MADS-box family of transcription factors. Eur J Biochem 229:1–13

8. Pellegrini L, Tan S, Richmond TJ (1995) Structure of serum response factor core bound to DNA. Nature 376:490–498

9. Treisman R (1994) Ternary complex factors: growth factor regulated transcriptional activators. Curr Opin Genet Dev 4:96–101

10. Affolter M, Montagne J, Walldorf U, Groppe J, Kloter U, LaRosa M, Gehring WJ (1994) The *Drosophila* SRF homologue is expressed in a subset of tracheal cells and maps within a genomic region required for tracheal development. Development (Camb) 120:743–753

11. Dolan JW, Fields S (1991) Cell-type-specific transcription in yeast. Biochim Biophys Acta 1088:155–169

12. Messenguy F, Duboin E (1993) Genetic evidence for a role for MCM1 in the regulation of arginine metabolism in *Saccharomyces cerevisiae*. Mol Cell Biol 13:2586–2592

13. Passmore S, Maine GT, Elble R, Christ C, Tye B-K (1988) *Saccharomyces cervisiae* protein involved in plasmid maintenance is necessary for mating of *MATa* Cells. J Mol Biol 204:593–606

14. Watanabe Y, Irie K, Matsumoto K (1995) Yeast RLM1 encodes a serum response factor-like protein that may function downstream of the Mpk1 (slt2) mitogen-activated protein kinase pathway. Mol Cell Biol 15:5740–5749

15. Ma H, Yanofsky MF, Meyerowitz EM (1991) *AGL1–AGL6*, an *Arabidopsis* gene family with similarity to floral homeotic and transcription factor genes. Genes Dev 5:484–495

16. Bowman JL, Smyth DR, Meyerowitz EM (1991) Genetic interactions among floral homeotic genes of *Arabidopsis*. Development (Camb) 112:1–20

17. Coen E, Meyerowitz EM (1991) The war of the whorls: genetic interactions controlling flower development. Nature 353:31–37

18. Weigel D, Meyerowitz EM (1994) The ABCs of floral homeotic genes. Cell 78:203–209

19. Colombo L, Franken J, Koetje E, van Went J, Dons HJM, Angenent GC, van Tunen AJ (1995) The petunia MADS box gene *FBP11* determines ovule identity. Plant Cell 7:1859–1868

20. Theißen G, Saedler H (1995) MADS-box genes in plant ontogeny and phylogeny: Haeckel's "biogenetic law" revisited. Curr Opini Genet Dev 5:628–639

21. Weigel D, Meyerowitz EM (1993) Activation of floral homeotic genes in *Arabidopsis*. Science 261:1723–1726

22. Doyle JJ (1994) Evolution of a plant homeotic multigene family: toward connecting molecular systematics and molecular developmental genetics. Syst Biol 43:307–328

23. Purugganan MD, Rounsley SD, Schmidt RJ, Yanofsky MF (1995) Molecular evolution of flower development: diversification of the plant MADS-box regulatory gene family. Genetics 140:345–356

24. Tandre K, Albert VA, Sundas A, Engström P (1995) Conifer homologues to genes that control floral development in angiosperms. Plant Mol Biol 27:69–78

25. Chase MW, Soltis DE, Olmstead RG, Morgan D, Les DH, Mishler BD, Duvall MR, Price RA, Hills HG, Qiu Y-L, Kron KA, Rettig JH, Conti E, Palmer JD, Manhart JR, Sytsma KJ, Michaels HJ, Kress WJ, Karol KG, Clark WD, Hedrén M, Gaut BS, Jansen RK, Kim K-J, Wimpee CF, Smith JF, Furnier GR, Strauss SH, Xiang Q-Y, Plunkett GM, Soltis PS, Swensen SM, Williams SE, Gadek PA, Quinn CJ, Equiarte LE, Golenberg E, Learn GH Jr, Graham SW, Barrett SCH, Dayanandan S, Albert

VA (1993) Phylogenetics of seed plants: an analysis of nucleotide sequences from the plastid gene *rbc*L. Ann MO Bot Gard 80:528–580

26. Huang H, Tudor M, Weiss CA, Hu Y, Ma H (1995) The *Arabidopsis* MADS-box gene *AGL3* is widely expressed and encodes a sequence-specific DNA-binding protein. Plant Mol Biol 28:549–567

27. Flanagan CA, Ma H (1994) Spatially and temporally regulated expression of the MADS-box gene *AGL2* in wild-type and mutant *Arabidopsis* flowers. Plant Mol Biol 26:581–595

28. Savidge B, Rounsley SD, Yanofsky MF (1995) Temporal relationship between the transcription of two *Arabidopsis* MADS box genes and the floral organ identity genes. Plant Cell 7:721–733

29. Dolan JW, Fields S (1991) Cell-type-specific transcription in yeast. Biochim Biophys Acta 1088:155–169

30. Pnueli L, Hareven D, Broday L, Hurwitz C, Lifschitz E (1994) The *TM5* MADS box gene mediates organ differentiation in the three inner whorls of tomato flowers. Plant Cell 6:175–186

31. Mandel MA, Gustafson-Brown C, Savidge B, Yanofsky MF (1992) Molecular characterization of the *Arabidopsis* floral homeotic gene *APETALA1*. Nature 360:273–277

32. Rounsley SD, Ditta GS, Yanofsky MF (1995) Diverse roles for MADS box genes in *Arabidopsis* development. Plant Cell 7:1259–1269

33. Mena M, Mandel MA, Lerner DR, Yanofsky MF, Schmidt RJ (1995) A characterization of the MADS-box gene family in maize. Plant J 8:845–854

34. Crane PR (1985) Phylogenetic analysis of seed plants and the origin of angiosperms. Ann MO Bot Gard 72:716–793

35. Doyle JA, Donoghue MJ (1986) Seed plant phylogeny and the origin of angiosperms: an experimental cladistic approach. Bot Rev 52:321–431

36. Loconte H, Stevenson W (1990) Cladistics of the Spermatophyta. Brittonia 42:197–211

37. Rothwell GW, Serbet R (1994) Lignophyte phylogeny and the evolution of spermatophytes: a numerical cladistic analysis. Syst Bot 19:443–482

38. Troitsky AV, Melekhovets YuF, Rakhimova GM, Bobrova VK, Valiejo-Roman KM, Antonov AS (1991) Angiosperm origin and early stages of seed plant evolution deduced from rRNA sequence comparisons. J Mol Evol 32:253–261

39 Chaw S-M, Zharkikh A, Sung H-M, Lau T-C, Li W-H (1996) Molecular phylogeny of gymnosperms and seed plant evolution: analysis of 18S rRNA sequences. Am J Bot 83(suppl):211

40. Hasebe M, Kofuji R, Ito M, Kato M, Iwatsuki K, Ueda K (1992) Phylogeny of gymnosperms inferred from *rbcL* gene sequences. Bot Mag Tokyo 105:673–679

41. Goremykin V, Bobrova V, Pahnke J, Troitsky A, Antonov A, Martin W (1996) Noncoding sequences from the slowly evolving chloroplast inverted repeat in addition to *rbcL* do not support Gnetalean affinities of angiosperms. Mol Biol Evol 13:383–396

42. Irish VF, Sussex IM (1990) Function of the *APETALA1* gene during *Arabidopsis* floral development. Plant Cell 2:741–753

43. Bowman JL, Alvarez J, Weigel D, Meyerowitz EM, Smyth DR (1993) Control of flower development in *Arabidopsis thaliana* by *APETALA1* and interacting genes. Development (Camb) 119:721–743

44. Mandel MA, Yanofsky MF (1995) A gene triggering flower formation in *Arabidopsis*. Nature 377:522–524

45. Kempin SA, Savidge B, Yanofsky MF (1995) Molecular basis of the *cauliflower* phenotype in *Arabidopsis*. Science 267:522–525
46. Huijser P, Klein J, Lönnig W-E, Meijer H, Saedler H, Sommer H (1992) Bracteomania, an inflorescence anomaly, is caused by the loss of function of the MADS-box gene *squamosa* in *Antirrhinum majus*. EMBO J 11:1239–1249
47. Hardenack S, Saedler DYH, Grant S (1994) Comparison of MADS box gene expression in developing male and female flowers of the dioecious plant white campion. Plant Cell 6:1775–1787
48. Mandel MA, Yanofsky MF (1995) The Arabidopsis *AGL8* MADS box gene is expressed in inflorescence meristems and is negatively regulated by *APETALA1*. Plant Cell 7:1763–1771
49. Schwartz-Sommer Z, Hue I, Huijser P, Flor JP, Hansen R, Tetens F, Lönnig W-E, Saedler H, Sommer H (1992) Characterization of the *Antirrhinum* floral homeotic MADS-box gene *deficiens*: evidence for DNA binding and autoregulation of its persistent expression throughout flower development. EMBO J 11:251–263
50. Tröbner W, Ramirez L, Motte P, Hue I, Huijser P, Lönnig W-E, Saedler H, Sommer H, Schwarz-Sommer Z (1992) *GLOBOSA*: a homeotic gene which interacts with *DEFICIENS* in the control of *Antirrhinum* floral organogenesis. EMBO J 11:4693–4704
51. Goto K, Meyerowitz EM (1994) Function and regulation of the *Arabidopsis* floral homeotic gene *PISTILLATA*. Genes Dev 8:1548–1560
52. Jack T, Brockman LL, Meyerowitz EM (1992) The homeotic gene *APETALA3* of *Arabidopsis thaliana* encodes a MADS box and is expressed in petals and stamens. Cell 68:683–697
53. Angenent GC, Busscher M, Franken J, Mol JNM, van Tunen AJ (1992) Differential expression of two MADS box genes in wild-type and mutant *Petunia* flowers. Plant J 4:983–993
54. Kush A, Brunelle A, Shevell D, Chua N-H (1993) The cDNA sequence of two MADS box proteins in *Petunia*. Plant Physiol (Bethesda) 102:1051–1052
55. van der Krol AR, Chua N-H (1993) Flower development in *Petunia*. Plant Cell 5:1195–1203
56. van der Krol AR, Brunelle A, Tsuchimoto S, Chua N-H (1993) Functional analysis of the petunia floral homeotic MADS-box gene *pMADS1*. Genes Dev 7:1219–1228
57. Angenent GC, Franken J, Busscher M, Colombo L, van Tunen AJ (1993) Petal and stamen formation in petunia is regulated by the homeotic gene *fbp1*. Plant J 4:101–112
58. Halfter U, Ali N, Stockhaus J, Ren L, Chua N-H (1994) Ectopic expression of a single homeotic gene, the *Petunia* gene *green petal*, is sufficient to convert sepals to petaloid organs. EMBO J 13:1443–1449
59. Angenent GC, Franken J, Busscher M, van Dijken A, van Went JL, Dons HJM, van Tunen AJ (1995) A novel class of MADS box genes is involved in ovule development in petunia. Plant Cell 7:1569–1582
60. Colombo L, Franken J, Koetje E, van Went J, Dons HJM, Angenent GC, van Tunen AJ (1995) The petunia MADS box gene *FBP11* determines ovule identity. Plant Cell 7:1859–1868
61. Yanofsky MF, Ma H, Bowman JL, Derws GN, Feldman KA, Meyerowitz EM (1990) The protein encoded by the *Arabidopsis* homeotic gene *agamous* resembles transcription factors. Nature 346:35–39
62. Bradley D, Carpenter R, Sommer H, Hartley N, Coen E (1993) Complementary floral homeotic phenotypes result from opposite orientations of a transposon at the *plena* locus of *Antirrhinum*. Cell 72:85–95

63. Bowman JL, Drews GN, Meyerowitz EM (1991) Expression of the *Arabidopsis* floral homeotic gene *AGAMOUS* is restricted to specific cell types late in flower development. Plant Cell 3:749–758

64. Bowman JL, Smyth DR, Meyerowitz EM (1991) Genetic interactions among floral homeotic genes of *Arabidopsis*. Development (Camb) 112:1–20

65. Jufuku KD, den Boer BGW, Montagu MV, Okamuro JK (1994) Control of *Arabidopsis* flower and seed development by the homeotic gene *APETALA2*. Plant Cell 6:1211–1225

66. Modrusan Z, Reiser L, Feldmann KA, Fischer RL, Haughn GW (1994) Homeotic transformation of ovules into carpel-like structures in *Arabidopsis*. Plant Cell 6:333–349

67. Sakai H, Medrano LJ, Meyerowitz ME (1995) Role of *SUPERMAN* in maintaining *Arabidopsis* floral whorl boundaries. Nature 378:199–203

68. Gaiser JC, Robinson-Beers K, Gasser CS (1995) The *Arabidopsis SUPERMAN* gene mediates asymmetric growth of the outer integument of ovules. Plant Cell 7:333–345

69. Ray A, Robinson-Beers K, Ray S, Baker SC, Lang JD, Preuss D, Milligan SB, Gasser CS (1994) *Arabidopsis* floral homeotic gene BELL (*BEL1*) controls ovule development through negative regulation of *AGAMOUS* gene (*AG*). Proc Natl Acad Sci USA 91:5761–5765

70. Reiser L, Modrusan Z, Margossian L, Samach A, Ohad N, Haughn GW, Fischer RL (1995) The *BELL1* gene encodes a homeodomain protein involved in pattern formation in the *Arabidopsis* ovule primordia. Cell 83:735–742

71. Léon-Kloosterziel KM, Keijzer CJ, Koornneef M (1994) A seed shape mutant of *Arabidopsis* that is affected in integument development. Plant Cell 6:385–392

72. Klucher KM, Chow H, Reiser L, Fischer RL (1996) The *ANINTEGUMENTA* gene of *Arabidopsis* required for ovule and female gemetophyte development is related to the floral homeotic gene *APETALA2*. Plant Cell 8:137–153

73. Elliott RC, Betzner AS, Huttner E, Oakes MP, Tucker WQL, Gerentes D, Perez P, Smyth DR (1996) *ANINTEGUMENTA*, an *APETALA2*-like gene of *Arabidopsis* with pleiotropic roles in ovule development and floral organ growth. Plant Cell 8:155–168

74 Nadeau JA, Zhang XS, Li J, O'Neill SD (1996) Ovule development: identification of stage-specific and tissue-specific cDNAs. Plant Cell 8:213–239

75. Heck GR, Perry SE, Nichols KW, Fernandez DE (1995) AGL15, a MADS domain protein expressed in developing embryos. Plant Cell 7:1271–1282

76. Mandel T, Lutziger I, Kuhlemeier C (1994) A ubiquitously expressed MADS-box gene from *Nicotiana tabacum*. Plant Mol Biol 25:319–321

77. Eberle J, Nemacheck J, Wen C-K, Hasebe M, Banks JA (1995) Ceratopteris: a model system for studying sex-determining mechanisms in plants. Int J Plant Sci 156:359–366

78. Cove DJ, Knight CD (1993) The moss *Physcomitrella patens*, a model system with potential for the study of plant reproduction. Plant Cell 5:1483–1488

79. Saitou N, Nei M (1987) The neighbor-joining method: a new method for reconstructing phylogenetic trees. Mol Biol Evol 4:406–425

80. Felsenstein J (1993) *PHYLIP (Phylogeny Inference Package)* version 3.5c. Available from the author. Department of Genetics, University of Washington, Seattle

81. Thompson JD, Higgins DG, Gibson, TJ (1994) CLUSTAL W: improving the sensitivity of progressive multiple sequence alignment through sequence weighting, positions-specific gap penalties and weight matrix choice. Nucleic Acids Res 22:4673–4680

CHAPTER 9

Palynological Approaches to the Origin and Early Diversification of Angiosperms

MASAMICHI TAKAHASHI

9.1 Introduction

Palynology deals with the walls of pollen grains and spores, which show great diversity in structure and sculpture [1]. Modern palynological studies consist of pollen morphology, pollen development, paleopalynology, and pollen analysis. Three international symposia have been held on palynology. The first symposium, on the evolutionary significance of the exine, organized by Ferguson and Muller, was held at the Linnean Society of London and the Royal Botanical Gardens, Kew in 1974 [2]. The second symposium, titled "Pollen and Spore: Form and Function," was held at the Linnean Society of London and the British Museum (Natural History) in 1985 [3]. The third symposium, "Pollen and Spores: Patterns of Diversification," was held at the Linnean Society of London and the Natural History Museum in 1990 [4]. These symposia demonstrated the exceptional opportunities afforded by pollen grains and spores for bringing together findings from studies of the origin and diversification of land plants. The systematic and paleontological applications of palynology exploit the complexity and diversity of organization present in pollen grains and spores. Recent advances in palynology, in association with the progress of electron microscopy, have had a remarkable impact on our understanding of the origin and diversification of the angiosperms. In this chapter, some distinguished contributions to paleopalynology and studies of pollen morphology and pollen development are shown, with special attention to the origin and diversification of angiosperms. Pollen analysis as a technique for reconstructing the vegetation and environment of the past is not dealt with in this review, but is shown in Faegri and Iversen [5].

Department of Biology, Faculty of Education, Kagawa University, 1-1 Saiwai-cho Takamatsu 760, Japan

9.2 Morphological Approach to the Diversification of Angiosperms

Comparative pollen morphology has a long history of more than 150 years, beginning with workers like Mohl [6] and Hassell [7]. A historical background of comparative pollen morphology is reviewed by Ferguson [8]. One of the outstanding palynologists is Erdtman [1], who used light microscopy to make elegant descriptions of the pollen morphology of all families in the angiosperms. He demonstrated a wide range of diversity and the systematic significance of pollen morphology in the angiosperms. Exine structure and sculpture had not, however, been described in detail with light microscopy.

The application of electron microscopy has served as an impetus to pollen morphological studies. Fernandez-Moran and Dahl [9] observed ultra-thin frozen sections of pollen grains using transmission electron microscopy. Rowley [10] applied transmission electron microscopy with carbon replica to a comparative study of Commelinaceae pollen. Thornhill et al. [11] first showed the use of the scanning electron microscope in pollen morphology.

The 1970s and 1980s saw notable progress in palynology with the improved quality of scanning and transmission electron microscopes. Many excellent palynological contributions were made to the systematics of the Berberidaceae [12], Centrospermae [13–16], Compositae [17–21], and Leguminosae [22–26]. These elegant studies showed that pollen grains demonstrate multiple characters significant in systematizing taxonomic categories. Takahashi [27–28] suggested that in *Trillium* the pollen sculpture would reflect infrageneric relationships. The infrageneric relationships thus suggested in *Trillium* were later supported by rbcL genetic analysis [29]. Pollen morphology provides significant and critical support for plant systematics.

Most angiosperms have pollen characterized by columellate exines, while some angiosperms have distinct pollen with non-columellate or granular exines that are highly developed in relation to those of the primitive angiosperms. *Degeneria* (Degeneriaceae) and *Eupomatia* (Eupomatiaceae) have pollen with structureless, atectate-amorphous exines [30–32]. Most of the Magnoliaceae and Annonaceae have monosulcate pollen characterized by granular exines [32]. Walker and Skvarla [31] examined exine structures in the ranalean complex and showed that the exine in some of the primitive families of angiosperms is structurally amorphous, without columellate or alveolar structure. They proposed that the alveolar structure of gymnosperm pollen and the columellate structure of angiosperm pollen have evolved from atectate, structureless pollen via a more or less similar granular stage.

On the other hand, the granular or beaded exines are also recognized in the Onagraceae and the Leguminosae [33–34]. Walker [32] suggested the possibility of reversal trends during which tectate-imperforate pollen may lose its columellar structure and become secondarily granular and atectate in some angiosperms. Zavada [33] hypothesized that the primitive tectate columellate wall structure gave rise to monocotyledonous atectate or granular wall and finally achieved extreme reduction of the exine.

Exine-less pollen is very common in aquatic plants [35–41]. The water plant *Ceratophyllum* has a reduced exine with spinules. The thin exinous layer in *Ceratophyllum* is regarded as equivalent to the endexine of other angiosperm pollen in structure and development [41]. The reduction of exines is regarded as an adaptation to the aquatic environment. However, reduced exines are also present in several unrelated terrestrial plants, e.g., the Lauraceae [42–44], the Cannaceae [45–47], and the Zingiberaceae [48–53]. The evolutionary trends of exine stratification in relation to the origin of angiosperms remain to be clarified.

9.3 Developmental Approach to the Homology of Pollen Wall Layers

The comparative ontogenetic approach may provide insight into the evolutionary relationships of primitive angiosperms and some gymnosperm groups and aid the determination of pollen character polarity among the seed plants [54]. Different interpretations of wall layer homologies in the angiosperms and gymnosperms have been proposed using developmental criteria [55–59]. The primexine scheme that was originated by Heslop-Harrison [60] had been widely accepted in many previous pollen developmental studies. However, a new model of exine pattern formation incorporated with the plasma membrane has been proposed for the angiosperms [61–65].

Hesse [66] reviewed recent advances in pollen development studies. The new scheme for exine development showed that the invaginated plasma membrane takes a tectum pattern before the columellae formation [61–63]. This scheme of exine pattern formation supported the hypothesis that the ektexine and the endexine in gymnosperm pollen are homologous with the ektexine and the endexine in the angiosperm pollen [57]. Rowley [58] further interpreted the endexines of the gymnosperms and angiosperms as being equivalent to the exospore of the pteridophytes. There is no fundamental difference between gymnosperms and angiosperms in the developmental process of exines.

There are, however, some exceptional developmental processes in supratectal elements. Takahashi and Kouchi [67] showed that the supratectal elements are formed during the free microspore stage in *Hibiscus* pollen. The supra-striate components in *Ipomopsis* pollen are also formed during the free microspore stage [65]. The later development of the supratectal elements clarified the difference between tectum and supratectum in ontogeny, suggesting the supratectal components are not homologous with tectum. On the contrary, the exinous spines of *Nuphar* pollen extend into the callose wall during the tetrad stage [68]. The exines of angiosperms are composed of elements constructed through different developmental processes.

There are some controversial aspects of the macromolecule structure of the exine substructure, as reviewed by Takahashi [62]. Rowley and his colleagues have proposed a helical coiled model of exine substructure [69–70]. Some previous works analyzed the chemical composition of the sporopollenin that constitutes exine [71–77]. Kawase and Takahashi [77] recently suggested that the sporopollenin is composed of an aliphatic main chain with an aromatic or conjugated side chain and organosilicon compounds. The chemical composition of sporopollenin and the exine substructure will be revealed in the near future. Further comparative chemical studies of sporopollenin and exine development will provide new evidence for clarifying pollen wall homology, the early diversification of land plants, and the origin of angiosperms.

9.4 Paleopalynological Approach to the Early Diversification of Angiosperms

Dispersed fossil pollen is an invaluable source of data for understanding the early diversification of angiosperms. Recently, the anthophyte hypothesis has stimulated renewed discussion of the possibility of pre-Cretaceous angiosperms [78–83]. Angiosperm-like pollen with tectate-columellate wall structure occurs in the *Crinopolles* group from the Upper Triassic [82–83]. The *Crinopolles* group, including 6 genera and 11 species, is defined by the reticulate-columellate exine structure, two or more furrows at the distal and equatorial sides of a grain. Cornet [83] suggested that the Triassic *Crinopolles* group probably evolved within the angiosperms rather than within a pre-angiospermous anthophytes group, and that the earliest angiosperms may have produced polyplicate inaperturate as well as tectate-granular monosulcate pollen. Doyle and Hotton [84] cast doubt on the angiospermous affinity of *Crinopolles* pollen, because the endexine of *Crinopolles* pollen is nonlaminated and uniform in thickness, which is different to exines of the angiosperms. The taxonomic position of the *Crinopolles* group is still controversial, although it remains possible that the angiosperm lineage diverged from related gymnosperm groups as early as the Triassic. It is expected that recovery of fossil reproductive organs with *Crinopolles* pollen will help elucidate the origin of angiosperms.

The early diversification of angiosperms in the Lower Cretaceous has been suggested by the fossil pollen evidence, as reviewed by Crane [85] and Traverse [86]. Some unequivocal angiosperm pollen grains are reported from the Hauterivian, and through the Hauterivian to Cenomanian angiosperm pollen and other reproductive structures increase dramatically in their variety and abundance in fossil assemblages [79,85,87]. Doyle [88] and Muller [89] noted that Magnoliaceae and related families (Magnoliales) have smooth monosulcate pollen with granular exine that is difficult to distinguish from pollen of the gymnosperms. Ward, Doyle, and Hotton [90] described fossil monosulcate pollen, *Lethomasites*, from the lower Potomac Group of Delaware and Virginia. The pollen is characterized by psilate, foveolate, or fossulate tectum and infratectal granular layer, and resembles pollen of the Magnoliales. The other representative angiosperm pollen grains from the Lower Cretaceous are monosulcate types, *Clavatipollenites*, *Retimonocolpites*, and *Liliacidites*, that are characterized by the reticulate tectum, columellae, and nexine structure. *Clavatipollenites* is considered to be included in the Chloranthaceae [91], and this systematic affinity is supported by fossil fruit and seed characters of *Couperites* [92]. Further, Taylor and Hickey [92] proposed a Paleoherb hypothesis that contrasts with the Magnolialean hypothesis. Although their cladistic analysis placed the nonmagnolialean

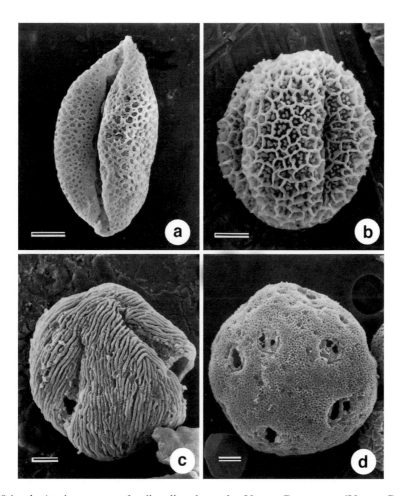

Fig. 9.1a–d. Angiospermous fossil pollen from the Upper Cretaceous (Upper Campanian) of Saghalin. Scanning electron micrographs (SEM). **a** Monocolpate pollen of *Liliacidites* sp. *Scale bar* = 5 μm. **b** Tricolporate pollen with reticulate exine. The reticulum includes distinct bacules in the lumina. Undescribed pollen. *Scale bar* = 3 μm. **c** Tricolporate pollen with striate exine. Undescribed pollen. *Scale bar* = 3 μm. **d** Pantoporate pollen with punctum. Undescribed pollen. *Scale bar* = 3 μm

Chloranthaceae at the bottom of the basal angiosperms [93], *Clavatipollenites* has ancestral angiosperm pollen grains from the Lower Cretaceous, but these are not of angiosperm origin. The Paleoherb concept [92] differs from the hypotheses of Doyle and Hotton [84] and Donoghue and Doyle [93], who recognized five major clades, Magnoliales, Laurales, Winteroids, Eudicots, and Paleoherbs, using a cladistic analysis. They considered the Paleoherb clade to consist of the Lactoridaceae, Aristolochiaceae, Piperaceae, Saururaceae, Nymphaeaceae, Cobambaceae, and the monocots. The Chloranthaceae is included in the Laurales [84,93]. Fossil pollen records from the Barremian and Aptian show that all the five clades had diverged early in the angiosperm evolution of the early Cretaceous. Although the number of dispersed grains of uncertain systematic affinities is still relatively large (Fig. 9.1), features of the pollen wall as observed with transmission electron microscopy provide additional characters of systematic importance, but so far relatively few ultrastructral studies of in situ pollen in Cretaceous angiosperm reproductive organs have been carried out. This has limited the exent to which dispersed pollen may be used to clarify the timing of evolutionary events and to test phylogenetic models hypothesized from studies of mesofossil and macrofossil remains [79].

Palynological studies, in association with studies of plant mesofossils [94], will play an important role in solving the problems of the origin and early diversification of angiosperms.

References

1. Erdtman G (1952) Pollen morphology and plant taxonomy. 1. Angiosperms. Almqvist & Wiksell, Stockholm
2. Ferguson IK, Muller J (eds) (1976) The evolutionary significance of the exine. Linnean Society of symposium series No. 1. Academic, London
3. Blackmore S, Ferguson IK (eds) (1986) Pollen and spore: form and function. Academic, London
4. Blackmore S, Barnes SH (eds) (1991) Pollen and spores: patterns of diversity, Systematic Association special vol 44. Clarendon, Oxford
5. Faegri K, Iversen J (1989) Textbook of pollen analysis. Wiley, Chichester
6. Mohl H (1835) Sur la structure et les formes des grains de pollen. Ann Sci Nat ser 2, 3:148–180
7. Hassell AH (1842) Observations on the structure of the pollen granule, considered principally in reference to its eligibility as a means of classification. Ann Mag Nat Hist 889:92–108, 544–573
8. Ferguson IK (1985) The role of pollen morphology in plant systematics. An Asoc Palinol Leng Esp 2:5–18
9. Fernandez-Moran H, Dahl AO (1952) Electron microscopy of ultra-thin frozen sections of pollen grains. Science 116:465–467
10. Rowley JR (1959) The fine structure of the pollen wall in the Commelinaceae. Grana 2:3–31
11. Thornhill JW, Matta RK, Wood W H (1965) Examining three-dimensional microstructures with the scanning electron microscope. Grana 6:3–6

12. Nowicke JW and Skvarla JJ (1981) Pollen morphology and the relationships of the Berberidaceae. Smithsonian Contrib Bot 50:1–83

13. Nowicke JW (1975) Pollen morphology in the order Centrospermae. Grana 15:51–77

14. Nowicke JW, Skvarla JJ (1977) Pollen morphology and the relationship of the Plumbaginaceae, Polygonaceae and Primulaceae to the order Centrospermae. Smithsonian Contr Bot 37:1–64

15. Nowicke JW, Skvarla JJ (1980) Pollen morphology: the potential influence in higher order systematics. Ann Missouri Bot Gard 66:633–700

16. Skvarla JJ, Nowicke JW (1976) Ultrastructure of pollen exine in centrospermous families. Pl Syst Evol 126:55–78

17. Skvarla JJ, Turner BL (1966) Systematic implications from electron microscopic studies of Compositae pollen: a review. Ann Missouri Bot Gard 53:220–244

18. Skvarla JJ, Turner BL, Patel VC, Tomb AS (1977) Pollen morphology in the Compositae and in morphologically related families. In: Heywood VH, Harborne JB, Turner BL (eds) The biology and chemistry of the compositae. Academic, London, pp 141–248

19. Blackmore S (1981) Palynology and intergeneric relationships in subtribe Hyoseridinae (Compositae: Lactuceae). Bot J Linn Soc 82:1–13

20. Blackmore S (1982) Palynology of subtribe Scorzonerinae (Compositae: Lactuceae) and its taxonomic significance. Grana 21:149–160

21. Blackmore S (1986) The identification and taxonomic significance of lophate pollen in the Compositae. Can J Bot 64:3101–3112

22. Ferguson IK (1981) The pollen morphology of Macrotyloma (Leguminosae: Papilionoideae; Phaseoleae). Kew Bull 36:455–461

23. Ferguson IK (1984) Pollen morphology and biosystematics of the subfamily Papilionoideae (Leguminosae). In: Grant WF (ed) Plant Biosystematics. Academic, London, pp 377–394

24. Ferguson IK, Schrire BD, Shepperson R (1994) Pollen morphology of the tribe Sophoreae and relationships between subfamilies Caesalpinioideae and Papilionoideae. In: Ferguson IK, Tucker S (eds) Advances in legume systematics 6: structural Botany, Royal Botanic Gardens, Kew, pp 53–96

25. Ferguson IK, Skvarla JJ (1981) The pollen morphology of the subfamily Papilionoideae (Leguminoseae). In: Polhill RM, Raven PH (eds) Advances in legume systematics. Royal Botanic Gardens, Kew, pp 859–896

26. Ferguson IK and Skvarla JJ (1982) Pollen morphology in relation to pollinators in Papilionoideae (Leguminosae). Bot J Linn Soc 83:183–193

27. Takahashi M (1982) Pollen morphology in North American species of *Trillium*. Am J Bot 69:1185–1195

28. Takahshi M 1983 Pollen morphology in Asiatic species of *Trillium*. Bot Mag Tokyo 96:377–384

29. Kato H, Terauchi R, Utech FH, Kawano S (1995) Molecular systematics of the Trilliaceae sensu lato as inferred from *rbcL* sequence data. Mol Phylogenet Evol 4:184–193

30. Dahl AO, Rowley JR (1965) Pollen of *Degeneria vitiensis*. J Arnold Arbor 50:1–35

31. Walker JW, and Skvarla JJ (1975) Primitively columellaless pollen: a new concept in the evolutionary morphology of angiosperms. Science 187:445–447

32. Walker JW (1976) Evolutionary significance of the exine in the pollen of primitive angiosperms. In: Ferguson IK, Muller J (eds) The evolutionary significance of the exine. Linn Soc Symp Ser No. 1, Academic, London, pp 251–308

33. Ferguson IK, Skvarla JJ (1983) The granular interstitium in the pollen of subfamily Papilionoideae (Leguminosae). Am J Bot 70:1401–1408
34. Takahashi M, Skvarla JJ (1990) Pollen development in *Oenothera biennis* (Onagraceae). Am J Bot 77:1142–1148
35. Zavada MS (1984) Angiosperm origins and evolution based on dispersed fossil pollen ultrastructure. Ann Missouri Bot Gard 71:444–463
36. Matinsson K (1993) The pollen of Swedish Callitriche (Callitrichaceae)—trends towards submergence. Grana 32:198–209
37. Pettitt JM (1976) Pollen wall and stigma surface in the marine angiosperms, *Thalassia* and *Thalassodendron*. Micron 7:21–32
38. Pettitt JM (1980) Reproduction in seagrasses: nature of the pollen and receptive surface of the stigma in the Hydrocharitaceae. Ann Bot 45:257–271
39. Pettitt JM, Jermy AC (1975) Pollen in hydrophilous angiosperms. Micron 5:377–405
40. Takahashi M (1994) Pollen development in a submerged plant, *Ottelia alismoides* (L.) Pers. (Hydrocharitaceae). J Plant Res 197:161–164
41. Takahashi M (1995) Development of structure-less pollen wall in *Ceratophyllum dermersum* L. (Ceratophyllaceae). J Plant Res 198:205–208
42. Hesse M, Kubitzki K (1983) The sporoderm ultrastructure in *Persea, Nectandra, Hernandia, Gomortega,* and some other lauralean genera. Pl Syst Evol 141:299–311
43. Raj B, Van der Werff H (1988) A contribution to the pollen morphology of neotropical Lauraceae. Ann Missouri Bot Gard 75:130–167
44. Rowley J R, Vasanthy G (1993) Exine development, structure, and resistance in pollen of *Cinnamomum* (Lauraceae). Grana (Suppl) 2:49–53
45. Skvarla JJ, Rowley JR (1970) The pollen wall of *Canna* and its similarity to the germinal apertures of other pollen. Am J Bot 57:519–529
46. Rowley JR, Skvarla JJ (1975) The glycocalyx and initiation of exine spinules on microspores of *Canna*. Am J Bot 62:479–485
47. Rowley JR, Skvarla JJ (1986) Development of the pollen grain wall in *Canna*. Nord J Bot 6:39–65
48. Kress WJ (1986) Exineless pollen and structure and pollination systems of tropical *Heliconia* (Heliconiaceae). In: Blackmore S, Ferguson IK (eds) Pollen and spores: form and function. Academic, London, pp 329–345
49. Kress WJ, Stone DE (1982) Nature of the sporoderm in monocotyledons, with special reference to the pollen grains of *Canna* and *Heliconia*. Grana 21:129–148
50. Kress WJ, Stone DE (1983) Morphology and phylogenetic significance of exine-less pollen of *Heliconia* (Heloconiaceae). Syst Bot 8:149–167
51. Kress WJ, Stone DE, Seller SC (1978) Ultrastructure of exine-less pollen: *Heliconia* (Heliconiaceae). Am J Bot 65:1064–1076
52. Stone DE, Seller SC, Kress WJ (1979) Ontogeny of exineless pollen in *Heliconia*, a banana relative. Ann Missouri Bot Gard 66:701–730
53. Stone DE, Sellers SC, and Kress WJ (1981) Ontogenetic and evolutionary implications of a neotenous exine in *Tapeinochilos* (Zingerales: Costaceae) pollen. Am J Bot 68:49–63
54. Zavada MS (1991) Determinating character polarities in pollen. In: Blackmore S, Barnes H (eds) Pollen and spores: patterns of diversification. Oxford University Press, London, pp 239–256
55. Blackmore S, Barnes SH (1987) Embryophyte spore walls: origin, development, and homologies. Cladistics 3:185–195

56. Kurmann MH (1989) Pollen wall formation in *Abies concolor* and a discussion on wall layer homologies. Can J Bot 67:2489–2504
57. Kurmann MH (1990) Development of the pollen wall in *Tsuga canadensis* (Pinaceae). Nord J Bot 10:63–78
58. Rowley JR (1995) Are the endexines of pteridophytes, gymnosperms and angiosperms structurally equivalent? Rev Palaeobot Palynol 85:13–34
59. Zavada MS, Gabarayeba N (1991) Comparative pollen wall development of *Welwitschia mirabilis* and selected primitive angiosperms. Bull Torrey Bot Club 118:292–302
60. Heslop-Harrison J (1963) An ultrastructural study of pollen wall ontogeny in *Silene pendula*. Grana 4:7–24
61. Takahashi M (1989) Pattern determination of the exine in *Caesalpinia japonica* (Leguminosae: Caesalpinioideae). Am J Bot 76:1615–1626
62. Takahashi M (1993) Exine initiation and substructure in pollen of *Caesalpinia japonica* (Leguminosae: Caesalpinioideae). Am J Bot 80:192–197
63. Takahashi M (1995) Three-dimensional aspects of exine initiatiation and development in *Lilium longiflorum* (Liliaceae). Am J Bot 82:847–854
64. Takahashi M, Skvarla JJ (1991a) Exine pattern formation by plasma membrane in *Bougainvillea spectabilea* Willd. (Nyctaginaceae). Am J Bot 78:1063–1069
65. Takahashi M, Skvarla JJ (1991b) Development of striate exine in *Ipomopsis rubura* (Polemoniaceae). Am J Bot 78:1724–1731
66. Hesse M (1995) Cytology and morphogenesis of pollen and spores. Prog Bot 56:33–55
67. Takahashi M, Kouchi J (1988) Ontogenetic development of spinous exine in *Hibiscus syriacus* (Malvaceae). Am J Bot 75:1549–1558
68. Takahshi M (1992) Development of spinous exine in *Nuphar japonicum* De Candolle (Nymphaeaceae). Rev Palaeobot Palynol 75:317–322
69. Rowley JR, Dahl AO, Sengupta S, Rowley JS (1981) A model of exine substructure based on dissection of pollen and spore exines. Palynology 5:107–152
70. Rowley JR (1990) The fundamental structure of the pollen exine. Plant Syst Evol Suppl 5:13–29
71. Brooks J, Shaw G (1968) Chemical structure of the exine of pollen walls and a new function for carotenoids in nature. Nature 219:532–533
72. Crang EE, May G (1974) Evidence for silicon as a prevalent elemental component in pollen wall structure. Can J Bot 52:2171–2174
73. Prahl AK, Rittscher M, Wiermann R (1986) New aspects of sporopollenin biosynthesis. In: Stumpf PF (ed) The biochemistry of plants—a comprehensive treaties. Springer, Berlin Heidelberg New York, pp 313–318
74. Prahl AK, Springstubbe H, Grumbach K, Wiermann R (1985) Studies on sporopollenin biosynthesis: the effect of inhibitiors of carotenoid biosynthesis on sporopollenin accumulation. Z Naturforschung 40:621–626
75. Schulze OK, Wiermann R (1987) Phenols as investigated compounds of sporopollenin from *Pinus* pollen. J Plant Physiol 131:5–15
76. Herminghaus S, Arendt S, Gubatz S, Rittscher M, Wiermann R (1988) Aspects of sporopollenin biosynthesis: phenols as integrated compounds of the biopolymer. In: Cresti M, Gori P, Pacini E (eds) Sexual reproduction in higher plants. Springer, Berlin Heidelberg New York, pp 169–174
77. Kawase M, Takahashi M (1996) Gas chromatography-mass spectrometric analysis of oxidative degradation products of sporopollnin in *Magnolia grandiflora* (Magnoliaceae) and *Hibiscus syriacus* (Malvaceae). J Plant Res 109:297–299

78. Crane PR (1985) Phylogenetic analysis of seed plants and the origin of angiosperms. Ann Missouri Bot Gard 72:716–793

79. Crane PR, Friis EM, Pedersen KR (1995) The origin and early diversification of angiosperms. Nature 374:27–33

80. Doyle JA, Donoghue MJ (1986) Seed plant phylogeny and the origin of angiosperms: an experimental cladistic approach. Bot Rev 52:321–431

81. Doyle JA and Donoghue MJ (1993) Phylogenies and angiosperm diversification. Paleobiology 19:141–167

82. Cornet B (1979) Angiosperm-like pollen with tectate-columellate wall structure from the Upper Triassic (and Jurassic) of the Newark Supergroup, USA. Palynology 3:281–282

83. Cornet B (1989) Late Triassic angiosperm-like pollen from the Richmond rift basin of Virginia, USA. Palaeontographica 213b:37–89

84. Doyle JA, Hotton CL (1991) Diversification of early angiosperm pollen. In: Blackmore S, Barnes SH (eds) Pollen and spores: patterns of diveristy, 169–95. Syst Assoc Spec vol 44, Clarendon, Oxford, pp 169–195

85. Crane PR (1987) Vegetational consequences of angiopserm diversification. In: Friis EM, Chaloner WG, Crane PR (eds) The orgins of angiosperms and their biological consequences. Cambridge University Press, Cambridge, pp 107–144

86. Traverse A (1988) Paleopalynology, Unwin Hyman, Boston

87. Crane PR, Lidgard S (1989) Angiosperm diversification and paleolatitudinal gradients in Cretaceous floristic diversity. Science 246:675–678

88. Doyle JA (1969) Cretaceous angiosperm pollen of the Atlantic Coastal Plain and its evolutionary significance. J Arnold Arbor 50:1–35

89. Muller J (1970) Palynological evidence on early differentiation of angiosperms. Biol Rev 45:417–450

90. Ward, JW, Doyle JA, Hotton CL (1989) Probable granular magnoliid angiosperm pollen from the Early Cretaceous. Pollen Spores 33:101–120

91. Walker JW, Walker AG (1984) Ultrastructure of Lower Cretaceous angiosperm pollen and the origin and early evolution of flowering plants. Ann Missouri Bot Gard 71:464–521

92. Taylor DW, Hickey LJ (1992) Phylogenetic evidence for the herbaceous origin of angiosperms. Pl Syst Ecol 180:137–156

93. Donoghue MJ, Doyle JA (1989) Phylogenetic analysis of the angiosperms and the relationships of the "Hamamelidae". In: Crane PR, Blackmore S (eds) Evolution, systematics and fossil history of the Hamamelidae, I. Introduction and "Lower" Hamamelidae. Clarendon, Oxford, pp 17–45

94. Crane PR, Herendeen PS (1996) Cretaceous floras containing angiosperm flowers and fruits from eastern North America. Rev Paleobot Palynol 90:319–337

Chromosomal Evolution of Angiosperms

HIROSHI OKADA

10.1 Introduction

At the beginning and developing stages of cytological studies, cytotaxonomists have discussed mainly the interrelationships among taxa based on the observations of somatic metaphase chromosomes or chromosome pairings in meiosis [1–4]. These ordinal cytological characters used are mainly the chromosome numbers and the chromosome shapes, which are very stable within taxa. On the other hand, the genome analysis originated by Kihara and his schools has clarified the relations among chromosome complements, chromosome behavior in meiosis and gamete fertility, based mainly on observations of *Triticum* [5]. These approaches are still effective methods to understand the biosystematic aspects. All the higher plants have to generate fertile offspring, which arise from mating between female and male gametes or from parthenogenesis, or they could not persist. In some cases, it is very difficult to understand the speciation mechanisms without cytogenetic analysis.

The analysis of morphological dynamics of chromatin through somatic cell division, namely karyomorphology, has proposed plant relationships in many groups [6,7]. At the present time, molecular biological methods are regarded as reliable tools to establish phylogenetic trees of all organisms [8]. It seems that the karyomorphological approach has completed its role in analyzing the phylogeny of angiosperms. Recent biochemical investigations concerning cell division or chromatin condensation mechanisms, however, indicate a new possible use of karyomorphology as a tool to understand the control of morphogenesis through development and to predict continuing evolutionary divergence. This chapter reviews recent knowledge of chromosome characters together with biochemical and molecular genetic information and then discusses their significance in the analysis of biosystematic aspects.

Botanical Gardens, Faculty of Science, Osaka City University, 2000 Kisaichi, Katano Osaka 576, Japan

10.2 Chromosome Numbers

Somatic chromosome numbers are the same as the double numbers of a linkage group. Genes on the same chromosomes move together through cell division, but genes on different chromosomes recombine at meiosis. If some new gene combination occurs on the same chromosome, and if such combination is well fitted to withstand selection pressures, it becomes directly and quickly conserved. The strategy of forming linkage groups is considered to be one of the main factors in the adaptive evolution of higher organisms. Further, somatic chromosome numbers are strictly equivalent to the numbers of centromere in almost all the angiosperms, with the exception of some groups, such as *Carex* and *Poa*, which have holocentric chromosomes. Chromosomes can regularly move to each pole at cell division only by means of the functions of the centromere and microtubules. The structure of the centromere is composed of MAP (microtuble attached protein) protein, INCENP (inner centromere protein), CENP-A, -B, -C, -D, and -E (centromere protein) proteins, and special repetitive DNA (CENP-B box DNA) [9–12]. Chromosome segmentation sometimes has occurred with some mutagens, but these phenomena do not directly increase chromosome numbers because chromosome segments without a centromere cannot move to the poles, and degeneration follows. Further, increase or decrease of the whole chromosome usually affects organisms as lethality or sterility because of abnormal and irregular genetic controls.

Chromosome numbers are extremely stable characters within organisms. If angiosperms have not demonstrated any disadvantages accompanying the change of chromosome numbers, they have had to establish certain systems to cancel such deleterious effects. In fact, there are a great many examples concerning variations of chromosome numbers in angiosperms [13], and therefore such variations have had to occur in the evolutionary process. Hence, we can use chromosome numbers for the analysis of the processes of evolution.

The most useful character for the consideration of evolutionary processes concerning chromosome numbers is the phenomena of polyploidy; 70%–80% of all angiosperms are of polyploid origin [14]. We easily understand that the evolutionary process occurred from lower ploidy levels to higher ones, such as in Compositae [15], in *Calamagrostis* [16], and so on. Parthenogenesis in angiosperms also accompanies polyploidization in many cases. There are no instances of reverse directions with rare exceptions in special cases, such as agamospermous or agamosporous polyploids [17,18]. The angiosperms sometimes exhibit reticulate evolution. In such cases, hybridization and following polyploidization (amphidiploidization) acts as a very important role for constant and stable supply of fertile gametes (mentioned later).

The aneuploid variations of basic chromosome number among allied groups also have an important evolutionary role in angiosperms [15,19], which sometimes occurred together with change of chromosome structure, i.e., translocation, and following chromosome elimination, centric fusion, or fission. These phenomena may correlate with the reorganization of linkage groups and following reproductive isolation.

The other interesting usage of cytogenetic characters concerning chromosome numbers is the analysis of numerical variation of B chromosomes. B chromosomes usually carry no genetic activities with major effects, or in greater numbers they depress fertility and reduce growth [20,21], and therefore they can be used as a parameter of neutral mutations. Hotta et al. [22] and Okada [23] have presented examples of the population diversity of *Schismatoglottis* (Araceae) using the proportion of B chromosomes in populations (Fig. 10.1). The populations are composed of various proportions of individuals with B chromosomes resulting from geographical isolation, but without any relation to environmental factors. If we can find B chromosomes in object species, it

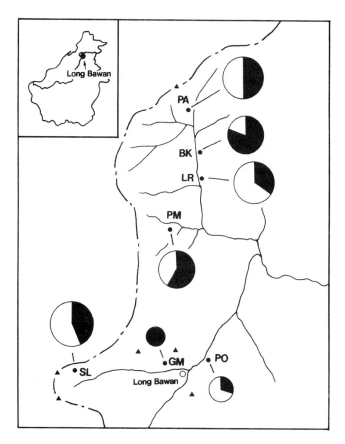

FIG. 10.1. Diversity among populations of *Schismatoglottis irrorata* detected by the proportions of individuals with B chromosomes (*solid parts* of pie graphs). (From [23], with permission)

is possible to use these for analysis of phenomena of population diversity. Thus we can analyze biosystematic aspects efficiently by means of variations of chromosome numbers.

10.3 Chromosome Shapes

The shape of a somatic chromosome is decided by the locations of some cytological markers, i.e., centromere, telomere, secondary constriction, or nucleolar organizing region (NOR), heterochromatic segments, banding patterns, etc. The meanings of these markers are analyzed biochemically and genetically (Table 10.1). For instance, according to classical studies of karyotype analysis, investigators usually paid attention to the shape of satellite chromosomes. The secondary constriction has the molecular biological meaning of being the location of repetitive rDNA, and its surrounding regions are composed of GC-rich DNA. Recent cytochemical advances offer new methods to identify the individuality of each chromosome, such as C-, G-, R-, and Q-banding patterns [24], DNA in situ hybridization by certain probes, i.e., MAP-specific DNA [25], T-DNA [26,27], 5S or 18S rDNA [28], and others [29–33]. These banding patterns, except for DNA in situ hybridization, are not yet characterized genetically in detail, but their positions are very stable within the individuals.

There are many examples using these characters that have genetic or molecular biological significance for analysis of aspects of population biology, i.e., variations of positions of centromere of autosomal chromosomes [34], positions of secondary constriction of satellite chromosomes [22], positions of the so-called heterochromatin with cold treatment [35], or C-banding patters [36].

TABLE 10.1. Cytological markers: their genetic meanings and molecular basis

Markers	Genetic meanings	Molecular basis
Centromere	Attached area with microtubles	Little is known; compound of MAP protein, INCENPs, CENP-A, -B, -C, -D, -E, and CENP-B box DNA
Telomere	Terminal of chromosome	Repetitive DNA, (ttaggg)n etc.
Secondary constriction	NOR, ribosome loci	Repetitive rDNA (GC-rich)
Heterochromatin	Unknown, usually inactive	AT-rich, GC-rich DNA
C-, G-, and R-band	Unknown	Unknown; usually corresponding to Q-band
Q-band (CMA)	(Ribosome loci)	GC-rich DNA
Q-band (DAPI)	Unknown	AT-rich DNA
In situ hybridization	Various	Various probes

NOR, nucleolar organizing region; CMA, chromomycin A_3; DAPI, 4′,6-diamidino-2-phenylindole dihydrochloride.

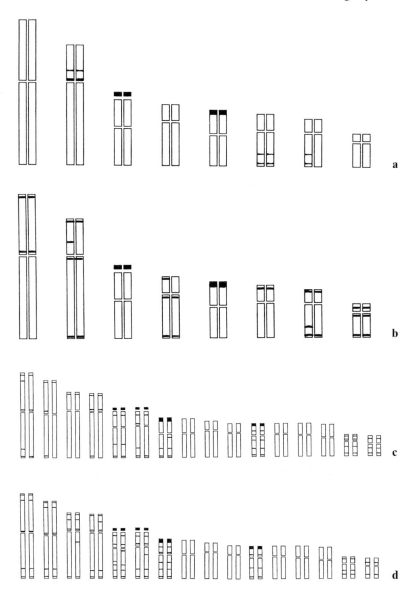

FIG. 10.2a–d. Ideograms of C-banding patterns of *Aconitum*. **a** *A. pterocaule* var. *glabrescens* (2n = 16). **b** *A. sanyoense* (2n = 16). **c** *A. gross-dentatum* (2n = 32). **d** *A. japonicum* var. *ibukiense* (2n = 32). Bands at chromosome pairs of 8th, 9th, 10th, 12th, 13th, and 14th of the latter two species (**c** and **d**) are not shown because these pairs are not distinguishable from the others

Figure 10.2 shows the differentiation of C-banding patterns at various taxo-nomic levels of the genus *Aconitum* (Okada, unpublished results). *Aconitum pterocaule* var. *glabrescens* belongs to the subgenus *Licoctonum* and the others to the subgenus *Aconitum*; *A. pterocaule* var. *glabrescens* and *A. sanyoense* are diploid species, while the others are tetraploid species. Diploid species have distinguishable chromosome complements displayed by banding patterns (Fig. 10.2a,b). The bigger bands at the satellite of the satellite chromosome and at the distal end of the short arm of the fifth pair are common to both species and are also common to tetraploid species. Some of the others are, however, peculiar. Further, *A. sanyoense* displays more complex banding patterns than *A. ptero-caule* var. *glabrescens* on the whole. The tetraploid-level species *A. gross-dentatum* and *A. japonicum* var. *ibukiense* showed more complex banding patterns than those of the diploid species *A. sanyoense* (Fig. 10.2c,d). The banding patterns of both tetraploid species are very similar to each other on the whole, although there exist some differences.

Furthermore, all the species exhibit many variations of banding patterns within and among populations. Based on banding patterns, identification of individuality is possible in *A. sanyoense* [37], in *A. gross-dentatum*, and in *A. japonicum* var. *ibukiense* (Okada, unpublished data). In this case, analysis of the banding patterns may explain the evolutionary processes of the genus more accurately than does the molecular evidence [38]. This example suggests that the structural changes of chromosomes, in other words, chromosomal evolution, has proceeded in conjunction with plant evolution. If there exist any effectual markers on chromosomes, we can analyze cytogenetically the aspect of population diversity.

10.4 Gamete Fertility and Chromosomes

Cytogenetic features can be used for analysis of the aspects of biosystematics or population biology. The change of chromosome numbers and chromosome struc-tures strikingly influences reproduction directly, that is, chromosome pairing at meiosis and following gamete formation. These phenomena affect population structures, the genetic component of populations, and reproductive isolation between different species or cytotypes. Among the factors of reproductive isola-tion, postmating isolation between allied species, populations, or individuals especially can be analyzed by cytogenetic observation [14]. For instances, *Ranun-culus sirelifolius* and *R. cantoniensis* exhibit polymorphism in chromosome shape [39,40]. Each cytotype is isolated reproductively [41,42]. The pollen grains of hybrids between cytotypes show low fertility, and as a result, seed sets become low.

Further, Okada [42] has presented a peculiar phenomenon concerning pollen grain fertility of hybrids. The origin of *R. cantoniensis* is considered to be from polyphyletic allopolyploid hybrids between *R. sirelifolius* and *R. chinensis*. The F_1 diploid hybrid between both species is almost sterile. Metaphase I of the pollen mother cells (PMCs) of the F_1 hybrid seems to be normal (Fig. 10.3a). Most chromosomes form bivalents, but seemingly normal tetrads (Fig. 10.3b) do not advance to microspore cell division. The nuclei of microspores of the F_1 hybrid soon degenerate (Fig. 10.3c), and normal pollen grains do not then develop, irregularly shaped sterile pollen grains being formed. When diploid F_1 hybrids were polyploidized artificially, artificial allopolyploid hybrids exhibited almost normal metaphase I PMCs (Fig. 10.3d), with one to two multivalents and normal tetrads (Fig. 10.3e), similar to the diploid F_1 hybrids. However, striking differences from diploid F_1 hybrids appeared in the following stages. Allopolyploid hybrids form almost fertile pollen grains through normal microspore cell division (Fig. 10.3f). The following offspring of artificial allopolyploid hybrids

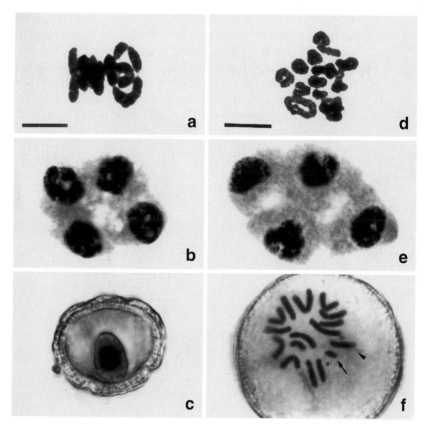

FIG. 10.3a–f. Microphotographs of metaphase I (**a, d**) and tetrad (**b, e**) of pollen mother cell (PMC) divisions and microspore divisions (**c, f**) of artificial hybrid between *Ranunculus silerifolius* and *R. chinensis*. Note the similarity of metaphase I and tetrad and the difference in microspore division between diploid (**a–c**) and tetraploid (**d–f**). *Scale bars =* indicate 10 μm. (Modified from [42])

exhibit pollen grains with 70%–90% fertility and aneuploid variations from $2n = 30$ to $2n = 35$ [43].

The process to establish the reproductive system to supply normal offspring is not clarified yet. It may require some modification of chromosome structure, such as translocation, for the stable formation of bivalents and subsequent completely fertile gametes. Thus, the origin and conservative persistence of *R. cantoniensis* could be traced by the cytogenetical approach. It might be difficult to trace the biological processes of such reticulate evolutionary processes with a molecular biological approach.

10.5 Karyomorphology

In addition to recent advances in study of the identification of individuality of chromosomes with banding patterns, analysis of chromatin condensation patterns through somatic cell division or cell differentiation, the so-called karyomorphology, has been carried out to understand the phylogenetic relationships among or within groups of angiosperms [6,7,44]. Figure 10.4 explains cytologically the possible evolutionary processes within the Chloranthaceae, that is, parallel with numerical changes including polyploidization and following aneuploidal reduction of basic chromosome numbers, karyomorphological variation occurred at least twice. On the other hand, recent advances of molecular biolog-

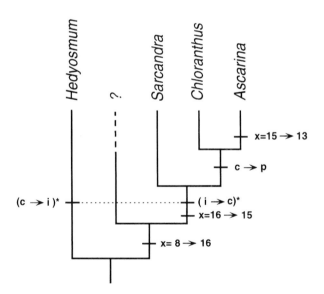

Fig. 10.4. Hypothesis of karyological relationships and postulated steps of karyological differentiation for the genera of Chloranthaceae. *c*, simple chromocenter type; *i*, intermediate type between diffuse type and complex chromocenter type; *p*, prochromosome type (From [7], with permission)

ical techniques offer a new approach to clarify the phylogeny of angiosperms [8]. At the moment, it is the common consensus that the molecular biological techniques will result in the most reliable phylogenetic trees. Although the results of karyomorphological analyses usually do not contradict the results of molecular biological analyses, karyomorphology might end its role in understanding the phylogeny of angiosperms. Based on the following facts, however, I consider that analysis of karyomorphological features still has an important meaning in understanding what has occurred in phylogenetic diversification.

Gene expressions occur only when the chromatin is diffused, such as the puffs of salivary gland chromosomes and the loops of lump brush chromosomes. The formation and the disappearance of puffs of salivary gland chromosomes proceed regularly during stages of development [45]. The interphase nuclei of *Drosophila* have no condensed bodies within the interphase nuclei at the beginning stage of development, while there are heterochromatic bodies at the later stages [46]. The nuclei of differentiated tissue cells show specific morphology in every tissue [47]. Miller et al. [48] demonstrated that a certain substance induces decondensation of "facultative heterochromatin" and that RNA synthesis follows. Recent biochemical evidence shows the correlation between chromatin condensation and the features of nuclear proteins. The phosphorylation and methylation of H1 histone [49–51] and of nonhistone protein [52] give rise to changes in the conformation of DNA or chromatin of differentiating cells. Gene groups aligned in tandem may be totally regulating their activation by the timing of chromatin condensation and decondensation. These facts suggest that the shift of timing of chromatin condensation and decondensation through development may dramatically affect its morphogenesis. It may be said that if karyomorphological features change to other ones with certain factors, an organism evolves its morphology as well as do gene mutations, as shown in *Arabidopsis* [53].

Even if karyomorphological analysis cannot contribute directly to delineating the phylogenetic tree of angiosperms, the attempt to analyze the relation between karyomorphological characteristics and biochemical or molecular genetic evidences will demonstrate some features of the evolutionary processes of angiosperms.

Acknowledgments. The author thanks an anonymous reviewer for constructive criticism of an earlier version of this manuscript.

References

1. Langlet O (1932) Uber Chromosomenverhaltnisse und Systematik der Ranunculaceae. Sven Bot Tidskr 26:381–400
2. Gregory WC (1941) Phylogenetic and cytological studies in the Ranunculaceae. Trans Am Philos Soc 31:443–521
3. Darlington CD (1956) Chromosome botany and the origins of cultivated plants. Hafner, New York

4. Stebbins GL (1971) Chromosomal evolution in higher plants. Arnold, London
5. Kihara H (1924) Cytologische und genetische Studien bei wichtigen Getreidearten mit besonderer Rucksicht auf das Verhalten der Chromosomen und die Sterilitat in den Bastarden. Mem Coll Sci Kyoto Imp Univ B 1:1–200
6. Tanaka R (1971) Types of resting nuclei in Orchidaceae. Bot Mag Tokyo 84:118–122
7. Okada H (1995) Karyological studies of four genera of the Chloranthaceae. Plant Syst Evol 195:177–185
8. Zurawski G, Clegg MT (1993) rbcL sequence data and phylogenetic reconstruction in seed plants: foreword. Ann Missouri Bot Gard 80:523–525
9. Earnshaw WC, Sullivan K, Machlin PS, Cooke CA, Kaiser DA, Pollard TD, Rothfield NF, Cleveland DW (1987) Molecular cloning of cDNA for CENP-B, the major human centromere autoantigen. J Cell Biol 104:817–829
10. Masumoto H, Matukata H, Moro Y, Nozaki N, Okazaki T (1989) A human centromere antigen (CENP-B) interacts with a short specific sequence in alphoid DNA, a human centromeric satellite. J Cell Biol 109:1963–1973
11. Willard HF (1990) Centromeres of mammalian chromosomes. Trends Genet 6:410
12. Earnshaw WC, Bernat RL (1991) Chromosomal passengers: toward an integrated view of mitosis. Chromosoma (Berl) 100:139–146
13. Grant V (1981) Plant speciation. Columbia University Press, New York
14. King M (1993) Species evolution. The role of chromosome change. Cambridge University Press, Cambridge
15. Shimotomai N (1933) Zur Karyogenetik der Gattung Chrysanthemum. J Sci Hiroshima Univ Ser B Div 2 (Bot) 2:1–100
16. Tateoka T (1986) Cytogeographical analysis of the grass genus Calamagrostis (Poaceae) in Japan. In: Iwatsuki K, Raven PH, Bock WJ (eds) Modern aspects of species. University of Tokyo Press, Tokyo, pp 107–123
17. Morita T, Sterk AA, Nijs JCM (1990) The significance of agamospermous triploid pollen donors in the sexual relationships between diploids and triploids in Taraxacum (Compositae). Plant Species Biol 5:167–176
18. Lin SJ, Kato M, Iwatsuki K (1992) Diploid and triploid offspring of triploid agamosporous fern Dryopteris pacifica. Bot Mag Tokyo 105:443–452
19. Babcock EB (1947) The genus Crepis, I and II. Univ Calif Publ Bot 21, 22:1–1030
20. Jones RN (1975) B-chromosome systems in flowering plants and animal species. Int Rev Cytol 40:1–100
21. Jones RN, Rees H (1982) B chromosomes. Academic Press, London
22. Hotta M, Okada H, Ito M (1985) Species diversity at wet tropical environment. I. Polymorphic variation and population structure of Schismatoglottis lancifolia (Araceae) in West Sumatra. Contrib Biol Lab Kyoto Univ 27:9–71
23. Okada H (1992) Population diversity of Schismatoglottis irrorata (Araceae) at Malesian wet tropics with reference to the distribution of B chromosome. Cytologia (Tokyo) 57:401–407
24. Bickmore WA, Sumner AT (1989) Mammalian chromosome banding: an expression of genome organization. Trends Genet 5:144–148
25. Rayburn AL, Gill BS (1985) Use of biotin-labeled probes to MAP specific DNA sequences on wheat chromosomes. J Hered 76:78–81
26. Ambros PF, Matzke MA, Matzke AJM (1989) Detection of a 17-kb unique sequence (T-DNA) in plant chromosomes by in situ hybridization. Chromosoma (Berl) 94:18
27. Blackburn EH (1991) Structure and function of telomeres. Nature 350:569–573

28. Hizume M (1994) Allodiploid nature of *Allium wakeqi* Araki revealed by genomic in situ hybridization and localization of 5S and 18S rDNAs. Jpn J Genet 69:407–415

29. Rayburn AL, Gill BS (1985) Molecular evidence for the origin and evolution of chromosome 4A in polyploid wheats. Can J Genet Cytol 27:246–250

30. Rayburn AL, Gill BS (1986) Molecular identification of the D-genome chromosomes of wheat. J Hered 77:253–255

31. Lapitan NLV, Sears RG, Rayburn AL, Gill BS (1986) Wheat-rye translocations. Detection of chromosome breakpoints by in situ hybridization with a biotin-labeled DNA probe. J Hered 77:415–419

32. Bergey DR, Stelly DM, Price HJ, McKnight TD (1989) In situ hybridization of biotinylated DNA probes to cotton meiotic chromosomes. Stain Technol 64:25–37

33. Wu HK, Chung MC, Wu T, Ning CN, Wu R (1991) Localization of specific repetitive DNA sequences in individual rice chromosomes. Chromosoma (Berl) 100:330–338

34. Haga T, Noda S (1976) Cytogenetics of the *Scilla scilloides* complex. I. Karyotype, genome, and population. Genetica (Dordr) 46:161–176

35. Fukuda I, Channell RB (1975) Distribution and evolutionary significance of chromosome variation in *Trillium ovatum*. Evolution 29:257–266

36. Noguchi J (1986) Geographical and ecological differentiation in the *Hemerocallis dumortierii* complex with special reference to its karyology. J Sci Hiroshima Univ Ser B Div 2 (Bot) 20:29–193

37. Okada H (1991) Correspondence of Giemsa C-band with DAPI/CMA fluorochrome staining pattern in *Aconitum sanyoense* (Ranunculaceae). Cytologia (Tokyo) 56:135–141

38. Kita Y, Ueda K, Kadota Y (1995) Molecular phylogeny and evolution of the Asian *Aconitum* subgenus *Aconitum* (Ranunculaceae). J Plant Res 108:429–442

39. Fujishima H, Kurita M (1974) Chromosome studies in Ranunculaceae. XXVI. Variation in karyotype of *Ranunculus ternatus* var. *qlaber*. Mem Ehime Univ Ser B (Biol) 7:62–68

40. Okada H, Tamura M (1977) Chromosome variations in *Ranunculus quelpaertensis* and its allied species. J Jpn Bot 52:360–369

41. Okada H (1981) On sexual isolation caused by karyotype variations in *Ranunculus silerifolius* Lev. J Jpn Bot 56:41–49

42. Okada H (1984) Polyphyletic allopolyploid origin of *Ranunculus cantoniensis* (4×) from *R. silerifolius* (2×) × *R. chinensis* (2×). Plant Syst Evol 148:89–102

43. Okada H (1989) Cytogenetical changes of offsprings from the induced tetraploid hybrid between *Ranunculus silerifolius* ($2n = 16$) and *R. chinensis* ($2n = 16$) (Ranunculaceae). Plant Syst Evol 167:129–136

44. Okada H (1975) Karyomorphological studies on woody Polycarpicae. J Sci Hiroshima Univ Ser B Div 2 (Bot) 15:115–200

45. Berendes HD (1965) Salivary gland function and chromosomal puffing patterns in *Drosophila hydei*. Chromosoma (Berl) 17:35–77

46. Cooper KW (1959) Cytogenetic analysis of major heterochromatic elements (especially XH and Y) in *Drosophila melanogaster* and the theory of "heterochromatin." Chromosoma (Berl) 10:535–588

47. Hirahara S (1980) Karyomorphological studies on somatic tissues in *Spiranthes sinensis*. J Sci Hiroshima Univ Ser B Div 2 (Bot) 17:9–49

48. Miller G, Berlowitz L, Regelson W (1971) Chromatin and histones in mealy bug explants: activation and decondensation of facultative heterochromatin by a synthetic polyanion. Chromosoma (Berl) 32:251–261

49. Holmgren P, Johanson T, Lambertsson A, Rasmuson B (1985) Content of histone H-1 and histone phosphorylation in relation to the higher order structures of chromatin in *Drosophila*. Chromosoma (Berl) 93:123–131

50. Lin R, Cook RG, Allis CD (1991) Proteolytic removal of core histone amino termini and dephosphorylation of histone H1 correlated with the formation of condensed chromatin and transcriptional silencing during *Tetrahmena* macronuclear development. Genes Dev 5:1601–1610

51. Roth SY, Allis CD (1992) Chromatin condensation: does histone H1 dephosphorylation play a role? Trends Biochem Sci 17:93–98

52. Haaf T, Dominguez-Steglich M, Schmid M (1990) Immunocytogenetics. VI. A nonhistone antigen is cell type specifically associated with constitutive heterochromatin and reveals condensation centers in metaphase chromosomes. Cytogenet Cell Genet 54:121–126

53. Tsukaya H (1995) The genetic control of morphogenesis in *Arabidopsis* and its relevance to the development of biodiversity. In: Arai R, Kato M, Doi Y (eds) Biodiversity and evolution. National Science Museum Foundation, Tokyo, pp 253–265

Speciation and Mechanisms
of Diversification

Mating Systems and Evolution in Flowering Plants

KENT E. HOLSINGER and JENNIFER E. STEINBACHS

From time immemorial, flowers have appealed to our highest thoughts, ideals, and emotions. We associate them with happiness, soft music, and sunshine; we send them to our mothers on Mothering Sunday and shower our lady singers with them after a successful aria. . . . The discovery that flowers possess sex and are actually designed as organs for a plant's sexual reproduction was, therefore, greeted with some surprise!

(Meeuse and Morris 1984, *The Sex Life of Flowers*. Facts on File, p 7 [1])

11.1 Introduction

Flowers are not only "designed as organs for a plant's sexual reproduction." They are, as Sprengel [2] pointed out more than two centuries ago, also designed to attract pollinators. Because plants are sessile, outcrossing among them requires that an external agent be recruited to move pollen from one to another. The enormous variety of external agents available—wind, water, bees, flies, bats, butterflies, and moths, to name only a few—is complemented by an equally enormous variety of floral forms. Pollination biologists have explained this variety by following the grand tradition established by Darwin [3–5] and the encyclopedists [6,7], i.e, they have explained this diversity as a result of natural selection to ensure outcrossing.

For almost as long as pollination biologists have studied the intricate relationships between flowering plants and their pollinators, plant evolutionists have studied the multifarious patterns of mating within plant populations and the consequences of these different patterns. Knight [8], for example, showed that hybrid varieties of garden peas are usually more vigorous than inbred strains, leading him to suggest that "Nature . . . abhors perpetual self-fertilization" (p 293). Indeed, the great diversity in floral form and function in flowering plants leads to an even greater diversity in mating patterns within populations of flow-

Department of Ecology and Evolutionary Biology, University of Connecticut, 75 N. Eagleville Road, U-43, Storrs, CT 06269-3043, USA

ering plants [9], providing plant evolutionists with the opportunity to study the evolutionary impact of mating system differences far greater than those found in animals with separate sexes.

Understanding the evolutionary forces responsible for producing this diversity of mating systems is clearly an important task for plant evolutionists. Of equal or greater importance, however, is understanding the impact that various reproductive modes have on patterns of variation within populations and on processes leading to divergence of populations. The mating system is, after all, arguably the most important influence on the genetic structure of plant populations [10–12]. We no longer describe those differences as a result of balancing short-term adaptation and long-term flexibility, as did Darlington [13], Mather [14], Grant [15], and Stebbins [16]. It is, nonetheless, reasonable to expect that plants differing in their mating system will also differ in the frequency, extent, and mode by which populations diverge from one another (cf. [17,18]). In short, evolutionary responses to changed reproductive conditions might also affect the evolutionary dynamics of traits unrelated to the mating system.

11.2 Evolution of Mating Systems in Flowering Plants

The gamete-producing parts of flowers may be arranged in many different ways. Organs producing male and female gametes may be part of the same flower, part of different flowers on the same plant, part of different flowers on different plants, or any combination of these; seed may be produced either sexually or asexually. As a result flowering plants also exhibit an enormous range in the the frequency of different reproductive events. Either sexual or asexual reproduction may be the sole means of reproduction, as may any combination of the two. Sexual reproduction involving fusion of gametes from different individuals (intercrossing) is the norm in many species, but in many others sexual reproduction occurs primarily through self-fertilization. Even in those species in which intercrossing is the norm, a significant proportion of mating events are likely to occur between relatives of some degree because of limited pollen and seed dispersal. These differences form three important axes along which reproductive systems can be classified (Table 11.1) (cf. [19]) and describe three important ways in which the genetic systems of flowering plants may differ from one another. In this chapter, we focus on the evolution of different modes of sexual reproduction. Although similar principles apply to interpreting the evolution and consequences of asexual reproduction, the differences are substantial enough that they warrant separate discussion.

11.2.1 Genetic Systems

Early in the evolutionary synthesis, Darlington [13] coined the term "genetic system" to refer to those aspects of an organism's biology that affect the extent of sexual recombination which occurs during reproduction and the amount of ge-

TABLE 11.1. Modes of reproduction in flowering plants

Mode	Description
Asexual	
Vegetative reproduction	Division of vegetative body into independent units
Agamospermy	Production of asexual seed
Adventitious embryony	Seed develops without development of gametophyte
Gametophytic apomixis	Seed develops from asexual gametophyte
Sexual	
Self-fertilization[a]	Uniting gametes produced by same sporophyte
Geitonogamy	Uniting gametes produced by different flowers on same sporophyte
Autogamy	Uniting gametes produced by same flower
Cleistogamy	Uniting gametes produced in structurally specialized flowers that never open
Intercrossing	Uniting gametes produced by different sporophytes
Outcrossing	Uniting gametes produced by unrelated sporophytes
Biparental inbreeding	Uniting gametes produced by different, but related, sporophytes

[a] See Lloyd and Schoen [136] for an exhaustive description of modes of self-pollination. Deposition of self pollen does not guarantee self-fertilization, but self-fertilization cannot occur without self-pollination. Thus, we expect a relationship between rates of self-pollination and rates of self-fertilization [60,70].

netic variation that will be found among that organism's progeny. He realized that not only does the genetic system affect the pattern of evolutionary responses within a population but that the genetic system itself may be subject to natural selection. He and Mather [14] explored the ways in which observed genetic systems might represent a compromise between long-term flexibility and immediate adaptation. Botanists were quick to pick up these ideas and to use them in trying to understand the reproductive systems of flowering plants. Grant [15] argued, for example, that members of the genus *Clarkia* (Onagraceae) are so distantly related to those of the genus *Gilia* (Polemoniaceae) that similarities in their genetic systems cannot be attributed to recent common ancestry. Rather, it seems reasonable to presume that "each type of genetic system . . . should display properties particularly suited for the conditions of life of the organism possessing it" (Grant [15], p 337).

Similarly, Stebbins [16] argued that long-lived perennials, particularly those of stable habitats, should exhibit high frequencies of outcrossed, sexual reproduction because of the large amounts of genetic variability that can be stored in such populations to respond to future environmental change. Plants occurring in more temporary habitats or with a shorter life cycle, he argued, should more often reproduce through selfing or agamospermy because they can ill afford the immediate cost of producing variable progeny, many of which may not be well suited to the severe conditions in which the plants are found (cf. [15]). Lloyd [20] was probably the first to notice that the arguments put forward by Darlington, Mather, Stebbins, and Grant depended on populational, not individual, differences in fitness. He pointed out that natural selection acting on individuals cannot balance immediate fitness and long-term flexibility. Differences in rates of speciation or extinction associated with different mating systems, if they exist, will have a large impact on taxonomic distribution and the frequency with which different mating systems are encountered. If selfers are more prone to extinction than outcrossers, for example, we would expect to find that most selfers are relatively recent derivatives of outcrossing ancestors and that diversification in a lineage with both selfers and outcrossers is primarily a result of divergence among the outcrossers. Nonetheless, to understand how natural selection has produced mating system differences among species we must first understand their immediate individual fitness consequences (cf. [21,22]). Understanding the evolutionary implications of these immediate fitness consequences has been the focus of experimental and theoretical work on plant mating system evolution for the past 30 years.

11.2.2 Evolution of Self-Fertilization

It has been apparent for at least 40 years that self-fertilizing species have evolved independently many different times in many different lineages of flowering plants [15,16,18,19,23,24]. In fact, Stebbins [25] suggested that the evolutionary pathway from predominant outcrossing to predominant selfing has been the most frequently followed pathway in flowering plant evolution. Moreover, changes in the rate of self-fertilization are often regarded as an important component of the selective forces affecting the evolution of other reproductive modes, e.g., heterostyly, dioecy, monoecy, and self-incompatibility. As a result, much of the experimental and theoretical work on plant mating systems in the past 30 years has been directed specifically toward identifying the evolutionary forces that determine why some species are predominantly outcrossing, others are predominantly selfing, and some reproduce by a mixture of the two modes. The surge of interest has been partly driven by two controversial suggestions:

1. Mixed mating systems are evolutionarily transient and only predominant selfing and predominant outcrossing are evolutionarily stable states [26–30].
2. Inbreeding depression is a poor predictor of when selfing is likely to evolve [31–36].

In spite of the controversy surrounding these two topics, it is clear that any immediate fitness advantage associated with selfing comes either from increased success as an ovule parent when seed production is pollen limited—i.e., reproductive assurance—or from increased success as a pollen parent when pollen devoted to selfing is more likely to fertilize ovules than pollen devoted to outcrossing—i.e., an automatic selection advantage [37].

Reproductive Assurance

Even if pollen availability rarely limits seed production (see [38–42] for reviews of the evidence), reproductive assurance may still explain why many plants are predominantly selfing [43,44]. Darwin [4] argued, for example, that selfing evolved in *Ophrys apifera* and several other orchid species because they now occur in areas where pollinators are not sufficiently abundant to ensure seed set. Similarly, Hagerup [45,46] argued that mechanisms to ensure self-pollination by rain evolved in *Ranunculus bulbosus*, *R. flammula*, *Caltha palustris*, and *Narthecium ossifragum* on the Faroe Islands because the insects that would normally be expected to pollinate them are not found there. Even if there is inbreeding depression, some degree of selfing will be evolutionarily favored whenever some ovules remain unfertilized [37,47]. "It is better, after all, to have reproduced by selfing than never to have reproduced at all" ([48], p iii).

Changes in floral morphology that allow more self pollen to be deposited autonomously, for example, provide substantial reproductive assurance and will be favored by natural selection, provided they do not significantly reduce opportunities for outcrossed reproduction [37]. Delayed self-fertilization, as when anthers brush by a receptive stigma during corolla dehiscence [49], is unconditionally favored provided that ovules remaining unfertilized after the opportunity for outcrossing has passed can be fertilized without reducing success as an outcross pollen parent.

Baker [50,51] pointed out one case in which reproductive assurance plays a particularly important role: the establishment of new populations after long-distance dispersal. Species that are self-compatible are more likely to be successful long-distance dispersers than those that are not. As a result, self-fertilization may be particularly common at the expanding margins of a species range or in species that have dispersed far from their center of origin. In both the genera *Armeria* and *Limonium* (Plumbaginaceae), for example, self-incompatible, heterostylous species are found primarily near the Eurasian center of distribution of the family, while North America and Australia contain chiefly homostylous, self-compatible species [52,53]. Successful establishment after long-distance dispersal might also explain the breakdown of tristyly in Jamaican populations of *Eichhornia paniculata*. Brazilian populations of this species commonly contain all three flower morphs and outcross at rates statistically indistinguishable from random mating, while Jamaican populations may consist of a single morph and self-fertilize at rates approaching 70% or more [53–55]. Stebbins [56] suggested that this "correlation . . . occurs so widely and has such great significance for studies of the origin and migration of genera of flowering plants . . . that it deserves recognition as Baker's law" (p 344).

Although reproductive assurance may provide a substantial immediate fitness advantage in many circumstances, it will not do so in all. Lloyd [57] pointed out that when self-pollination requires insect visits the extent of reproductive assurance that selfing can provide may be quite limited. The very thing that would make reproductive assurance beneficial, a lack of pollinator visits, is also the thing needed for any reproductive assurance to be provided. Under these circumstances, pollinator-mediated self-fertilization can increase seed set only when pollinators are present but the extent of cross-pollination is insufficient to ensure full seed set. Even then the extent of reproductive assurance provided may be quite small.

Automatic Selection Advantage

Reproductive assurance provides an advantage to self-fertilization through increased success as an ovule parent. Fisher [58] showed that self-fertilization might also have an advantage because of increased success as a pollen parent. Selfers produce more successful gametes, on average, than outcrossers. A selfer serves as both pollen and ovule parent to its own selfed progeny, and it may serve as an outcross pollen parent to the outcrossed progeny of other individuals. An outcrosser, on the other hand, will serve only as an ovule parent to its own outcrossed progeny and as an outcross pollen parent to the outcrossed progeny of other individuals. Thus, a selfer has three successful gametes for every two successful gametes of an outcrosser. This 3:2 excess of successful gametes is the source of what Jain [43] called the "automatic selection advantage" of selfers [22,59,60]. Notice, however, that this advantage exists only to the extent that selfers are able to produce their own selfed progeny without reducing their ability to serve as outcross pollen parents to the outcrossed progeny of others [61–63].

Holsinger et al. [63] introduced the discounting rate as a measure of the proportional reduction in success as an outcross pollen parent associated with a particular selfing rate. Although less than 1% of the pollen a plant produces typically reaches the stigmas of other plants [9], it is not the amount of pollen used in selfing but the functional relationship between selfing rates and discounting rates that determines whether an automatic selection advantage will exist. When seed set does not depend on the amount of pollen received, a genotype with a higher selfing rate will have an advantage over one with a lower selfing rate only if the additional success the selfer has as a self pollen parent to its own progeny more than compensates for lost opportunities to serve as an outcross pollen parent to the progeny of other individuals [37]. In the absence of pollen discounting, selfers have the complete 3:2 advantage attributed to them in Fisher's [58] model. With complete discounting, selfers have no advantage at all.

There have only been a few direct attempts to measure the functional relationship between selfing rates and discounting rates. Ritland [64] found complete discounting in mixed populations of selfing and outcrossing members of the *Mimulus guttatus* complex. In contrast, Rausher et al. [65] found no evidence of pollen discounting in *Ipomoea purpurea*, while Schoen and Clegg [66] presented evidence that the white-flowered form of *I. purpurea*, which selfs at a higher rate than other color morphs, is actually more successful as an outcross pollen parent than other color morphs in some populations. Similarly, elongation of one or more short-level stamens in the mid-style morph of tristylous *Eichhornia paniculata* apparently increases both the amount of self-pollen deposited on stigmas and the amount of pollen successfully transported to mid-level stigmas of other plants [67–69].

An alternative to the direct experimental approach is to build explicit models of the pollination and fertilization processes and to use these models to infer the functional relationship between selfing rates and discounting rates. The recently proposed mass-action approach [70,71] can be used for just this purpose (see also [72]). Models developed so far have shown these results [70]:

1. Outcrossing may be evolutionarily stable even in the absence of inbreeding depression if pollen devoted to outcrossing is more likely to participate in fertilization than pollen devoted to selfing.
2. Complete selfing is evolutionarily stable only when reproduction is pollen limited.
3. Mixed mating systems are evolutionarily stable whenever selfing can evolve.

More importantly, such an approach helps to focus theoretical attention on the same variables that pollination biologists have studied for more than a century, e.g., the intricate variety of positions in which floral parts are held, the timing and sequence in which floral parts mature, and the functional relationships among floral morphology, the length and number of pollinator visits, the amount of self pollen deposited, the amount of outcross pollen received, and the amount of pollen exported to other plants. Because each of these variables affects patterns of pollen dispersal, they in turn affect mating opportunities and the frequency with which selfed progeny are produced. The time is now ripe for ecological and evolutionary studies that explore the relationship between variation in pollination systems and variation in mating systems [9,37].

11.2.3 Evolution of Self-Incompatibility

The preceding discussion emphasized the roles of reproductive assurance and pollen transfer dynamics in determining the evolutionary forces acting on plant mating systems. In understanding the evolution of self-incompatibility, however, differences in patterns of pollen transfer have not yet been explicitly considered, although differences in pollen performance matter greatly. A self-incompatibility allele, after all, reduces the reproductive success of a plant carrying it to the extent that self pollen had previously played a role in reproduction. Thus, self-incompatibility can arise only when the additional reproductive success as an ovule parent that results from excluding self pollen exceeds the reproductive success lost as a self pollen parent. In short, self-incompatibility can evolve only when there is some inbreeding depression.

Evolutionary Dynamics of Self-Incompatibility

Charlesworth and Charlesworth [73] presented the first explicit model for the evolution of self-incompatibility. In their model a single locus controls all aspects of the self-incompatibility reaction, although they did consider the possibility that allelic interactions within the pollen or stigma could influence the outcome of the evolutionary process. As might be expected from the results previously presented for the evolution of selfing, alleles promoting self-incompatibility are incorporated into populations only when inbreeding depression is greater than one-half, regardless of whether the reaction is gametophytically or sporophytically determined. Interestingly, self-compatibility alleles are not eliminated from populations if they are initially present, even when new self-incompatibility alleles are strongly favored.

Their results [73] also help to understand the evolutionary dynamics associated with the loss of self-incompatibility. A population with a gametophytic self-incompatibility system is more susceptible to invasion by variants restoring self-compatibility than those with sporophytic systems, because variants restoring self-compatibility have a greater reproductive advantage in gametophytic systems than in sporophytic ones. In gametophytic systems, pollen carrying a newly arisen self-compatibility allele is able to germinate on stigmas of any plant where it is deposited. In sporophytic systems the compatibility type of pollen is determined by the diploid genotype of the plant producing it. Thus, pollen carrying a newly arisen self-compatibility allele is able to germinate only on stigmas of plants not sharing a compatibility allele with the plant that produced the pollen. Unless the number of compatibility groups in a population is very large, therefore, variants conferring self-compatibility have fewer mating opportunities in a population with a sporophytic self-incompatibility system than in one with a gametophytic system. This effect is particularly marked if the self-compatibility allele in a sporophytic system is recessive; under these conditions, its effect will not be apparent until individuals homozygous for the self-compatibility allele are formed [74].

Localized pollen and seed dispersal not only contributes to local variation in allele frequencies but also produces genetic correlations between uniting gametes (see the discussion on impacts of isolation by distance that follows). The correlation between uniting gametes may be thought of as one form of biparental inbreeding. Uyenoyama [75] examined the effect of differing frequencies of biparental inbreeding on the evolution of gametophytic self-incompatibility. Her results show that biparental inbreeding alone is unlikely to cause the evolution of self-incompatibility; only when there is some degree of self-fertilization are variants causing self-incompatibility able to become established. Furthermore, the threshold level of inbreeding depression necessary for invasion of a variant inducing self-incompatibility is greater for dominant than for recessive variants.

Recent molecular studies of gametophytic self-incompatibility suggest that it is a result of coordinated expression of several distinct loci [76]. Because it is more faithful to these molecular details, Uyenoyama's [77] three-locus model is a better tool for understanding the evolution of this coordinated response than the simpler, one-locus models that preceded it. One of the three loci in her model is an overdominant viability locus that determines offspring viability. Inbreeding depression in the model is a result of the higher frequency of homozygous progeny associated with self-fertilization. A second locus is an antigen locus that determines specificities in a self-incompatibility reaction, and the third locus is an unlinked modifier locus which influences the degree of expression of the incompatibility, perhaps by regulating the level at which the antigens are produced. Uyenoyama showed that when a mutation increasing the strength of the self-incompatibility reaction arises at the modifier locus, associations among the three loci develop immediately. The coordinated expression of the three gene loci evolves naturally because of the way in which these interlocus associations develop. Furthermore, these associations promote the establishment of self-incompatibility. Even though the level of inbreeding depression never reaches one-half, which is the threshold for evolution of self-incompatibility found in previous models, self-incompatibility will still become established in the population. New developments in the molecular biology of sporophytic self-incompatibility show that its molecular basis is quite different from that of gametophytic self-incompatibility [78], but models which reflect the structure of the locus, similar to that of Uyenoyama [77], have yet to be considered.

Phylogeny of Self-Incompatibility Systems

The evolutionary relationship between self-incompatibility systems in different families of flowering plants has long been a topic of interest. Similarities in the developmental trajectories of species with different systems have suggested to some that sporophytic systems arose from gametophytic systems through a simple change in the time at which the S locus is expressed during pollen development [76,79–82]. Physiological differences in the mode of gene action and the enormous sequence differences between genes with different modes of action suggest, however, that many instances of self-incompatibility have arisen independently.

Examining the phylogenetic distribution of different types of self-incompatibility further strengthens the hypothesis that self-incompatibility systems have been evolved independently many times within flowering plants (Fig. 11.1). Although there are several clades in which all reported instances of self-incompatibility are gametophytic (all monocot clades with self-incompatibility, ranunculids, and rosid IV), every clade from which sporophytic self-incompatibility has been reported also has members with gametophytic self-incompatibility. Even closely related pairs of families may exhibit different incompatibility systems, e.g., Solanaceae (gametophytic) versus Convolvulaceae (sporophytic) and Goodeniaceae (gametophytic) versus Asteraceae (sporophytic). Moreover, the genetic and physiological mechanisms underlying the incompatibility reactions in those groups in which this has been examined in detail (Brassicaceae, Papaveraceae, Poaceae, and Solanaceae) are markedly different from one another, and the genes involved may not even be homologous to one another [76].

It remains an important task for future research to determine which, if any, self-incompatibility systems of different plant families are homologous to one another, but some preliminary clues are available. Comparative analysis of nucleotide sequences from S alleles of Solanaceae (asterid I), *Antirrhinum* (Scrophu-

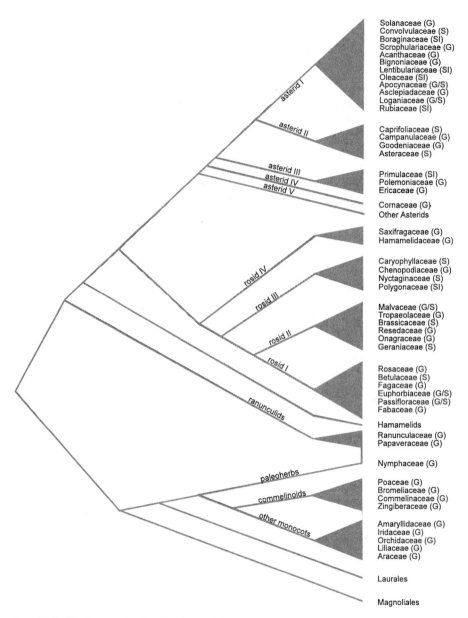

Fig. 11.1. Phylogenetic distribution of homomorphic self-incompatibility systems in flowering plants. Symbols in parentheses indicate the type of incompatibility system present. *S*, families with sporophytic self-incompatibility; *G*, families with gametophytic self-incompatibility; *G/S*, groups in which incompatibility type depends on the type of pollen produced [79]; *SI*, families from which self-incompatibility is reported but the type of incompatibility is unknown. Groups lacking a symbol have no reported instances of self-incompatibility. The phylogeny is simplified from Chase et al. [133]; the information on self-incompatibility types is derived from Fryxell [19], Pandey [79], de Nettancourt [80], Charlesworth [134], and Mau et al. [135]

lariaceae; asterid I; [83]) and *Malus* (Rosaceae; rosid I; [84]) shows that *S* alleles of each family form distinct lineages. Nonetheless, *S* alleles resemble one another more closely than other members of this multigene family [83], suggesting that gametophytic self-incompatibility might have been present in the common ancestor of the rosid and asterid clades. Analyses of *S* allele sequences from members of the remaining rosid and asterid clades with gametophytic self-incompatibility are necessary, however, before we can confidently exclude the possibility that similar genes were recruited independently from members of the *S*-like ribonuclease gene family.

11.3 Population Structure and Mating Systems

The genetic system approach to describing reproductive systems [13,14] was attractive partly because it emphasized the synergistic effects that reproductive systems have on the entire evolutionary process. The evolution of a new reproductive mode affects the distribution and abundance of genetic variation in populations. Because any response to natural selection depends on the distribution and abundance of genetic variation, the evolution of a new reproductive system may also affect the way in which a population responds to new selective challenges. Indeed, evolution of a new reproductive system, perhaps more than evolution in any other feature, has consequences reaching far beyond the immediate change in the structures it affects to impacts on almost any other evolutionary response. Moreover, changes in the reproductive system may limit the opportunities for evolution of new reproductive modes. If, for example, selfers purge themselves of inbreeding depression [26,85], a genetic variant promoting outcrossing can invade only if its increased success as an outcross pollen parent more than compensates for any decreased success as a self pollen parent [37].

11.3.1 Isolation by Distance

Botanists have long recognized that populations of widely distributed species vary markedly across their range. As Epling and Dobzhansky [86] described it: "Some [races] reflect an adaptive response of the species to the local environment. [Others] display no apparent relation to a particular environment . . ." (p 317).

Wright [87] showed that genetic differentiation among geographically distant individuals is the inevitable result of localized mating. Even in a continuously distributed population subject to uniform selection pressures, widely separated individuals will evolve separately from one another and will become genetically distinct if more than one homozygous genotype can give rise to the same phenotype [88]. Wright coined the term "isolation by distance" to refer to this phenomenon, and it reflects the possibility that genetic drift, local mating, and mutation may be sufficient to explain genetic differentiation within a species. As Epling and Dobzhansky suggested, variation need "display no apparent relation to a particular environment," and it need not be adaptive.

Pollen and seed are rarely dispersed far in natural populations. Grant [15], for example, cited a variety of studies dating to the 1940s suggesting that mating rarely occurs at distances of more than 100–200 m, even in trees as large as Coulter pine (*Pinus coulteri*). He also noted that in bee-pollinated plants matings rarely occur at distances of more than 10–20 m. More recent studies have suggested that matings are commonly even more local than Grant supposed (see, for example, [89–93]; but see [94] for an example suggesting that gene flow may be more extensive in small populations of self-incompatible plants). Even such localized dispersal may lead to positive genetic correlations across a surprisingly broad area. With a mutation rate of 10^{-3} per locus per generation, for example, the genetic correlation between individuals approaches zero at 30 times the root mean squared dispersal distance. With a more realistic mutation rate of 10^{-5}, the genetic correlation does not approach zero until well beyond 100 times the root mean squared dispersal distance [95,96]. These distances correspond to between 300 m and more than 2 km for bee-pollinated plants and between 3 and 20 km for wind-pollinated plants.

Although numerous examples of small-scale differentiation within plant populations exist [97–99], it is unclear whether isolation by distance or local adaptive differentiation is the primary contributor to such differentiation [96]. Epling and Dobzhansky [86], and Wright [87], for example, initially concluded that isolation by distance alone could account for the pattern of color polymorphism they found in *Linanthus parryae*. Subsequent studies have shed doubt on this interpretation [100]. The theoretical results cited previously do suggest, however, that for traits in which isolation by distance plays an important role, the scale of genetic differentiation will be measured at least in tens of meters if not more.

11.3.2 Selfing

Wright [87] and Malecot [95] explained isolation by distance as the result of genetic drift in small, genetically isolated neighborhoods. Of course, Wright [101] pointed out that continued mating in small populations leads to an increase in homozygosity analogous to that seen with selfing and other forms of inbreeding. The analogy between genetic drift and inbreeding has been widely exploited ever since, even in the derivation of formulas for effective population size [102,103]. Not surprisingly, differences in population size may either accentuate or diminish differences between species in population structure associated with differences in their mating system (cf. [43,104]). There is, however, one important difference. Homozygosity increases with inbreeding without a change in allele frequency. Homozygosity increases with genetic drift, in contrast, because of the gradual loss of allelic diversity [102]. As a result, patterns of genetic diversity within and among plant populations are strongly affected both by mating system and population size.

Allozyme and Nucleotide Sequence Variation

The mating system, in fact, is arguably the most important influence on the genetic structure of plant populations. It has been commonplace for more than 15 years to note that the balance between selfing and outcrossing has an enormous effect both on the abundance of variation at allozyme loci and on the partitioning of that variation within and among loci [10,11,105]. Selfers tend to harbor less electrophoretic diversity than outcrossers, and populations of selfers tend to be more genetically distinct from one another than those of outcrossers.

Plant species that reproduce predominantly by self-fertilization have, on average, only one-third as many loci with electrophoretically detectable variation as outcrossers, and the mean heterozygosity in selfers is less than 0.1% while it is nearly 9% in outcrossers [11]. Some reduction in diversity is expected as a result of the reduced in effective population size of selfers, but the variance effective size of a partially selfing population is never less than half that of an outcrossing population [106,107]. Thus, much of this reduction must be a result either of the greater variance in effective population size among populations of selfers [108] or of background selection at other loci [109]. More than 50% of the total electrophoretically detectable variation is a result of among-population differences in selfers, while only about 12% of the genetic variation in outcrossers is a result of among-population differences [12,110]. Indeed, Hamrick and Godt [12] found that mating system (selfed, mixed mating, outcrossed) and life form (annual, short-lived perennial, long-lived perennial) account for nearly 40% of among-species differences in partitioning of genetic variation within and among populations.

Similar data for nucleotide sequences or restriction fragment polymorphisms are not readily available for plant populations. Fenster and Ritland [111] reported a fourfold reduction in chloroplast DNA (cpDNA) nucleotide diversity in the highly selfing *Mimulus micranthus* relative to its outcrossing congener *M. guttatus*, as inferred from restriction fragment polymorphisms. This result is initially surprising because cpDNA in these species is maternally transmitted. Thus, the mating system has no effect on its transmission, and it might be expected to have no impact on nucleotide diversity in cpDNA. The observed reduction in the highly selfing species is apparently a manifestation of background selection at other loci [109]. The higher the selfing rate in a population, the more that selection at any one locus in that population affects patterns of diversity at all other loci in that population, even nonnuclear loci.

Unfortunately, the sampling design Fenster and Ritland [111] employed does not allow patterns of among-population differentiation to be compared in these species. If the results of Mason-Gamer et al. [112] and Holsinger and Mason-Gamer [113] are any guide, however, future analyses of cpDNA variation may show much greater levels of among-population differentiation in cpDNA than are found at allozyme loci in the same species. As much as 90% of cpDNA nucleotide diversity in the outcrossing *Coreopsis grandiflora* is a result of among-population differences, especially between populations in Georgia and

Arkansas, while earlier allozyme analyses revealed little among-population differentiation [114,115]. If the pattern in *C. grandiflora* is typical of other species, it may be that mating systems, while they have an enormous effect on the amount of cpDNA diversity, have only a minor effect on the distribution of that diversity.

Polygenic Variation

Although much is known about the influence of mating system differences on the distribution and abundance of allozyme and nucleotide sequence variation in plant populations, little is known about the extent to which such variation follows patterns similar to that of variation in traits with more immediate ecological significance [116]. Traits such as growth rate, size, leaf shape, and reproductive effort, which are likely to be more directly related to survival and reproduction than allozyme variants, are generally influenced by allelic variation at many loci. Unfortunately, little is known about the distribution and abundance of polygenic variation within and among plant populations (cf. [116]). Even less is known about the impact of mating system differences on the partitioning of genetic variation among families and populations. Stebbins [56], for example, argued that populations of selfing species commonly consist of a variety of morphological types, each represented by many similar individuals. Consistent with the assumption that these biotypes, as he called them, correspond to homozygous genotypes, seed collected from a single mother tends to breed true. In contrast, the study by Knowles [117] of *Bromus mollis* suggested that each maternal family is genetically distinct from every other family, and Allard et al. [104] cited this result approvingly, arguing that "no single genotype is represented by more than a few individuals and perhaps every individual in the population differs genotypically from every other individual" (pp 98–99).

For traits not subject to natural selection, it is straightforward to use results from Crow and Kimura [102] to gain further insight into what the impact of mating system differences is likely to be. The equilibrium genetic variance in a trait increases as the selfing rate increases, provided that heterozygous genotypes have an intermediate phenotype. More importantly, the partitioning of the variance present changes dramatically (Fig. 11.2). In outbred populations, between 20% and 25% of the total genetic variance is a result of differences among maternal families; in selfing populations, all the genetic variance is a result of differences among maternal families. The proportion of genetic variance attributable to among-family differences increases roughly linearly with the selfing rate, and it is little affected by the degree of dominance within loci or the allele frequencies at those loci. Only if there are strong epistatic interactions among alleles at different loci may the pattern be markedly different.

In populations subject to selection, the situation is much more complex. The amount of genetic variation present in a population depends strongly on underlying allele frequencies, and allele frequencies in a partially selfing population are generally quite different from those in an outcrossing population subject to

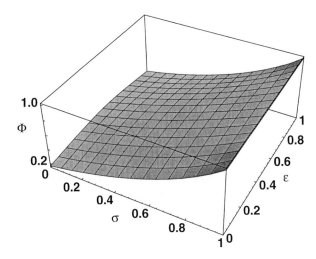

Fɪɢ. 11.2. Proportion of total genetic variation at a single locus with two alleles resulting from differences among maternal families: phi (ϕ) as a function of selfing rate, sigma (σ), and degree of dominance, epsilon (ε). (Epsilon = 0 corresponds to complete dominance of one allele; epsilon = 1, to complete dominance of the other allele; epsilon = 0.5, to additivity)

identical selection pressures [104,118]. As a result, equilibrium genetic variance may decline with an increase in selfing rate in traits subject to selection while it is expected to increase with selfing rate in traits not subject to selection. Charlesworth and Charlesworth [116] showed, for example, that the equilibrium genetic variance at a mutation–selection balance in a completely selfing population is only one-fourth that for a randomly mating population when heterozygotes are intermediate. Nonetheless, our numerical studies of the partitioning of genetic variation within and among families show that the proportion of genetic variation attributable to differences among families is little affected by underlying allele frequencies.

It is likely, therefore, that the relationship depicted in Fig. 11.2 applies, at least qualitatively, to all polygenic traits, whether subject to selection or not. In general, therefore, we expect populations with high selfing rates to harbor less polygenic variation than those with low selfing rates, and we expect more of the variation present in populations with high selfing rates to be the result of among-family differences than in populations with low selfing rates. Charlesworth and Charlesworth [116] noted that too few data are currently available for reliable empirical conclusions about the relationship between selfing rates and the amount of polygenic variation in plant populations. The same is true for empirical conclusions about the relationship between selfing rates and the proportion of genetic variation resulting from among-family differences, but the limited data available appear to be inconsistent with theoretical predictions.

Clay and Levin [118] have described the result of analyses partitioning pheno-typic variation in two *Phlox* species into within-family, among-family within-population, and among-population components. They found, contrary to expectation, more among-family variation in 11 of 20 characters measured in the predominantly outcrossing *P. drummondii* than in the predominantly selfing *P. cuspidata*. Similarly, Charlesworth and Mayer [119] failed to find a relationship between genetically estimated selfing rates and the proportion of among-family variation in four populations of *Collinsia heterophylla*. The reason for these discrepancies is not clear. Nongenetic variation within families might artificially inflate the importance of this variance component, meaning that the results are artefactual, or strong epistatic interactions might lead to more complex relation-ships between selfing rate and partitioning of genetic variance than those we have outlined. In either case, experiments in which within-family variation reflects only genetic differences among siblings are clearly necessary for a more complete understanding of the impact of mating system differences on the genetic structure of plant populations.

11.3.3 Self-Incompatibility

Besides affecting the distribution of genetic variation within and among popula-tions through their impact on the mating system, the pattern of genetic diversity at loci enforcing self-incompatibility is a fascinating topic in its own right. Wright [120] developed some of the original theory for the population genetic consequences of gametophytic self-incompatibility, based on Emerson's work involving *Oenothera organensis* [121,122]. He pointed out that when a self-incompatibility allele arises, it has an automatic reproductive advantage. In a species with gametophytic self-incompatibility, for example, a newly arisen allele will be able to fertilize ovules of all plants in the population except the plant in which it has arisen. Resident alleles are able to fertilize ovules only of plants with which they do not share an allele. As a result of this frequency-dependent advantage, rare mutations in the self-incompatibility alleles are less likely to be lost than newly arisen neutral alleles, and the strength of this advantage increases as the number of alleles segregating in the population decreases. Furthermore, population substructuring may also have an important role in maintaining allelic diversity.

Early empirical studies of self-incompatibility focused on attempts to under-stand their evolutionary dynamics within natural populations. In the past decade, however, studies have tended to focus more on the molecular details of the way in which self-incompatibility is expressed than on the distribution and frequency of self-incompatibility alleles. Relatively few studies report the distribution and frequencies of self-incompatibility alleles, and those that do have tended to examine species with a gametophytic self-incompatibility system (e.g., *Solanum carolinense* [123]; *Lolium perenne* [124]; *Papaver rhoeas* [125]). We were unable to locate any studies documenting or concerning allele frequencies in groups with sporophytic self-incompatibility, although there has been substantial interest in transspecific polymorphisms in groups with both types of incompatibility systems (e.g., *Brassica* [126]; Solanaceae [123,127]).

Population genetic studies on *Papaver rhoeas* far outnumber those on any other species. Lawrence et al. [128] reported a substantial degree of overlap in the complement of alleles present in different populations. In fact, populations contain almost the same set of alleles [125]. These results parallel Emerson's classic results on the distribution of self-incompatibility alleles in *Oenothera organensis* [121,122]. Of the 34 alleles in the 1937 collection, for example, 7 are found in all of the three most thoroughly studied canyon populations; 14 are found in two populations, and only 13 are restricted to a single population. Richman et al. [123] also found that populations of *Solanum carolinense* from North Carolina and Tennessee share many self-incompatibility alleles. The relatively low number of self-incompatibility alleles present in this species and geographic distance between the study populations, however, suggests that alleles are shared because populations of this weedy species have recently expanded into their current range, not because of contemporary gene flow between them. More surprising than the broad overlap in allelic constitution in populations of *P. rhoeas*, therefore, was the observation that alleles were present in very unequal frequencies in each study population, in spite of the fact that the populations had been established long before the investigations occurred. Two explanations for the uneven allele frequencies are possible.

First, variation at the *S* locus in *Papaver rhoeas* might be under the influence of an effect of selection beyond that associated with self-incompatibility. A series of controlled crosses showed that self-incompatibility alleles are associated with alleles at one or more loci controlling seed dormancy [129,130], as might be be expected from Uyenoyama's demonstration that associations between self-incompatibility loci and loci affecting the level of inbreeding depression form rapidly during the evolution of self-incompatibility [77]. Second, Brooks et al. [131] showed that localized pollen and seed dispersal could lead to spatial aggregations with very different allele frequencies. Neither seed nor pollen dispersal alone seemed to influence the equilibrium allele frequencies, but when both types of dispersal are considered together with variation in plant size, alleles tended to form clusters, resulting in very different equilibrium frequencies in simulated populations.

Studies of *Lolium perenne* have focused on the number of alleles present, not on their frequencies (e.g., [124,132]). Devey et al. [132], for example, examined the alleles present in a cultivar established by five genotypes. The five genotypes originally present could have carried at most 10 self-incompatibility alleles at each of the two loci determining self-incompatibility in grasses. After open pollination for three to five generations, however, more than 17 alleles were found at the *S* locus and 13 alleles were found at the *Z* locus. Because the presumed mutation rate of self-incompatibility alleles is far too low to explain this discrepancy, it appears that substantial migration of pollen into this population occurred. These results are similar to those of Devlin and Ellstrand [94], who found that 6%–7% of the progeny in two populations of self-incompatible *Raphanus sativus* near Riverside, California, were the result of pollen migration. These results suggest that substantial between-population pollen flow may be a common phenomenon in small populations of self-incompatible plant species.

In addition to counting the number of alleles in populations of *Lolium perenne*, Fearon et al. [124] also estimated the frequency with which each allele occurred. Because the self-incompatibility reaction in grasses is determined by the genotype at two, unlinked loci, the *S* locus and the *Z* locus, an enormous number of different self-incompatibility types are possible. Any population is likely to contain only a subset of the possible genotypes, unless it is very large. In addition, the large number of genotypes possible means that the effect of selection on allelic diversity is much weaker than the effect of selection at a single self-incompatibility locus. In short, it is reasonable to expect that drift will have a much larger impact on the dynamics of the self-incompatibility polymorphism in grasses than in groups with a single-locus self-incompatibility system. Perhaps for this reason (and the relatively small number of generations for which the population had existed), the very uneven allele frequencies found at both the *S* locus and the *Z* locus [124] are not unexpected.

Because of the strong frequency-dependent selection to maintain polymorphism at self-incompatibility loci, it should not be surprising to discover that these polymorphisms may be quite ancient. Polymorphism at the *S* locus in Solanaceae, for example, predates many speciation events within this family [123,127]. In fact, certain alleles from *Nicotiana tabacum* and *Solanum chacoense*, which belong to different subfamilies, share a more recent common ancestor with one another than other alleles within *N. tabacum* or *S. chacoense*. Similarly, Uyenoyama [126] estimated that some *S* alleles in the genus *Brassica* diverged more that four times as long ago as the species within which they are currently found. Given these patterns, it would be of great interest to analyze closely related families with the same self-incompatibility system to see whether these transspecific and transgeneric polymorphisms might also be transfamilial polymorphisms.

11.4 Conclusions

Much has been learned in the last century about the evolution of flowering plant mating systems. It is now clear that any fitness advantage to be found in self-fertilization will be found in its immediate impact on differences in individual survival and reproductive success [20], not in a balance between immediate fitness and long-term flexibility [13,14]. Even the source of that advantage is clear: either self-fertilization increases reproductive success as a result of increased seed set, or it increases reproductive success as a result of increased success as a pollen parent [37].

Investigations of the functional relationship between particular floral morphologies, patterns of pollen transfer, and rates of self-fertilization are, however, still in their infancy. As Harder and Barrett [9] pointed out, pollination biologists focus their attention on rates of pollinator visitation, patterns of pollen export, and seed production by individual plants. "They typically pay little attention to who has mated with whom" (p 142). Population geneticists, on the other hand, have focused primarily on the relative frequency of selfing and outcrossing while tending to ignore the features of floral morphology and pollinator behavior responsible for the observed pattern of matings. It should be clear by now, however, that future progress in understanding the evolution of plant mating systems is likely to emerge primarily from efforts to integrate observations and theory from these two fields.

Just as much has been learned about the evolution of plant mating systems, so also has much been learned about the evolutionary impact of mating system differences on the distribution and abundance of genetic variation in populations. It is clear that the localized pollen and seed dispersal typical of many plant populations may result in evolutionarily independent units on a scale as small as tens or hundreds of meters, even in a continuously distributed population [87,88,96]. The degree to which genetic differentiation reflects isolation by distance rather than local adaptation remains uncertain, although theoretical analyses suggest that isolation by distance alone is unlikely to account for differences observed on scales of a meter or less.

Hamrick and Godt [12] summarized results from hundreds of allozyme analyses of plant populations in which the impact of mating system differences on the partitioning of that variation within and among populations is both clear and dramatic. Selfers tend to harbor much less allozyme variation than closely related outcrossers, and about four times more of the variation they do contain is a result of differences among populations than in outcrossers. Unfortunately, little is known about patterns of genetic variation in characters that may be of more immediate ecological relevance, characters typically subject to polygenic control. The theoretical expectations are, however, clear. (1) Levels of variation within populations are expected to decline with an increase in the selfing rate if polygenic variation is maintained by a mutation–selection balance [116]. (2) More of the genetic variation present is expected to be a result of among-family differences in populations with a high selfing rate than in those with a low selfing rate.

Not only has the mating system been among the most labile characters in flowering plant evolution [25], mating system differences are arguably the most important influence on the genetic structure of plant populations. The extent to which patterns of variation in polygenic traits or in chloroplast-encoded genes mimic those so extensively documented for allozyme loci is not known. The results summarized here, however, continue to suggest that the mating system will be the predominant influence on those patterns, even if they are different from those found at allozyme loci. Clearly, understanding why so many independent lineages of angiosperms have undergone shifts in their mating system and understanding the further evolutionary consequences of those shifts remains one of the most important tasks facing plant evolutionists.

Acknowledgments. We have benefited from many informal discussions about plant mating systems and self-incompatibility with Greg Anderson, John Gabel, Don Les, Carl Schlichting, and Pati Vitt. Their influence will be found throughout this paper, although they cannot be held responsible for any of the ideas expressed here (unless they want to be). Tetsukazu Yahara provided helpful comments on an earlier version of this paper. The work described here on evolution of selfing would not have been possible without the continuous support of the University of Connecticut Research Foundation and of two grants from the National Science Foundation (BSR-9107330 and DEB-9509006).

References

1. Meeuse B, Morris S (1984) The sex life of flowers. Facts on File, New York
2. Sprengel CK (1793) Das entdeckte Geheimnis der Natur im Bau un in der Befruchtung der Blumen. I. Vieweg, Berlin
3. Darwin C (1859) On the origin of species by means of natural selection or the preservation of favoured races in the struggle for life. Murray, London
4. Darwin C (1862) The various contrivances by which orchids are fertilised by insects. Murray, London
5. Darwin C (1876) The effects of cross- and self-fertilization in the vegetable kingdom. Murray, London
6. Muller (1883) The fertilization of flowers. Macmillan, London
7. Knuth P (1899–1905) Handbuch der Blutenbiologie, vol I–III. Engelmann, Leipzig
8. Knight T (1799) Experiments on the fecundation of vegetables. Philos Trans R Soc Lond 89:195–204
9. Harder LD, Barrett SCH (1996) Pollen dispersal and mating patterns in animal-pollinated plants. In: Lloyd DG, Barrett SCH (eds) Floral biology: studies on floral evolution in animal-pollinated plants. Chapman & Hall, New York, pp 140–190
10. Brown AHD (1979) Enzyme polymorphism in plant populations. Theor Popul Biol 15:1–42
11. Gottlieb LD (1981) Electrophoretic evidence and plant populations. In: Reinhold L, Harborne JB, Swain T (eds) Progress in phytochemistry, vol 7. Pergamon, Oxford, pp 1–46
12. Hamrick JL, Godt MJW (1990) Allozyme diversity in plant species. In: Brown AHD, Clegg MT, Kahler AT, Weir BS (eds) Plant population genetics, breeding, and genetic resources. Sinauer, Sunderland, MA, pp 43–63
13. Darlington CD (1939) Evolution of genetic systems. Cambridge University Press, Cambridge
14. Mather K (1943) Polygenic inheritance and natural selection. Biol Rev 18:32–64
15. Grant V (1958) The regulation of recombination in plants. Cold Spring Harbor Symp Quant Biol 23:337–363
16. Stebbins GL (1958) Longevity, habitat, and the release of genetic variability in the higher plants. Cold Spring Harbor Symp Quant Biol 23:365–378
17. Grant V (1956) The influence of breeding habit on the outcome of natural hybridization in plants. Am Nat 90:319–322
18. Baker HG (1959) Reproductive methods as factors in speciation in flowering plants. Cold Spring Harbor Symp Quant Biol 24:177–190
19. Fryxell PA (1957) Mode of reproduction in higher plants. Bot Rev 23:135–233

20. Lloyd DG (1965) Evolution of self-compatibility and racial differentiation in *Leaven-worthia* (Cruciferae). Contrib Gray Herb Harv Univ 195:3–133
21. Williams GC (1975) Sex and evolution. Princeton University Press, Princeton
22. Maynard Smith J (1978) The evolution of sex. Cambridge University Press, Cambridge
23. Stebbins GL (1950) Variation and evolution in higher plants. Columbia University Press, New York
24. Lewis H (1953) The mechanism of evolution in the genus *Clarkia*. Evolution 7:1–20
25. Stebbins GL (1974) Flowering plants: evolution above the species level. Harvard University Press, Cambridge
26. Lande R, Schemske DW (1985) The evolution of self-fertilization and inbreeding depression. I. Genetic models. Evolution 39:24–40
27. Schemske DW, Lande R (1985) The evolution of self-fertilization and inbreeding depression in plants. II. Empirical observations. Evolution 37:523–539
28. Aide TM (1986) The influence of wind and animal pollination on variation in outcrossing rates. Evolution 40:434–435
29. Waller DM (1986) Is there disruptive selection for self-fertilization? Am Nat 128:421–426
30. Barrett SCH, Eckert CG (1990) Variation and evolution of mating systems in seed plants. In: Kawano S (ed) Biological approaches and evolutionary trends in plants. Academic, London, pp 229–254
31. Holsinger KE (1988) Inbreeding depression doesn't matter: the genetic basis of mating system evolution. Evolution 42:1235–1244
32. Charlesworth D, Charlesworth B (1990) Inbreeding depression with heterozygote advantage and its effect on selection for modifiers changing the outcrossing rate. Evolution 44:870–888
33. Charlesworth D, Morgan MT, Charlesworth B (1990) Inbreeding depression, genetic load, and the evolution of outcrossing rates in a multilocus system with no linkage. Evolution 44:1469–1489
34. Uyenoyama MK, Waller DM (1991) Coevolution of self-fertilization and inbreeding depression. I. Genetic modification in response to mutation-selection balance at one and two loci. Theor Popul Biol 40:14–46
35. Uyenoyama MK, Waller DM (1991) Coevolution of self-fertilization and inbreeding depression. II. Symmetric overdominance in viability. Theor Popul Biol 40:47–77
36. Uyenoyama MK, Waller DM (1991) Coevolution of self-fertilization and inbreeding depression. III. Homozygous lethal mutations at multiple loci. Theor Popul Biol 40:173–210
37. Holsinger KE (1996) Pollination biology and the evolution of mating systems in flowering plants. Evol Biol 29:107–149
38. Bierzychudek P (1981) Pollinator limitation of plant reproductive effort. Am Nat 117:838–840
39. Bruneau A, Anderson GJ (1988) Reproductive biology of diploid and triploid *Apios americana* (Leguminosae). Am J Bot 75:1876–1883
40. Zimmerman M (1988) Nectar production, flowering phenology, and strategies for pollination. In: Lovett-Doust J, Lovett-Doust L (eds) Plant reproductive biology: patterns and strategies. Oxford University Press, New York, pp 157–178
41. Young HJ, Young TP (1992) Alternative outcomes of natural and experimental high pollen loads. Ecology (Washington DC) 73:639–647

42. Wilson P, Thomson JD, Stanton ML, Rigney LP (1994) Beyond floral Batemania: gender biases in selection for pollination success. Am Nat 143:283–296
43. Jain SK (1976) The evolution of inbreeding in plants. Annu Rev Ecol Syst 7:469–495
44. Lloyd DG (1980) Demographic factors and mating patterns in angiosperms. In: Solbrig OT (ed) Demography and evolution in plant populations. Blackwell, Oxford, pp 67–88
45. Hagerup O (1950) Rain pollination. K Dan Vidensk Selsk Biol Medd 18:1–19
46. Hagerup O (1951) Pollination in the Faroes—in spite of rain and poverty in insects. K Dan Vidensk Selsk Biol Medd 18:3–48
47. Lloyd DG (1979) Some reproductive factors affecting the selection of self-fertilization in plants. Am Nat 113:67–79
48. Holsinger KE (1992) Comment: functional aspects of mating system evolution in plants. Int J Plant Sci 153(3):iii–v
49. Dole JA (1992) Reproductive assurance mechanisms in three taxa of the *Mimulus guttatus* complex (Scrophulariaceae). Am J Bot 79:650–659
50. Baker HG (1955) Self-compatibility and establishment after "long-distance" dispersal. Evolution 9:347–348
51. Baker HG (1948) Dimorphism and monomorphism in the Plumbaginaceae. I. A survey of the family. Ann Bot 12:207–219
52. Baker HG (1953) Dimorphism and monomorphism in the Plumbaginaceae. III. Correlation of geographical distribution patterns with dimorphism and monomorphism. Ann Bot 17:615–627
53. Barrett SCH (1979) The evolutionary breakdown of tristyly in *Eichhornia crassipes* (Mart.) Solms (water hyacinth). Evolution 33:499–510
54. Glover DE, Barrett SCH (1986) Variation in the mating system of *Eichhornia paniculata* (Spreng.) Solms (Pontederiaceae). Evolution 40:1122–1131
55. Barrett SCH, Morgan MT, Husband BC (1989) The dissolution of a complex genetic polymorphism: the evolution of self-fertilization in tristylous *Eichhornia paniculata* (Pontederiaceae). Evolution 43:274–311
56. Stebbins GL (1957) Self-fertilization and population variability in the higher plants. Am Nat 91:337–354
57. Lloyd DG (1992) Self- and cross-fertilization in plants. II. The selection of self-fertilization. Int J Plant Sci 153:370–380
58. Fisher RA (1941) Average excess and average effect of a gene substitution. Ann Eugen 11:53–63
59. Kimura M (1959) Conflict between self-fertilization and outbreeding in plants. Ann Rep Natl Inst Genet Japan 9:87–88
60. Holsinger HG (1992) Ecological models of plant mating systems and the evolutionary stability of mixed mating systems. In: Wyatt RW (ed) Ecology and evolution of plant reproductive systems. Chapman & Hall, New York, pp 169–191
61. Nagylaki T (1976) A model for the evolution of self-fertilization and vegetative reproduction. J Theor Biol 58:55–58
62. Charlesworth B (1980) The cost of sex in relation to mating system. J Theor Biol 84:655–671
63. Holsinger KE, Feldman MW, Christiansen FB (1984) The evolution of self-fertilization in plants. Am Nat 124:446–453
64. Ritland K (1991) A genetic approach to measuring pollen discounting in natural plant populations. Am Nat 138:1049–1057

65. Rausher MD, Augustine D, VanderKooi A (1993) Absence of pollen discounting in a genotype of *Ipomoea purpurea* exhibiting increased selfing. Evolution 47:1668–1695

66. Schoen DJ, Clegg MT (1985) The influence of flower color on outcrossing rate and male reproductive success in *Ipomoea purpurea*. Evolution 39:1242–1249

67. Kohn JR, Barrett SCH (1992) Floral manipulations reveal the cause of male fitness variation in experimental populations of *Eichhornia paniculata* (Pontederiaceae). Funct Ecol 6:590–595

68. Kohn JR, Barrett SCH (1992) Experimental studies on the functional significance of heterostyly. Evolution 46:43–55

69. Kohn JR, Barrett SCH (1994) Pollen discounting and the spread of a selfing variant in tristylous *Eichhornia paniculata*: evidence from experimental populations. Evolution 48:1576–1594

70. Holsinger KE (1991) Mass-action models of plant mating systems: the evolutionary stability of mixed mating systems. Am Nat 138:606–622

71. Holsinger KE, Thomson JD (1994) Pollen discounting in *Erythronium grandiflorum*: mass-action estimates from pollen transfer dynamics. Am Nat 144:799–812

72. Gregorius HR, Ziehe M, Ross MD (1987) Selection caused by self-fertilization. I. Four measures of self-fertilization and their effects on fitness. Theor Popul Biol 31:91–115

73. Charlesworth D, Charlesworth B (1979) The evolution and breakdown of *S*-allele systems. Heredity 43:41–55

74. Charlesworth D (1988) Evolution of homomorpic sporophytic self-incompatibility. Heredity 60:445–453

75. Uyenoyama MK (1988) On the evolution of genetic incompatibility systems. II. Initial increase of strong gametophytic self-incompatibility under partial selfing and half-sib mating. Am Nat 131:700–722

76. Matton DP, Nass N, Clarke AE, Newbigin E (1994) Self-incompatibility: how plants avoid illegitimate offspring. Proc Natl Acad Sci USA 91:1992–1997

77. Uyenoyama MK (1991) On the evolution of genetic incompatibility systems. VI. A three-locus modifier model for the origin of gametophytic self-incompatbility. Genetics 128:453–469

78. Nasrallah JB, Stein JC, Kandasamy MK, Nasrallah ME (1994) Signaling the arrest of pollen tube development in self-incompatible plants. Science 266:1505–1508

79. Pandey KK (1960) Evolution of gametophytic and sporophytic systems of self-incompatibility in angiosperms. Evolution 14:98–115

80. de Nettancourt D (1977) Incompatibility in angiosperms. Springer, Berlin Heidelberg New York

81. Dzelzkalns VA, Nasrallah JB, Nasrallah ME (1992) Cell-cell communication in plants: self-incompatibility in flower development. Dev Biol 153:70–82

82. Newbigin E, Anderson MA, Clarke AE (1993) Gametophytic self-incompatibility systems. Plant Cell 5:1315–1324

83. Xue Y, Carpenter R, Dickinson HG, Coen E (1996) Origin of allelic diversity in *Antirrhinum S* locus RNases. Plant Cell 8:805–814

84. Broothaerts W, Janssens GA, Proost P, Broekaert W (1995) cDNA cloning and molecular analysis of two self-incompatibility alleles from apple. Plant Mol Biol 27:499–511

85. Barrett SCH, Charlesworth D (1991) Effects of a change in the level of inbreeding on the genetic load. Nature 352:522–524

86. Epling C, Dobzhansky D (1942) Genetics of natural populations. VI. Microgeographic races in *Linanthus parryae*. Genetics 27:317–332
87. Wright S (1943) An analysis of local variability of flower color in *Linanthus parryae*. Genetics 28:139–156
88. Goldstein DB, Holsinger KE (1992) Maintenance of polygenic variation in spatially structured populations: roles for local mating and genetic redundancy. Evolution 46:412–429
89. Levin DA, Kerster HW (1974) Gene flow in seed plants. Evol Biol 7:139–220
90. Levin DA (1981) Dispersal versus gene flow in plants. Ann MO Bot Gard 68:233–253
91. Levin DA (1984) Immigration in plants: an exercise in the subjunctive. In: Dirzo R, Sarukhan J (eds) Perspectives on plant population ecology. Sinauer, Sunderland, MA, pp 242–260
92. Adams WT, Birks DS (1991) Estimating mating patterns in forest tree populations. In: Fineschi S, Malvolti ME, Cannata F, Hattemer HH (eds) Biochemical markers in the population genetics of forest trees. SPB Academic, The Hague, Netherlands, pp 157–172
93. Adams WT, Griffin AR, Moran, GF (1992) Using paternity analysis to measure effective pollen dispersal in plant populations. Am Nat 140:762–780
94. Devlin B, Ellstrand NC (1990) The development and application of a refined method for estimating gene flow from angiosperm paternity analysis. Evolution 44:248–259
95. Malecot G (1969) The mathematics of heredity. Freeman, San Francisco
96. Holsinger KE (1992) The evolutionary dynamics of fragmented plant populations. In: Kareiva P, Kingsolver J, Huey R (eds) Biotic interactions and global change. Sinauer, Sunderland, pp 198–216
97. Schaal BA (1975) Population structure and local differentiation in *Liatris cylindracea*. Am Nat 109:511–528
98. Epperson BK, Clegg MT (1986) Spatial autocorrelation analysis of flower color polymorphisms within substructured populations of morning glory (*Ipomoea purpurea*). Am Nat 128:840–858
99. Campbell DR, Dooley JL (1992) The spatial scale of genetic differentiation in a hummingbird-pollinated plant: comparison with models of isolation by distance. Am Nat 139:735–748
100. Epling C, Lewis H, Ball FM (1960) The breeding group and seed storage: a study in population dynamics. Evolution 14:238–255
101. Wright S (1931) Evolution in Mendelian populations. Genetics 16:97–159
102. Crow JF, Kimura M (1970) An introduction to population genetics theory. Burgess, Minneapolis
103. Ewens WJ (1979) Mathematical population genetics. Springer, Berlin Heidelberg New York
104. Allard RW, Jain SK, Workman PL (1968) The genetics of inbreeding populations. Adv Genet 14:55–131
105. Hamrick JL, Linhart YB, Mitton JB (1979) Relationships between life history characteristics and electrophoretically detectable genetic variation in plants. Annu Rev Ecol Syst 10:173–200
106. Pollak E (1987) On the theory of paritally inbreeding finite populations. I. Partial selfing. Genetics 117:353–360
107. Maruyama K, Tachida H (1992) Genetic variability and geographical structure in partially selfing populations. Jpn J Genet 67:39–51

108. Schoen DJ, Brown AHD (1991) Intraspecific variation in population gene diversity and effective population size correlates with the mating system in plants. Proc Natl Acad Sci USA 88:4494–4497

109. Charlesworth B, Morgan M, Charlesworth D (1993) The effect of deleterious mutations on neutral molecular variation. Genetics 134:1289–1303

110. Loveless MD, Hamrick JL (1984) Ecological determinants of genetic structure in plant populations. Annu Rev Ecol Syst 15:65–95

111. Fenster CB, Ritland K (1992) Chloroplast DNA and isozyme diversity in two *Mimulus* species (Scrophulariaceae) with contrasting mating systems. Am J Bot 79:1440–1447

112. Mason-Gamer RJ, Holsinger KE, Jansen RK (1995) Chloroplast DNA haplotype variation within and among populations of *Coreopsis grandiflora* (Asteraceae). Mol Biol Evol 12:371–381

113. Holsinger KE, Mason-Gamer RJ (1996) Hierarchical analysis of nucleotide diversity in geographically structured populations. Genetics 142:629–639

114. Crawford DJ, Smith EB (1984) Allozyme divergence and interspecific variation in *Coreopsis grandiflora* (Compositae). Syst Bot 9:219–225

115. Cosner MB, Crawford DJ (1990) Allozyme variation in *Coreopsis* sect. *Coreopsis* (Asteraceae). Syst Bot 15:256–265

116. Charlesworth D, Charlesworth B (1995) Quantitative genetics in plants: the effect of the breeding system on genetic variability. Evolution 49:911–920

116a. Holsinger KE (1991) Conservation of genetic diversity in rare and endangered plants. In: Dudley EC (ed) The unity of evolutionary biology. Dioscorides Press, Portland, pp 626–633

117. Knowles PF (1943) Improving an annual bromegrass. J Am Soc Agron 35:584–594

118. Clay K, Levin DA (1989) Quantitative variation in *Phlox*: comparison of selfing and outcrossing species. Am J Bot 76:577–588

119. Charlesworth D, Mayer S (1995) Genetic variability of plant characters in the partial inbreeder *Collinsia heterophylla* (Scrophulariaceae). Am J Bot 82:112–120

120. Wright S (1939) The distribution of self-sterility alleles in populations. Genetics 24:538–552

121. Emerson S (1938) The genetics of self-incompatibility in *Oenothera organensis*. Genetics 23:190–202

122. Emerson S (1939) A preliminary survey of the *Oenothera organensis* population. Genetics 24:524–537

123. Richman AD, Kao TH, Schaeffer SW, Uyenoyama MK (1995) S-allele sequence diversity in natural populations of *Solanum carolinense* (horsenettle). Heredity 75:405–415

124. Fearon CH, Cornish MA, Hayward MD, Lawrence MJ (1994) Self-incompatibility in ryegrass. X. Number and frequency of alleles in a natural population of *Lolium perenne* L. Heredity 73:254–261

125. O'Donnell S, Lawrence MJ, Lane MD (1993) The population genetics of the self-incompatibility polymorphism in *Papaver rhoeas*. VI. Estimation of the overlap between the allelic complements of a pair of populations. Heredity 71:591–595

126. Uyenoyama MK (1995) A generalized least-squares estimate for the origin of sporophytic self-incompatibility. Genetics 139:975–992

127. Ioerger TR, Clark AG, Kao TH (1990) Polymorphism at the self-incompatibility locus in Solanaceae predates speciation. Proc Natl Acad Sci USA 87:9732–9735

128. Lawrence MJ, Lane MD, O'Donnell S, Franklin-Tong VE (1993) The population genetics of the self-incompatibility polymorphism in *Papaver rhoeas*. V. Cross-classification of the *S*-alleles of samples from three natural populations. Heredity 71:581–590

129. Lawrence MJ, Franklin-Tong VE (1994) The population genetics of the self-incompatibility polymorphism in *Papaver rhoeas*. IX. Evidence of an extra effect of selection acting on the *S*-locus. Heredity 72:363–364

130. Lane MD, Lawrence MJ (1995) The population genetics of the self-incompatibility polymorphism in *Papaver rhoeas*. X. An association between incompatibility geno-type and seed dormancy. Heredity 75:92–97

131. Brooks RJ, Tobias AM, Lawrence MJ (1996) The population genetics of the self-incompatibility polymorphism in *Papaver rhoeas*. XI. The effects of limited pollen and seed dispersal, overlapping generations and variation in plant size on the variance of *S*-allele frequencies in populations at equilibrium. Heredity 76:367–376

132. Devey F, Fearon CH, Hayward MD, Lawrence MJ (1994) Self-incompatibility in ryegrass. XI. Number and frequency of alleles in a cultivar of *Lolium perenne* L. Heredity 73:262–264

133. Chase MW, Soltis DE, Olmstead RG, Morgan D, Les DH, Mishler BD, Duvall MR, Price RA, Hills HG, Qiu YL, Kron KA, Rettig JH, Conti E, Palmer JD, Manhart JR, Sytsma KJ, Michaels HJ, Kress MJ, Karol KG, Clark WD, Hedren M, Gaut BS, Jansen RK, Kim KJ, Wimpee CF, Smith JF, Furnier GR, Strauss SH, Xiang QY, Plunkett GM, Soltis PS, Swensen SM, Williams SE, Gadek PA, Quinn CJ, Eguiarte LE, Golenberg E, Learn GH Jr, Graham SW, Barrett SCH, Dayanandan S, Albert VA (1993) Phylogenetics of seed plants: an analysis of nucleotide sequences from the plastid gene *rbcL*. Ann MO Bot Gard 80:528–580

134. Charlesworth D (1985) Distribution of dioecy and self-incompatibility in an-giosperms. In: Greenwood PJ, Harvey PH, Slatkin M (eds) Evolution: essays in honor of John Maynard Smith. Cambridge University Press, Cambridge

135. Mau SL, Anderson MA, Heisler M, Haring V, McClure BA, Clarke AE (1991) Molecular and evolutionary aspects of self-incompatibility in flowering plants. In: Jenkins GI, Schuch W (eds) Symposia of the Society for Experimental Biology, vol 45. Molecular biology of plant development. The Company of Biologists, Depart-ment of Zoology, University of Cambridge, Cambridge

136. Lloyd DG, Schoen DJ (1992) Self- and cross-fertilization in plants. I. Functional dimensions. Int J Plant Sci 153:358–369

CHAPTER 12

Plant Speciation on Oceanic Islands

DANIEL J. CRAWFORD[1] and TOD F. STUESSY[2]

12.1 Introduction

Few would argue that one of the important questions in evolutionary biology is how morphologically distinguishable groups, commonly recognized as species, evolve or develop through time. Although the process of speciation has been viewed as unknowable because it usually occurs gradually, and one must infer past processes from current observations, there has been renewed interest in speciation during the past decade (Otte and Endler [1], Coyne [2,3], Levin [4], Mayr [5], as examples). Any discussion of systematic evolutionary research to the year 2000 and beyond should give high priority to intensive studies of plant speciation. Attempts to classify modes of speciation and to apply definitions have received considerable attention in recent years (Baum and Donoghue [6], Davis [7] Luckow [8], Mc Dade [9], Olmstead [10], Levin [4] Rieseberg and Brouillet [11], Mallet [12], Gavrilets and Hastings [13]). We will not engage in this discussion in any detail, but will consider the issue when it becomes necessary within the context of the central foci of the chapter. As for the definition of speciation used in this chapter, slight modification and elaboration of the one given by Levin [4] will suffice. Severe reduction in effective gene flow must occur between populations and it becomes permanently disadvantageous for migrants of one species to occupy the same niche as another.

Plants endemic to oceanic islands have long attracted the attention of botanists and naturalists because they are sometimes very different morphologically from continental relatives; indeed they are sometimes so distinct that it is difficult to identify their ancestors (see excellent general discussions in Carlquist [14,15]). In this paper we will argue that insular endemics are of interest beyond documenting the diversity of these curious and marvelous products of the evolutionary process. The purposes of the present paper are to discuss the use of plants

[1] Department of Plant Biology, The Ohio State University, 1735 Neil Avenue, Columbus OH 43210, USA
[2] Institut für Botanik, Universität Wien, Rennweg 14, A-1030, Vienna, Austria

endemic to oceanic islands as model systems for the study of plant speciation, to discuss insights and questions about speciation provided by studies from insular endemics, and to suggest some directions for future studies.

12.2 Studying Plant Speciation

The two general methods for studying speciation are experimental studies and comparative studies of two or more closely related species. The latter approach, which will be discussed in this chapter, is more widely employed and involves determining those features by which two taxa differ, and then inferring differences associated with speciation. As observed by Coyne [2], the process of speciation can be studied (and inferred) by the "patterns it leaves behind." To study these patterns, it is critical that the proper congeners be compared, that is, those taxa related as progenitor-derivative or as sister species evolved from a common ancestor; the former situation is more desirable than the latter and outstanding examples include *Stephanomeria* by Gottlieb [16], *Clarkia* by Lewis [17], and *Mimulus* by MacNair and Cumbes [18].

The number of tools available to determine most closely related species has increased with the routine use of molecular data to infer phylogenetic relationships [19]. Molecular data have an advantage over morphological characters in phylogenetic reconstruction (particularly when morphological features may be selected in radiating lineages on islands) because the molecular features often employed in such studies are largely neutral. Once robust phylogenies have been produced, preferably from more than one molecular or other data set, then species comparisons can be made with some confidence that the correct taxa are being studied.

An important issue in species comparisons is distinguishing differences associated with speciation as opposed to those differences accumulated subsequent to speciation [20]. This is a difficult problem, and it can never be known with certainty when particular differences between taxa evolved. The magnitude of the problem, however, can be minimized by studying recently diverged species; as time passes after speciation differences will accumulate and these either have nothing to do with the initial divergence or, more rarely, may involve "fine tuning" of the two lineages. Once differences between species have been identified, a next step is to determine the genetic basis of the differences. This requires that the species be interfertile so that crosses can be made between them. Also, it is highly desirable that the plants be self-compatible so that the F_1 generation can be selfed to produce the segregating F_2 generation.

12.3 Insular Endemic Plants as Model Systems for Studying Speciation

Congeneric insular endemic species have many of the desirable features discussed in the preceding section. Many endemics appear to be relatively recent in origin compared to most continental species (although there are exceptions and caution must be exercised); two lines of evidence argue for this generalization. First, a number of studies from a variety of different genera using several molecular techniques such as allozymes, chloroplast DNA restriction sites, and various sequences show low divergence between species. Several representative examples include *Argyranthemum* (Asteraceae) from the Macaronesian Islands [21–23], *Dendroseris* (Asteraceae) from the Juan Fernandez Islands [24,25], and *Dubautia* (Asteraceae) species with $n = 13$ chromosome number [26,27] from the Hawaiian Islands. Although there are exceptions, a general picture of low molecular divergence is emerging for insular congeners. While there is considerable debate about the accuracy of a "molecular clock" [19,28], there is little question that molecular divergence increases with time, and thus low divergence is an indication of recent speciation in insular endemics.

Other evidence for the youth of endemic species is the young ages of the islands on which they occur. The basic assumption is that the species evolved on the islands where they now occur, and as argued by Carlquist [29], this seems a fair assumption in most instances. There is the possibility that, in archipelagos such as Hawaii where there is a series or chain of earlier islands now below sea level, extant species may have originated on these older islands and then migrated to the high islands. Cases have been made for this happening in certain genera [30], and there seems little doubt that some species may be older than present high islands. The critical question is whether there was biological continuity between the older and younger islands, and this can never be known with certainty. In the absence of data to the contrary it, is probably best to assume that extant species originated on islands now in existence [29].

Crossing studies in several different genera indicate that congeners are cross-compatible and interfertile despite their differences in morphology and ecology. Genera that have been studied in some detail include. *Argyranthemum* in Macaronesia [31], *Bidens* in Hawaii [32], *Dubautia* and two other genera of the silversword alliance in Hawaii [33,34], and *Tetramolopium* in Hawaii [35]. The level of fertility can vary, but in many instances fertility between species is comparable to within species. The ability to cross different species to produce fertile F_1 and F_2 generations is critical to the study of genetic differences between taxa. Very similar species may be intersterile because of genetic or chromosomal factors [4], making it difficult to study inheritance of species differences. Insular congeners are unusual in differing by a variety of features yet being interfertile. The topic of determining the genetic basis of the morphological differences is discussed in more detail below. The lack of (or weak) internal barriers to gene flow between endemic congeners may be taken as additional evidence of their recent divergence. This is so because chromosomal-genetic differences accumulate with time after initial divergence and speciation. Elegant evidence for this scenario in island plants was provided by Carr and Kyhos [34], who demonstrated that plants of the silversword alliance with chromosome numbers of $n = 14$ occur on the older islands of the Hawaiian archipelago whereas the derived number of $n = 13$ is restricted to younger islands. It appears that, with sufficient time, chromosome mutations occur (there are other chromosome differences in the silversword alliance in addition to aneuploid reduction). The allozyme data of Witter and

Carr [26] are concordant with the cytogenetic data in suggesting that the $n = 14$ species of *Dubautia* are more highly divergent (and thus older) than the presumably derived $n = 13$ species.

The availability of molecular markers for mapping makes it possible to infer the genetic basis of morphological and other characters in a more refined way than has been possible in the past. The so-called QTLs or quantitative trait loci are mapped with molecular markers and can then be used to infer the genetic basis of characters by one of two general methods. One method determines whether each of the segregating QTLs is significantly correlated with a quantitative trait. A second, more refined method is called interval mapping. A discussion of these methods is beyond the present discussion; very brief general introductions are given by Finchan [36] and Doebley [37], and examples of the use of QTLs may be found in Doebley and Stec [38,39] and Vlof et al. [40]. Most available examples of the use of QTLs involve domesticated plants, and studies of naturally occurring species are needed. Such work, using endemic species of *Tetramolopium* from the Hawaiian Islands, is now in progress (T.K. Lowrey and R. Whitkus, personal communication).

12.4 Speciation in Insular Endemics: Special Cases or Typical of Flowering Plants?

If island endemics are to be used as model systems for the study of plant speciation, then it must be assumed that the process in them is typical of speciation in flowering plants in general. Several processes such as polyploidy and chromosomal evolution appear to be rare phenomena in island plants, and will be treated in the next section. In this section, the reasons for considering the situation in islands typical of many other flowering plants will be presented. One reason islands appear at first glance like unusual situations is that the process of speciation is telescoped in space and time compared to many continental situations.

Levin [4] has correctly argued that speciation is usually at the local level because there is no easy mechanism to explain the transformation of entire geographical races into species. In other words, subspecies are not incipient species in the sense that they evolve en masse into species with time. Rather, the locus of speciation is at the local population (or a metapopulation), with subsequent dispersal and colonization. Also, it is possible that dispersal from the local population could result in speciation, founder effect speciation being one possible means by which this occurs. The model put forth by Levin [4] is applicable to speciation on oceanic islands because of the presence of many well-defined species occupying very different and, in many cases, adjacent habitats. Interisland colonization would likewise be compatible with the local speciation hypothesis because dispersal from one population can provide the ancestor for a species.

Consider next the factors leading to the initial divergence from the local population. One possibility is the dispersal of diaspores from the population to a distance where gene flow via pollen or seed becomes low or nonexistent. This new locality would presumably provide a niche or habitat distinct from that of the parental population. In many island situations this is quite feasible because a variety of different habitats is often found within a short distance of one another. Divergence through time could then produce a morphologically stable population adapted to its new habitat. This could involve both selection on alleles at various loci or various recombinant types. If populations are small, as they likely would be on islands, then stochastic factors could also play an important role in the establishment of novel types [41].

Studies of insular endemics show clearly that isolation between species is the result of spatial separation, with species often occurring in different habitats. Available evidence for the existence of internal isolating barriers is rare to nonexistent for island groups, with the exception of the aforementioned silversword alliance [33,34]. It seems obvious that the primary isolating factors associated with plant speciation on islands are those that reduce or eliminate gene flow between the parental and the "new" population. These factors could be adaptive or stochastic. In the silversword alliance, it seems likely that the chromosomal differences do not represent the primary isolating factors for the species, but rather that habitat, spatial separation, or both have been critical; the chromosomal differences likely have accumulated with time. Also, the group to which the silverswords belong is known to undergo chromosomal restructuring during evolution, and this propensity was probably brought to the islands by their ancestors.

Congeneric plant species in continental situations display a wide range of factors that prevent or greatly reduce their ability to exchange genes [42,43]. The important question is the feature or features responsible for the initial isolation versus characters that evolved after isolation. One issue has been the role of chromosomal restructuring as a primary isolating mechanism. Perhaps the best example of this is the elegant work of Harlan Lewis and collaborators [17,44–48] with the genus *Clarkia* showing that morphologically similar species often differ by structural differences in their chromosomes so that their experimental hybrids are highly sterile. Experimental studies by Lewis [48] showed that when the two very closely related progenitor-derivative species pair *Clarkia biloba–C. lingulata* were grown together in mixed populations that the species present in lower frequency would not survive because of gametic wastage in highly sterile F_1 hybrids. Lewis [48] concluded that restricted gene flow between populations of the two species is necessary for both to occur in close proximity. In other words, the chromosomal differences were not the primary isolating factor leading to speciation. Suffice it to say that there is little empirical or experimental support for chromosomal change as a mechanism of speciation; Howard [41] presents a balanced discussion of the issue for animals, and most of his points apply equally well to plants. In general, the differences seen between the most recently diverged species of insular endemics are similar to the initial stages of speciation in many continental situations.

12.5 Factors Limiting Use of Insular Endemics for Speciation Studies

12.5.1 Practical and Biological Limitations

Although island species offer several important advantages for studying speciation, there are also some limitations. One potential difficulty is obtaining the plants because they may be nearly inaccessible on remote islands. This is less of a problem on archipelagos such as Hawaii, but there may be a real limitation on, for example, the Juan Fernandez Islands. Many species often are rare and occur in a few small populations; locating the plants, therefore, may not be easy.

Systematists have been fascinated with the morphological differences between the insular species and have often treated them as different genera, and if species are viewed as distantly related they are not attractive subjects for study. More refined understanding of the high genetic similarity between species has enhanced their perceived value for speciation studies. For example, the detailed biosystematic work of Lowrey [35] on *Tetramolopium* in Hawaii elucidated the potential of the group for speciation studies.

Essentially all island endemics are perennial plants, some of which have the palmiform or rosette tree growth forms typical of islands. Genera such as *Dendroseris*, *Robinsonia*, and *Centaurodendron* of the Asteraceae on the Juan Fernandez Islands are typical examples, and one may appreciate examples from other archipelagos by consulting the excellent drawings and photographs in Carlquist [14,15]. A problem with quite a number of these unusual endemics is the inability to cultivate them to flowering or the time required to reach flowering in cultivation. Notable exceptions to this generalization include *Argyrantherum* from the Macaronesian Islands and *Tetramolopium* from Hawaii; species of both genera may be grown to flowering in a matter of several months.

Chromosome numbers must also be considered in selecting insular plants for study; diploids are desirable, yet some outstanding examples of endemic genera such as the silversword alliance and *Bidens* in Hawaii are polyploids, as are *Dendroseris* and *Robinsonia*. Excellent examples of extensive diversification at the diploid level include *Argyranthemum* and *Sonchus* in Macaronesia, and *Tetramolopium* in Hawaii. Polyploidy will be discussed in another context in the next section.

12.5.2 Processes Uncommon in the Evolution of Insular Species

Polyploidy is a common mechanism of speciation in flowering plants [42,49,50] and has often been referred to as the one good example of instantaneous speciation. Recent incorporation of molecular data into the study of polyploid formation has produced renewed interest in polyploidy, and there is now evidence that autopolyploidy is more common than traditionally believed. A comprehensive review of these topics may be found in Soltis and Soltis [51]. There are almost no reports of polyploid speciation in oceanic islands. Sahuquillo and Lumaret [52] present allozyme data indicating that polyploids have originated in *Dactylis* (Poaceae) in the Macaronesian Islands. There is also the possibility that a tetraploid species of *Tolpis* (Asteraceae) may have originated in situ on the island of Tenerife in the Canary archipelago (J. Francisco-Ortega, personal communication). The apparent rarity of polyploidy as a mechanism of speciation within oceanic archipelagos obviously precludes the use of insular endemics in the study of the origin, evolution, and dynamics of polyploids. The reasons for the absence of in situ polyploidy in islands are not known, but there are several possibilities. One reason could be that the youth of the species has not provided adequate time for the evolution of polyploids. This seems rather unlikely, however, because there is evidence that the "same" allopolyploids can arise multiple times and spread in a matter of decades in *Tragopogon* [53–56] and can likewise originate more than once in the fern genus *Asplenium* [57–59]. Given that many polyploids are allopolyploid [42,49,50], the rarity of hybridization between well differentiated populations, often recognized as species in some plant groups on oceanic islands, may be a factor. Another possibility may be the lack of niches for polyploids to become established where they are not in competition with their parental diploids; this could be true of both auto- and allopolyploids. Suffice it to say that, regardless of the reasons, polyploid speciation is rare to nonexistent on oceanic islands and insular endemics are not appropriate systems for studying the process.

The role of chromosomal mutation (other than polyploidy) in speciation is an issue of continuing discussion [60], and Kyhos and Carr [61] have recently considered the possible role of chromosomal rearrangements in adaptive evolution. Island plants offer little promise for studying the role of chromosomal evolution in speciation and in species isolation. With the exception of the aforementioned work in Hawaii [33,34], there is little evidence for chromosomal differences between species on islands. This may be due in part to the limited biosystematic work done, but it seems more likely that gross chromosomal changes are not common in island plants.

12.6 Insular Species, Hybridization, and Hybrid Speciation

In the past several years there has been renewed interest in the role of hybridization and introgression in plant speciation and evolution [62–65]; the availability of various molecular markers has been an important factor stimulating this new interest [62]. Also, hybridization is now evaluated more as a process that may be important in evolution rather than something that only produces complications for the taxonomist. In considering hybridization in island plants, it is important to make two distinctions. First, the distinction should be made between hybridization occurring early in radiating lineages and contemporary or recent gene exchange between well-differentiated species. Concern here will be with the latter situation. Secondly, introgression and the movement of genes from one species into another must be distinguished from the stabilization of hybrid recombinants, that is, hybrid speciation. The frequency of interspecific hybridization between insular endemics has been discussed; one view [66] is that it is infrequent whereas another [67] suggests that hybridization is much more common. Available evidence indicates that in some groups such as the silversword alliance [33,61] and *Wikstroemia* (Thymelaeaceae) of Hawaii [68], *Argyranthenum* [69–72] in the Canary Islands, and *Gunnera* [73] in the Juan Fernandez Islands interspecific hybridization occurs, to cite several of the available examples. By contrast, hybridization appears to be quite rare in other island groups such as *Dendroseris* and *Robinsonia* in the Juan Fernandez Islands [74] and *Bidens* [75] and *Tetramolopium* [35] on Hawaii.

The present lack of evidence for intertaxon hybridization should not be taken as evidence that this process was not of consequence in the evolution of a group. Carlquist [15,76] provides compelling arguments that hybridization was probably significant during the early radiation of island lineages because it provided for the segregation and recombination of the genetic variation taken to the island by the ancestor. This could be crucial to the successful radiation of a group when genetic variation is low. In *Tetramolopium* in Hawaii there is strong evidence that one taxon is of hybrid origin despite no evidence of contemporary gene flow between taxa [35] (T.K. Lowrey, personal communication).

Island plants offer certain advantages for studying speciation via the stabilization of interspecific hybrid derivatives. Because the species are often quite distinct morphologically, it is relatively easy to detect hybrids in the field; a particularly striking example of this is shown by Carr [77] for an intergeneric hybrid in the silversword alliance. In many instances, species occur in different habitats, yet may be spatially very close to one another; this not only makes it feasible for gene exchange to occur, but also the hybrid individuals may be found in this narrow area. If the hybrids have become established in niches not occupied by either parent, then it may be reasonably easy to discern this as compared to the study of more diffuse hybrid zones. At present, the role of hybridization in speciation on oceanic islands is not known with any degree of precision, but given the interfertility of many congeners, there seems little doubt that ample opportunities for gene exchange occur in certain groups. The evolutionary consequences of gene exchange need to be evaluated more critically.

12.7 The Origin and Radiation of Island Plants: More Specific Issues

One generalization that has been widely accepted is that the ancestor(s) of an endemic lineage on oceanic islands consisted of one to few diaspores, and thus little of the genetic diversity in the source population was dispersed to the island. Studies of allozyme diversity in species of a variety of genera have shown very low diversity compared to many continental taxa, although there are a number of notable exceptions. A review is in Crawford [78], and more recent results and discussion are given by Crawford et al. [79,80], de Joode and Wendel [81], and

Weller et al. [82]. Because lowered allozyme diversity is often associated with factors that otherwise lower genetic variation within populations, e.g., high or obligate selfing [83], it has often been tacitly assumed that allozyme loci are somewhat "representative" of genome diversity. This would seem to make sense for insular plants because the lowered allozyme diversity is explainable by loss associated with dispersal to the island, and presumably this would not be limited to allozyme loci. If this is indeed the situation, then genetically depauperate ancestors have given rise to an array of taxa that are more diverse morphologically and ecologically than their continental ancestors. This at first glance appears to be an anomalous situation, but there are several possible hypotheses to explain it. First, it is possible that the allozyme loci routinely surveyed in electrophoretic studies do not reflect the diversity present at or among gene loci controlling morphological, physiological, and other differences distinguishing endemic congeners. There is both experimental and theoretical evidence that this is the situation in some animals [84] despite the aforementioned correlations between allozyme diversity and other biological attributes affecting (or correlated with) population level diversity.

Another factor considered by Crawford et al. [66] is that some of the striking differences between insular species may have a more simple genetic basis than would be inferred based on experience with continental taxa. More specifically, the characters associated with speciation may be based on very few loci with major effects [20]. Whether this is as likely to occur as the accumulation of differences at many loci with small effects has been a matter of considerable discussion and it has centered on various aspects of the founder principle and founder effect speciation [84–87]. Provine [88], in particular, provides a discussion of the original concept of Mayr [89] and its modifications. A detailed consideration of the complex topic of the founder principle is beyond the scope of the present paper, but suffice it to say that the number of loci involved in speciation is quite speculative and subject to debate. It is profitable, however, to consider real data from plants. The comments by Doebley [37] on work done by him and co-workers on maize and its likely ancestor teosinte provide a balanced view of the situation based on real data rather than theory or speculation. The data from maize-teosinte indicate that the control of traits present in maize ranges from one major locus with several modifiers to a number of loci with rather small effects. Doebley [37] argues that this is what one might expect given that evolution is opportunistic and will "work" with whatever genetic variation or mutations are available. One important point from the maize research is that the view of Gottlieb [90], that a major locus with large effects may control differences between congeneric species, is quite reasonable; also, experimental results by Ford and Gottlieb [91] demonstrate that mutants with large effects need not have strong negative pleiotropic effects. It is possible, therefore, that these major loci may play a substantial role in divergence during speciation. If this is the case, then reduced diversity may not be a strong barrier to divergence.

Another factor that may allow more genetic variation to be carried by ancestral propagules to islands is polyploidy, and this seems particularly true in the Asteraceae. For example, *Bidens* [32] and the silversword alliance in Hawaii [33] are both polyploids, as are *Dendroseris* and *Robinsonia* on the Juan Fernandez [92] and *Scalesia* [93] from the Galapagos Islands. The presence of duplicated gene loci in polyploids could allow more alleles and higher heterozygosity to be carried to the islands. Electrophoretic studies of *Bidens* [94], the silversword alliance [26,27], *Dendroseris* [24], and *Robinsonia* [79] showed the expression of "extra" isozymes relative to the number expected in diploid flowering plants [95,96]. This demonstrates that these plants have not become diploidized, at least at a variety of isozyme loci, and the same may be true at many other loci. Reports by Carr [97] for various Hawaiian endemic taxa are notable for the high proportion with numbers indicating polyploidy. Lammers [98] provided counts for thirteen species in six genera of the Lobelioideae in Hawaii and concluded that the gametic number of $n = 14$ represents the polyploid condition.

Caution should be exercised, however, in generalizing about the significance of polyploidy in sequestering genetic variation in the founders of insular lineages. First, many groups have radiated successfully at the diploid level, including members of the Asteraceae such as *Argyranthemum* [22,23] and *Sonchus* and related genera in the Macaronesian Islands [99], and *Tetramolopium* [35] in Hawaii. Another reason for caution in interpreting the significance of polyploidy in the founding of new lineages is the possibility that polyploidy occurred after dispersal to the island. For example, Baldwin and Robichaux [100] noted the problem in identifying an appropriate extant polyploid ancestor for the silversword alliance; it is possible that the ancestor is extinct or that polyploidy occurred after dispersal. Each case must be evaluated individually with regard to the probable significance of polyploidy in the radiation of a lineage.

Ecological factors appear to isolate many congeneric insular endemics; indeed, the habitat diversity found in such groups as the silversword alliance [33,100], *Bidens* [68], *Tetramolopium* [35,101], and *Schiedea* and *Alsinidendron* [102] in Hawaii is truly remarkable. Despite these differences, very little research has been done either on the physiological ecology of the species or the genetic basis of features that appear highly adaptive in different species; a notable exception is the work of Robichaux and collaborators [103,104] on the silversword alliance in Hawaii. Related to the question of the genetic basis of seemingly adaptive features in species is how strongly species are adapted to the habitats in which they are now found. This question is worthy of study because one hypothesis that has been put forth for the morphological differences between congeneric endemics is that various recombinant types were originally able to become established because of the open habitats available on volcanic islands with lack of selecting agents such as competitors and pathogens [66]. This concept suggests that insular endemics evolved under relaxed selection, but the question remains largely unanswered as to how "fine-tuned" they are to the niches they now occupy.

Stochastic processes could operate in small founder populations such that the morphological and physiological features now seen in some species are not the result of strong selection (i.e., are not highly adaptive). This seems all the more possible because of the relaxed selection of the open habitats. There is no reason to believe that the situation is the same in all cases, but hard data are lacking. Some simple transplant experiments would be useful first steps, and given the dramatic differences in the habitats of various congeners, the experiments could be relatively straight-forward.

12.8 A Perspective and Future Studies

The renewed interest in speciation during the past 10–15 years has also produced considerable debate and discussion of species definitions [6–12] as well as theoretical discussions of speciation processes [13,105] and the role of experimental studies in understanding speciation [106,107]. These all represent topics worthy of discussion as part of the intellectual vitality of evolutionary biology. However, the focus of this chapter and the recommendations for future work involve extensive comparative studies of extant plant species as a means of understanding the process. The footprints are no doubt there if the proper taxa are chosen and the proper methods applied. As an example, recent dramatic progress in the study of recombinational speciation offers ample evidence that new analytical methods can provide compelling data on a process. Grant [108,109] used extensive experimental crossing studies to document that a new species could be synthesized via the recombination of features found in the parental taxa, and produced a model for the process. The question remained as to whether this process was feasible under natural conditions. Heiser [110,111], working with the genus *Helianthus*, proposed the hybrid origin of several taxa with the mechanism apparently following the Grant model. Rieseberg and collaborators [112–116] have provided elegant data supporting some of Heiser's hypotheses for recombinational speciation, and more importantly, genome mapping using molecular markers has confirmed the presence of parental chromosomal segments in the hybrid species, which supports Grant's model. Remarkably, one of the hybrid species has been reproduced experimentally [116]. The feasibility of recombinational speciation could be debated forever, but there is now experimental data that it has occurred in nature. How common is it and under what conditions does it occur? Only real data from real organisms will provide the answer.

Consider next the kinds of studies needed on island plants if speciation as a process is to be better understood, i.e., what should be done in the next century? In general, more extensive comparative studies of species pairs or groups of species are needed; among the best model studies of plants now available are those of Gottlieb and collaborators on *Stephanomeria* [16,117–120] and MacNair and others on *Mimulus* [121–123]. But even these exemplary studies fall somewhat short of what may be needed if refined insights are to be obtained.

The first part of any study must be selection of an appropriate group, and several desirable features were enumerated earlier. These include a diploid chromosome number, cultivation to flowering in several months, interfertility of interspecific hybrids, F_1 hybrids self-compatible, species occurring in contrasting habitats; and species differing by a variety of morphological features. Clearly, not all groups will possess all of these features, and trade-offs may be necessary. However, given the amount of effort going into such studies, it is important that the "nearly perfect" group be found.

Obtaining a robust phylogeny for the taxa is critical to the eventual success of the study. As indicated earlier, molecular data have proven useful for generating robust phylogenies for plants [19], and examples exist of the utility of DNA for resolving relationships among endemic species [30,100,124,125]. Data from several different molecular data sets as well as from morphological and anatomical characters should be used individually and in combination to arrive at the best phylogeny possible. Incongruences in data sets might in themselves be interesting in providing possible insights into processes such as hybridization [72].

Once the species have been selected, morphological and anatomical differences must be determined with care and precision. It should not be assumed a priori that some features are more "important" than others. These are the characters that will be studied genetically in the segregating F_2 hybrids.

The ecology of the two species should be studied in as much detail as possible (this, of course, depends on the time, money, and human resources available). There are general aspects of the ecology that should be investigated. First, the physical attributes of the habitat such as edaphic factors, precipitation amounts and patterns, elevation, and solar radiation must be examined. Secondly, the physiological diversity of the species themselves must be studied in great detail, and any discussion of this topic is beyond the scope of the present chapter. Suffice it to say that specific aspects of such general topics as water balance and photosynthetic rates must be examined. The genetic basis of these physiological differences should be examined in the same way as morphological and anatomical differences are studied. If the species appear to have specific pollinators, they need to be studied, particularly as they relate to differences in floral morphology.

The need for simple transplant experiments was mentioned earlier in the chapter, when it was suggested that plants of the two species could be reciprocally transplanted. The same experiments should be done except F_2 segregates could be planted in the two parental habitats to study survival of plants with different morphological and physiological characters.

Ultimately, it would be desirable to know not only the number of loci controlling differences between species but also the molecular-developmental basis of differences between species in morphological and physiological features. Ultimately, the striking morphological differences between insular endemics are the results of spatial and temporal changes in developmental events. These critical processes are not always easy to study, and this helps explain why so little is known about them [90,126]. One example of this is the MADS gene family which are floral homeotic genes controlling various aspects of floral development; discussion may be found in Doyle [127] and Purugganan et al. [128]. Molecular data were discussed earlier as suitable for generating phylogenetic hypotheses because the evolution of the DNA examined is likely neutral or nearly neutral, thus reducing the possibility of confounding common ancestry, and parallel evolution. At this later stage in the study of speciation, molecular-developmental data may be used to examine the basis of adaptive evolution. One aspect of these studies is to elucidate mechanisms at the molecular level that are the basis of morphological differences. This in turn entails examining both the structural and regulatory evolution of morphological differences between species. Insular endemics should be ideal systems for these studies because of their obvious and numerous morphological differences, and, because of their recent divergence, "tracks" should be present at the molecular-developmental level.

It is apparent from the present description of model speciation studies that one laboratory or one worker cannot possibly carry out the research. Rather, it must be a team effort with systematists, geneticists, ecologists, developmental and molecular biologists, and others involved. These will be long-term studies, and all who are involved must share the conviction that the results are of fundamental importance. Island plants offer some of the best possibilities for such studies, and it is only through such endeavors that major refinements can be made in understanding how such a marvelous and diverse array of plant species came into existence.

Acknowledgment. Research on plants of the Juan Fernandez Islands have been supported by the National Science Foundation through grants INT-7721637; BSR-8306436 and DEB-9500499 to the authors.

References

1. Otte D, Endler JA (eds) (1989) Speciation and its consequences. Sinauer, Sunderland
2. Coyne JA (1992) Genetics and speciation. Nature 355:511–515
3. Coyne JA (1994) Ernst Mayr and the origin of species. Evolution 48:19–30
4. Levin DA (1993) Local speciation in plants: the rule not the exception. Syst Bot 18:197–208
5. Mayr E (1993) Fifty years of progress in research on species and speciation. Proc Calif Acad Sci 48:131–140
6. Baum DA, Donoghue MJ (1995) Choosing among alternative "phylogenetic" species concepts. Syst Bot 20:560–573
7. Davis JI (1995) Species concepts and phylogenetic analysis—introduction. Syst Bot 20:555–559
8. Luckow M (1995) Species concepts: assumptions, methods and applications. Syst Bot 20:589–605
9. McDade LA (1995) Species concepts and problems in practice: insight from botanical monographs. Syst Bot 20:606–622
10. Olmstead RG (1995) Species concepts and pleisiomorphic species. Syst Bot 20:623–630
11. Rieseberg LH, Brouillet L (1994) Are many plant species paraphyletic? Taxon 43:1–32

12. Mallet J (1995) A species definition for the Modern Synthesis. Tr Ecol Evol 10:294–299

13. Gavrilets S, Hastings A (1996) Founder effect speciation: a theoretical reassessment. Am Nat 147:466–491

14. Carlquist S (1965) Island life—a natural history of the islands of the world. The Natural History Press, Garden City, New York

15. Carlquist S (1974) Island biology. Columbia University Press, New York

16. Gottlieb LD (1973) Genetic differentiation, sympatric speciation and the origin of a diploid species of *Stephanomeria*. Am J Bot 65:970–982

17. Lewis H (1973) The origin of diploid neospecies in *Clarkia*. Am Nat 107:161–170

18. MacNair MR, Cumbes QT (1989) The genetic architecture of interspecific variation in *Mimulus*. Genetics 122:211–222

19. Soltis PS, Soltis DE (1995) Plant molecular systematics: inferences of phylogeny and evolutionary processes. In: Hecht MK, Macintyre RJ, Clegg MT (eds) Evol Biol 28:139–194

20. Templeton AR (1981) Mechanisms of speciation—a population genetic approach. Ann Rev Ecol Syst 12:23–48

21. Francisco-Ortega J, Jansen RK, Crawford DJ, Santos-Guerra A (1995) Chloroplast DNA evidence for intergeneric relationships of the Macaronesian endemic genus *Argyranthemum* (Asteraceae). Syst Bot 20:413–422

22. Francisco-Ortega J, Crawford DJ, Santos-Guerra A, Sa-Fontinha S (1995) Genetic divergence among Mediterranean and Macaronesian genera of the subtribe Chrysantheminae (Asteraceae). Am J Bot 82:1321–1328

23. Francisco-Ortega J, Crawford DJ, Santos-Guerra A, Carvalho JA (1996) Isozyme differentiation in the endemic genus *Argyranthemum* (Asteraceae: Anthemideae) in the Macaronesian Islands. Pl Syst Evol 202:137–152

24. Crawford DJ, Stuessy TF, Silva O.M (1987) Allozyme divergence and the evolution of *Dendroseris* (Compositae: Lactuceae) on the Juan Fernandez Islands. Syst Bot 12:435–443

25. Crawford DJ, Stuessy TF, Cosner MB, Haines DW, Silva O.M, Baeza M (1992) Evolution of the genus *Dendroseris* (Asteraceae: Lactuceae) in the Juan Fernandez Islands: evidence from chloroplast and ribosomal DNA. Syst Bot 17:676–682

26. Witter MS, Carr GD (1988) Adaptive radiation and genetic differentiation in the Hawaiian silversword alliance (Composite: Madiinae). Evolution 42:1278–1287

27. Witter MA (1988) Duplicate expression of biochemical gene markers in the Hawaiian silversword alliance (Madiinae: Compositae). Biochem Syst Ecol 16:381–392

28. Li W-H (1993) So, what about the molecular clock hypothesis? Curr Opin Genet Devel 8:896–901

29. Carlquist S (1995) Introduction. In: Wagner WL, Funk VA (eds) Hawaiian biogeography—evolution on a hot spot archipelago. Smithsonian Institution, Washington, p 1

30. Givnish TK, Systma KJ, Smith JF, Hahn WJ (1995) Molecular evolution, adaptive radiation, and geographic speciation in *Cyanea* (Campanulaceae, Lobelioideae). In: Wagner WL, Funk VA (eds) Hawaiian biogeography—evolution on a hot spot archipelago. Smithsonian Institution, Washington, p 288

31. Humphries CJ (1979) Endemism and evolution in Macaronesia. In: Bramwell D (ed) Plants and islands. Academic, London, p 171

32. Gillett GW, Lim EKS (1970) An experimental study of the genus *Bidens* in the Hawaiian Islands. Univ Calif Publ Bot 56:1–63
33. Carr GD (1985) Monograph of the Hawaiian Madiinae (Asteraceae): *Argyroxiphium*, *Dubautia*, and *Wilkesia*. Allertonia. 4:1–123
34. Carr GD, Kyhos DW (1986) Adaptive radiation in the Hawaiian silversword alliance (Compositae: Madiinae). II. Cytogenetics of artificial and natural hybrids. Evolution 40:959–976
35. Lowrey TK (1986) A biosystematic revision of Hawaiian *Tetramolopium* (Compositae: Astereae). Allertonia 4:203–265
36. Fincham JR (1994) Genetic analysis—principles, scope and objectives. Blackwell Science, Oxford
37. Doebley J (1995) Genetics, development, and the morphological evolution of maize. In: Hoch PC, Stephenson AG (eds) Experimental and molecular approaches to plant biosystematics, pp 57–70. Monographs Syst Bot Missouri Bot Gard vol 53
38. Doebley J, Stec A (1991) Genetic analysis of the morphological differences between maize and teosinte. Genetics 129:285–295
39. Doebley JA, Stec A (1993) Inheritance of morphological differences between maize and teosinte: comparison of results for two F_2 populations. Genetics 134:559–570
40. Vlof EC, Van Houten WH, Mauthe S, Bachmann K (1992) Genetic and nongenetic factors influencing deviations from five pappus parts in a hybrid between *Microseris douglasii* and *M. bigelovii* (Asteraceae, Lactuceae). In J Pl Sci 153:89–97
41. Howard DJ (1993) Small populations, inbreeding, and speciation. In: Thornhill NW (ed) The natural history of inbreeding and outbreeding (theoretical and empirical perspectives). University of Chicago Press, Chicago, p 118
42. Grant V (1981) Plant speciation, 2nd edn. Columbia University Press, New York
43. Levin DA (1978) The origin of isolating mechanisms in flowering plants In: Hecht MK, Steere WC, Wallace B (eds), Evol Biol 11:185–317
44. Lewis H (1953) The mechanism of evolution in the genus *Clarkia*. Evolution 7:1–20
45. Lewis H, Roberts MR (1956) The origin of *Clarkia lingulata*. Evolution 10:126–138
46. Lewis H, Raven PH (1958) Rapid evolution in *Clarkia*. Evolution 12:319–336
47. Lewis H (1961) Experimental sympatric populations of *Clarkia*. Am Nat 95:155–168
48. Lewis H (1962) Catastrophic selection as a factor in speciation. Evolution 16:257–271
49. Stebbins GL (1950) Variation and evolution in plants. Columbia University Press, New York
50. Stebbins GL (1971) Chromosome evolution in higher plants. Arnold, London
51. Soltis DE, Soltis PS (1993) Molecular data and the dynamic nature of polyploidy. Crit Rev Pl Sci 12:243–273
52. Sahuquillo E, Lumaret R (1995) Variation in the subtropical group of *Dactylis glomerata* L. 1. evidence from allozyme polymorphism. Biochem Syst Ecol 23:407–418
53. Ownbey M (1950) Natural hybridization and amphiploidy in the genus *Tragopogon*. Am J Bot 37:487–499
54. Soltis DE, Soltis PS (1989) Allopolyploid speciation in *Tragopogon*: insights from chloroplast DNA. Am J Bot 76:1119–1124
55. Soltis PS, Soltis DE (1991) Multiple origins of the allotetraploid *Tragapogon mirus* (Compositae): rDNA evidence. Syst Bot 16:407–413

56. Soltis PS, Doyle JJ, Soltis DE (1992) Molecular and polyploid evolution in plants. In: Soltis PS, Soltis DE, Doyle JJ (eds) Molecular systematics of plants. Chapman and Hall, New York, p 177
57. Wagner WH (1954) Reticulate evolution in the Appalachian aspleniums. Evolution 8:103–118
58. Werth CR, Guttman SI, Eshbaugh WH (1985) Electrophoretic evidence for reticulate evolution in the Appalachian *Asplenium* complex. Syst Bot 10:184–192
59. Werth CR, Guttman SI, Eshbaugh WH (1985) Recurring origin of allopolyploid species in *Asplenium*. Science 228:731–733
60. King M (1993) Species evolution: the role of chromosome change. Cambridge University Press, Cambridge
61. Kyhos DW, Carr GD (1994) Chromosome stability and lability in plants. Evol Theor 10:227–248
62. Rieseberg LH, Ellstrand NC (1993) What can molecular and morphological markers tell us about plant hybridization? Crit Rev Pl Sci 12:213–241
63. Rieseberg LH, Wendel JF (1993) Introgression and its consequences in plants In: Harrison RG (ed) Hybrid zones and the evolutionary process. Oxford University Press, New York, p 70
64. Arnold ML, Bennett BD (1993) Natural hybridization in Louisiana irises: genetic variation and ecological determinants. In: Harrison RG (ed) Hybrid zones and the evolutionary process. Oxford University Press, New York, p 115
65. Arnold ML (1992) Natural hybridization as an evolutionary process. Ann Rev Ecol Syst 23:237–261
66. Crawford DJ, Whitkus R, Stuessy TF (1987) Plant evolution and speciation on oceanic islands. In: Urbanska K (ed) Differentiation patterns in higher plants. Academic, London, p 183
67. Carr GD (1987) Beggar's ticks and tarweeds: masters of adaptive radiation. Tr Ecol Evol 2:192–195
68. Mayer SS (1991) Artificial hybridization in *Wikstroemia* (Thymelaeaceae). Am J Bot 78:122–130
69. Borgen L (1976) Analysis of a hybrid swarm between *Argyranthemun adaucturn* and *A. fififolium* in the Canary Islands. Norweg J Bot 19:149–170
70. Brochmann C (1984) Hybridization and distribution of *Argyranthemun adauctum* (Asteraceae—Anthemideae) in the Canary Islands. Nord J Bot 4:729–736
71. Brochmann C (1987) Evaluation of some methods for hybrid analysis, exemplified by hybridization in *Argyranthemum* (Asteraceae). Nord J Bot 7:609–630
72. Francisco-Ortega F, Jansen RK, Santos-Guerra A (1996) Chloroplast DNA evidence of colonization, adaptive radiation, and hybridization in the evolution of the Macaronesian flora. Proc Natl Acad Sci USA 93:4085–4090
73. Pacheco P, Stuessy TF, Crawford DJ (1991) Natural hybridization in *Gunnera* (Gunneraceae) of the Juan Fernandez Islands, Chile. Pac Sci 4:389–399
74. Sanders RW, Stuessy TF, Marticorena C, Silva O.M (1987) Phytogeography and evolution of *Dendroseris* and *Robinsonia*, tree-composite of the Juan Fernandez Islands. Opera Bot 92:195–215
75. Ganders FR, Nagata KM (1984) The role of hybridization in the evolution of *Bidens* on the Hawaiian Islands. In: Grant WF (ed) Plant biosystematics. Academic, Toronto, p 179
76. Carlquist S (1980) Hawaii—a natural history. National Tropical Botanical Garden, Lawa, Hawaii

77. Carr GD (1995) A fully fertile intergeneric hybrid derivative from *Argyroxiphium sandwicense* ssp. *macrocephalum* × *Dubautia menziesii* (Asteraceae) and its relevance to plant evolution in the Hawaiian Islands. Am J Bot 82:1574–1581

78. Crawford DJ (1990) Plant molecular systematics: macromolecular approaches. John Wiley, New York

79. Crawford DJ, Stuessy TF, Haines DW, Cosner MB, Silva O.M, Lopez P (1992) Allozyme diversity within and divergence among four species of *Robinsonia* (Asteraceae: Senecioneae), a genus endemic to the Juan Fernandez Islands, Chile. Am J Bot 79:962–966

80. Crawford DJ, Stuessy TF, Rodriguez R, Rondinelli M (1993) Genetic diversity in *Rhaphithamnus venustus* (Verbenaceae), a species endemic to the Juan Fernandez Islands. Bull Torrey Bot Club 120:23–28

81. de Joode DR, Wendel JF (1992) Genetic diversity and origin of the Hawaiian Islands cotton, *Gossypium tomentosum*. Am J Bot 79:3911–3919

82. Weller SG, Sakai AK, Staub C (1996) Allozyme diversity and genetic identity in *Schiedea* and *Alsinidendron* (Caryophyllaceae: Alsinoideae) in the Hawaiian Islands. Evolution 50:23–34

83. Hamrick JL, Godt MJW (1989) Allozyme diversity in plant species. In: Brown AHD, Clegg MT, Kahler AL, Weir BS (eds) Plant population genetics, breeding, and genetic resources. Sinauer, Sunderland, p 43

84. Carson, HL (1990) Increased genetic variance after a population bottleneck. Tr Ecol Evol 5:228–230

85. Barton NH, Charlesworth B (1984) Genetic revolutions founder effects, and speciation. Ann Rev Ecol Syst 15:133–164

86. Barton NH (1989) Founder effect speciation. In: Otte D, Endler JA (eds) Speciation and its consequences. Sinauer, Sunderland, pp 229–256

87. Carson HL, Templeton AR (1984) Genetic revolutions in relation to speciation phenomena: the founding of new populations. Ann Rev Ecol Syst 15:97–131

88. Provine W (1989) Founder effects and genetic revolutions in microevolution and speciation: a historical perspective. In: Giddings LV, Kaneshiro KY, Anderson WW (eds) Genetics, speciation and the founder principle. Oxford University Press, New York, p 43

89. Mayr E (1942) Systematics and the origin of species. Columbia University Press, New York

90. Gottlieb LD (1984) Genetics and morphological evolution in plants. Am Nat 123:681–709

91. Ford VS, Gottlieb LD (1992) *bicalyx* is a natural homeotic floral variant. Nature 358:671–673

92. Sanders RW, Stuessy TF, Rodriguez R (1983) Chromosome numbers from the flora of the Juan Fernandez Islands. Am J Bot 70:799–810

93. Eliasson U (1974) Studies in Galapagos plants. XIV. The genus *Scalesia*. Arn Opera Botanica 36:1–117

94. Helenurm K, Ganders FR (1985) Adaptive radiation and genetic differentiation in Hawaiian *Bidens*. Evolution 39:753–765

95. Gottlieb LD (1982) Conservation and duplication of isozmes in plants. Science 216:373–380

96. Weeden NF, Wendel JF (1990) Genetics of plant isozymes. In: Soltis DE, Soltis PS (eds) Isozymes in plant biology. Dioscorides, Portland, p 46

97. Carr GD (1978) Chromosome numbers of Hawaiian flowering plants and the signifi-
    cance of cytology in selected taxa. Am J Bot 65:236–242
98. Lammers TG (1988) Chromosome numbers and their systematic implications in
    Hawaiian Lobelioideae (Companulaceae). Am J Bot 75:1130–1134
99. Kim S-C, Crawford DJ, Francisco-Ortega J, Santos-Guerra F (1996) A common
    origin for woody *Sonchus* and five related genera in the Macaronesian Islands:
    molecular evidence for extensive radiation. Proc Natl Acad Sci USA 93:7743–7748
100. Baldwin BG, Robichaux RH (1995) Historical biogeography and ecology of the
    Hawaiian silversword alliance (Asteraceae)—new molecular phylogenetic perspec-
    tives. In: Wagner WL, Funk VA (eds) Hawaiian biogeography—evolution on a hot
    spot archipelago. Smithsonian Institution, Washington, p 285
101. Lowrey TK (1995) Phylogeny, adaptive radiation, and biogeography of Hawaiian
    Tetramolopium (Asteraceae: Astereae). In: Wagner WL, Funk VA (eds) Hawaiian
    biogeography—evolution on a hot spot archipelago. Smithsonian Institution, Wash-
    ington, p 195
102. Wagner WL, Weller SG, Sakai AK (1995) Phylogeny and biogeography in *Schiedea*
    and *Alsinidendron* (Caryophyllaceae). In: WL Wagner, Funk VA (eds) Hawaiian
    biogeography-evolution on a hot spot archipelago. Smithsonian Institution, Wash-
    ington, p 221
103. Robichaux RH (1984) Variation in the tissue water relations of two sympatric Ha-
    waiian *Dubautia* species and their natural hybrid. Oceologia 65:75–81
104. Robichaux RH, Carr GD, Liebman M, Percy RW (1990) Adaptive radiation of the
    silversword alliance (Compositae: Madiinae): ecological, morphological, and physio-
    logical diversity. Ann Missouri Bot Gard 77:64–72
105. McCarthy EM, Asmussen MA, Anderson WW (1995) A theoretical assessment of
    recombinational speciation. Heredity 74:502–509
106. Rice WR, Hostert EE (1993) Laboratory experiments on speciation: what have we
    learned in 40 years? Evolution 47:1637–1653
107. Templeton AR (1996) Experimental evidence for the genetic transilience model of
    speciation. Evolution 50:909–915
108. Grant V (1966) Selection for vigor and fertility in the progeny of a highly sterile
    species hybrid in *Gilia*. Genetics 53:757–775
109. Grant V (1966) The origin of a new species of *Gilia* in a hybridization experiment.
    Genetics 54:1189–1199
110. Heiser CB (1947) Hybridization between the sunflower species *Helianthus annuus*
    and *H. petiolaris*. Evolution 1:249–262
111. Heiser CB (1958) Three new annual sunflowers (*Helianthus*) from the southwestern
    United States. Rhodora 60:272–283
112. Rieseberg LH (1991) Homoploid reticulate evolution in *Helianthus*: evidence from
    ribosomal genes. Am J Bot 78:1218–1237
113. Rieseberg LH, Beckstrom-Sternberg S, Doan K (1990) *Helianthus annuus* ssp. *texa-
    nus* has chloroplast DNA and nuclear ribosomal RNA genes of *Helianthus debilis* ssp
    cucumerifolius. Proc Natl Acad Sci USA 87:593–597
114. Rieseberg LH, Carter R, Zona S (1990) Molecular tests of the hypothesized hybrid
    origin of two diploid *Helianthus* species. Evolution 44:1498–1511
115. Rieseberg LH, Choi H, Chan R, Spore C (1993) Genomic map of a diploid hybrid
    species. Heredity 70:285–293
116. Rieseberg LH, Sinervo B, Linder CR, Ungerer MC, Arias DM (1996) Role of gene
    interactions in hybrid speciation: evidence from ancient and experimental hybrids.
    Science 272:741–745

117. Gottlieb LD (1977) Phenotypic variation in *Stephanomeria exigua* ssp. *coronaria* (Compositae) and its recent derivative species "Malheurensis". Am J Bot 64:873–880
118. Gottlieb LD (1979) The origin of phenotype in a recently evolved species. In: Solbrig OT, Jain S, Johnson GB, Raven PH (eds) Topics in plant population biology. Columbia University Press, New York, p 264
119. Gottlieb LD, Bennett JP (1983) Interference between individuals in pure and mixed cultures of *Stephanomeria malheurensis* and its progenitor. Am J Bot 70:276–284
120. Brauner S, Gottlieb LD (1989) Response to selection for time of bolting in *Stephanomeria exigua* ssp. *coronaria* and implications for the origin of *S. malheureusis* (Asteraceae). Sys Bot 14:516–524
121. MacNair MR (1983) The genetic control of copper tolerance in the yellow monkey flower, *Mimulus guttatus*. Heredity 50:283–293
122. Christie P, MacNair MR (1984) Complementary lethal factors in two North American populations of the yellow monkey flower. J Hered 75:510–511
123. Roberston AW, Diaz A, MacNair MR (1994) The quantificative genetics of floral characters in *Mimulus guttatus*. Heredity 72:300–311
124. Sang T, Crawford DJ, Stuessy TF, Silva O.M (1995) ITS sequences and the phylogeny of the genus *Robinsonia* (Asteraceae). Syst Bot 20:55–64
125. Sang T, Crawford DJ, Kim S-C, Stuessy TF (1994) Radiation of the endemic genus *Dendroseris* (Asteraceae) on the Juan Fernandez Islands: evidence from sequences of the ITS region of the nuclear ribosomal DNA. Am J Bot 81:1494–1501
126. Coyne J, Orr HA (1992) The genetics of adaptation—a reassessment. Am Nat 140:725–742
127. Doyle JJ (1994) Evolution of a plant homectic multigene family: toward connecting molecular systematics and molecular developmental genetics. Syst Biol 43:307–328
128. Purugganan MD, Rounsley SD, Schmidt RJ, Yanofsky MF (1995) Molecular evolution of flower development: diversification of the plant MADS-box regulatory gene family. Genetics 140:345–356

Relations of Environmental Change to Angiosperm Evolution During the Late Cretaceous and Tertiary

JACK A. WOLFE

13.1 Introduction

That morphology of the vegetative body of plants has been—and is continuing to be—shaped by environmental factors is a generally accepted concept. If so, then environmental change must be a significant factor in morphological change in plants. Morphological (used here in the broad sense to include anatomical) change can be equated to evolutionary change, at least relative to the vegetative body of plants.

Reproductive strategy, as mirrored in reproductive morphology, however, may be partially (or perhaps even largely) unrelated to environment. Cronquist [1] repeatedly emphasized that a given major taxonomic grouping of angiosperms occurs in a wide variety of environments; from this could be inferred that environment has played a small, or perhaps even no, role in plant evolution that led to divergences representing major taxonomic groupings, all of which are based on reproductive morphology.

Our understanding, however, of plant evolution may be strongly biased by the emphasis on reproductive strategy for delineation of higher taxonomic groupings. If, for example, the fossil record of major angiosperm categories is examined (Fig. 13.1), the conclusion might be drawn that the major evolutionary divergences in the angiosperms had all occurred by the end of the Cretaceous at about 65 million years (myr) ago and that evolutionary innovation has not occurred since that time. While the major divergences (at least from our present perspective) had indeed occurred by the end of the Cretaceous, innovations would continue.

How environmental, especially climatic, changes are related both to the major angiosperm divergences during the Late Cretaceous (~65 to 95 myr) and to the later innovations, especially during the Tertiary (~2 to 65 myr), is the major aim of this discussion; inextricably entwined, moreover, is historical biogeography of the angiosperms during the Late Cretaceous and Tertiary. Ideally, we would like to have floras represented by co-occurring organs (e.g., both vegetative and reproductive organs), floras from many different latitudes and continental posi-

Department of Geosciences, University of Arizona, Tucson, AZ 85721, USA

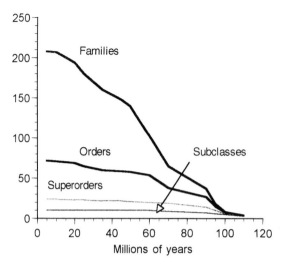

FIG. 13.1. Appearances of angiosperm taxa at various higher taxonomic levels from the mid-Cretaceous through the Tertiary. Compilation is based on both megafossil and microfossil citations in the literature; the taxonomic categories follow Takhtajan [87]

tions, and many floras in stratigraphic superposition. These ideal conditions are not present, and thus much of the discussion is necessarily speculative.

A number of environmental events have undoubtedly markedly affected the distribution of various angiosperm groups and hence their evolution. Among these are the separation of South America and Africa, which started during the Albian (~105 myr) and continues to the present day; the closing of the North American north-south epicontinental seaway near the end of the Cretaceous (~70 myr); the closing of the Eurasian north-south epicontinental seaway in western Siberia near the end of the Eocene (~35 myr); the temperature deterioration, also near the end of the Eocene; the separation of Australia from Antarctica during the Late Cretaceous and its significant northward drift during the early Tertiary; the floundering of the North Atlantic land bridge during the Eocene (~50 myr); the docking of India with Asia during the Oligocene (25–30 myr); and the emergence of the Panamanian isthmus near the end of the Miocene (~5 myr). Except for the temperature deterioration (see following), these events were, however, largely regional in scope.

13.1.1 Inferring Paleoenvironments

Correlation of climatic changes with evolutionary changes requires isolating inferences of one set of changes from the other. That is, if paleoenvironment is inferred from the taxonomic composition of a fossil assemblage, then we should not be inferring how the taxa in the fossil assemblage are related to environment; we have already assumed that the present taxon–environment relations were present during the time the fossil assemblage lived so as to arrive at the estimate of paleoenvironment. If Cronquist's [1] assertion of the apparent lack of taxon–environment relations at especially higher taxonomic levels is valid, then inferring paleoenvironments from taxa that are extinct, or only related to extant taxa at suprageneric levels, becomes fraught with pitfalls.

To avoid these pitfalls, neobotanists [2,3] have urged that paleobotanists, when attempting to infer paleoclimate, use character states of fossils that are related to environment [4]. Such character states are found in the vegetative parts of plants, especially in leaves (leaf physiognomy) and wood, and paleoclimatic inferences have been based on these organs [5–8]. Certainly evidence of general climatic regimes can be inferred from various character states, but any particular character state might result from moisture factors, temperature factors, or various combinations of both sets of factors. What is needed is a methodology that evaluates a given assemblage of fossils relative to modern assemblages that can be related to environmental factors with some degree of accuracy (validity); the methodology should also be precise (repeatable).

CLAMP (the Climate-Leaf Analysis Multivariate Program) is such a methodology [6,9]. Leaves of woody dicotyledons growing proximal to meteorological stations were sampled in a wide range of modern environments. A total of 31 rigorously defined character states of physiognomy were tabulated for a minimum of 20 species; these character states include form (i.e., lobed vs. unlobed), margin (toothed, and if so, characters of teeth), size, apical and basal configurations, length-to-width ratio, and general shape. At the present time, the database of the 150 modern samples is composed of only Northern Hemisphere, especially North American, samples; recently, however, K. Uemura has collected an extensive series of samples from Japan for addition to the CLAMP database.

TABLE 13.1. Inferred temperature parameters for some North American Late Cretaceous leaf assemblages

Flora	No. of species	Paleo-latitude °N	Age	MAT (°C)	CMMT (°C)	MART (°C)
Lance	35	55	MA	13.9	6.1	15.6
Medicine Bow	43	53	MA	18.8	14.0	9.6
Cooper Pit	36	42	MA	18.1	11.6	13.0
Perry Place	49	42	MA	20.7	14.6	12.2
Middendorf	47	38	CA	20.5	14.3	12.4
Gay Head	71	42	CA	22.5	18.3	8.4
North Slope	25	75	CO	12.5	5.7	13.6
Kamchatka	28	72	CO	9.0	0.0	18.0
Tuscaloosa	43	37	CO	21.5	17.3	8.4
Kamchatka	21	72	TU	6.9	−3.8	21.4
Novaya Sibir	20	82	TU	9.0	0.0	18.0
Woodbridge	56	36	CE	20.4	15.4	10.0
Woodbine	61	34	CE	21.4	16.5	9.8

MAT, mean annual temperature; CMMT, cold-month mean temperature; MART, mean annual range of temperature; CE, Cenomanian; TU, Turonian; CO, Coniacian; CA, Campanian; MA, Maastrichtian.
Data on Arctic assemblages after Herman and Spicer [12].

The sample tabulations are subjected to canonical correspondence analysis (CCA) [10,11], which is an ordination methodology widely used in ecology. A variety of meteorological data [e.g., mean annual temperature (MAT), cold-month mean temperature (CMMT), mean growing-season precipitation, relative humidity] are supplied to the analysis, which constrains the principal ordination axes with these parameters and yields vectors for each of the supplied parameters. Consequently, the various samples are approximately ranked by orthogonally projecting the sample plots to the vectors. The various parameters are predicted for the modern samples with varying degrees of accuracy; MAT and CMMT, for example, can be predicted with accuracy of less than 2°C and 3°C, respectively. Fossil samples, which have unknown meteorological parameters, are treated as passive samples and their ordination scores reflect positioning based on the 31 character states; projection of the plot of a given fossil sample to a meteorological parameter vector allows estimation of that parameter for the fossil assemblage. Because leaf physiognomy, as Herman and Spicer [12] aptly stated ". . . is controlled primarily by the physical laws of gas diffusion, fluid transport, and evaporation," CLAMP should be valid when applied to leaf assemblages of Tertiary and late Cretaceous age, in which leaves of woody dicotyledons are the dominant type of fossil. CLAMP is independent of the assignment to a higher taxonomic level of a given fossil species.

13.1.2 Geographic Coverage

To include all environmental changes that have impacted the angiosperms and to include paleobotanical data from all areas of the Earth here would inflate this discussion beyond reason. In many areas, especially those of the Southern Hemisphere, Late Cretaceous and Tertiary paleobotanical data are much more abundant than they were 10–20 years ago, but these data are still of less geographic coverage and of less stratigraphic continuity than Northern Hemisphere data. In the Northern Hemisphere, where is the best paleobotanical data coverage, I have concentrated on North America, partly because this area is most familiar to me but also because more Late Cretaceous and Tertiary floras have been recently analyzed using foliar physiognomy. Environmental change and its evolutionary impact can thus be better understood than in most other regions, although I have also relied, especially relative to the diversification of terrestrial herbs in the post-Eocene, on the Eurasian megafossil record.

13.2 Late Cretaceous

13.2.1 Late Cretaceous Diversification

Whatever the time of divergence of the angiosperms from their gymnosperm sister-group [13], the angiosperms first enter the fossil record in any abundance during the late part of the Early Cretaceous (~100–110 myr) and then only at low to middle latitudes. The spread of the angiosperms poleward, a phenomenon first hypothesized by Axelrod [14], occurred during the medial Cretaceous, and their assumption of dominance at high latitudes occurred during the Late Cretaceous [15–17].

Although many of the Early Cretaceous dicotyledonous leaves were attributed to extant families and genera by paleobotanists working in the nineteenth century, comparative studies involving extant dicotyledonous foliage indicate that these Early Cretaceous leaves represent a low evolutionary (unspecialized) grade now largely confined to Magnoliidae and especially the least specialized members of the magnoliids [18–20]. Because dicotyledons were not, insofar as the leaf record is concerned, diverse at the species level, analysis with CLAMP is not undertaken.

I consider much of the Cretaceous pollen record of putative extant taxa to be of dubious value. Although Muller [21] attempted to evaluate critically the Cretaceous pollen record, even in instances (e.g., *Pandaniidites*, which was thought by Muller to be definitely indicative of Pandanaceae) where relationships appeared to be well founded, further work by Stockey et al. [22] has shown the taxon to represent an extinct and unrelated clade of aquatics. Unless a particular Cretaceous palynomorph taxon can be demonstrated to be continuous both morphologically and geographically with an extant taxon, I have not accepted the Cretaceous occurrence as valid. For example, Muller [21] accepted as Leguminosae the pollen type called *Loranthacites* from the Cretaceous of Siberia; within the present range of the putatively similar *Sindora*, pollen of the extant type does not appear until the Neogene (~25 myr ago). These Cretaceous pollen grains of *Loranthacites* are much more likely modified from, and related to, the extinct triprojectate *Aquilapollenites*.

By the end of the Cenomanian (~91 myr), the dicotyledons were abundant and probably dominant in vegetation at middle paleolatitudes. Analyses (see Fig. 13.1) of two late Cenomanian leaf assemblages, the Woodbine from central Texas [23] and the Woodbridge from New Jersey [24], suggest that climate at about 35°N in eastern North America was very similar to present-day climate at about 30°N (e.g., central and northern parts of peninsular Florida) in both temperature and precipitation. That the Texas and New Jersey estimates are very similar is attributable to the rotation of North America; during the Cenomanian, New Jersey was only slightly north of central Texas [25]. This type of climate continued through the Turonian (~90 myr), as indicated by the Tuscaloosa assemblage from Alabama. Analyses of middle-, and especially high-, latitude leaf assemblages indicate very mild (i.e., low MART) climates.

The paleoclimatic inferences just discussed are partly different from those of Wolfe and Upchurch [26], who inferred that during the Late Cretaceous at middle latitudes of North America the climate was subhumid but precipitation was well distributed throughout the year. These earlier inferences were based on (1) analogy to present-day vegetation for Fiji and New Caledonia and (2) single-character physiognomic analyses (e.g., considering leaf margin type alone or leaf size alone). Further, we relied on the absence of ring porosity in associated fossilized dicotyledonous woods; this absence may be a reflection only of wood evolution (J. Chapman, oral communication, 1995).

The middle-latitude North American leaf assemblages of Cenomanian and Turonian age contain abundant and apparently diverse Magnoliidae [27,28] but are especially significant relative to origin and diversification of the Normapolles complex [29]. At least some of this complex includes the ancestors of at least some of the "higher Hamamelididae" or "Amentiferae," such as Juglandales as shown by Friis [30] and as inferred earlier [31]. Like those of their descendants, the Normapolles reproductive structures display specializations for wind pollination [30].

The Normapolles, which were characteristic of both eastern North America and Europe during much of the Late Cretaceous, were of low diversity and generally low abundance from the Cenomanian until the Santonian (~85 myr). From available data, the Normapolles originated in and achieved moderate diversification during a period of paratropical (MAT 20°–25°C) wet climate, a kind of climate that is typically characterized by closed-canopy forest. This type of climate is today occupied by woody plants that are primarily insect pollinated, although some wind-pollinated trees occur in the canopy, as emergents above the canopy, or in openings such as along rivers. For example, trees of the juglandaceous *Pterocarya* occur abundantly along the Mekong River in Laos, where the river flows through a region of broad-leaved evergreen paratropical rain forest.

The climates that are inferred to have supported closed-canopy, broad-leaved evergreen forest in eastern North America were replaced during the Campanian by vegetation that grew under climates that, from CLAMP analysis, had a marked dry season. Concomitant with this climatic change, the "simpler" Normapolles (i.e., the pollen type that now characterizes many of the higher Hamamelididae) replaced the older types of Normapolles [32]. By the early Maastrichtian (~75 myr), analysis of leaf assemblages such as those of the Ripley Formation (Perry Place and Cooper Pit) suggests vegetation very much like that of the "short deciduous forest" of present-day southern Sonora and Sinaloa; this does not imply, however, that the early Maastrichtian and present-day Sinaloan floras were similar but only that the short, open-canopy forests were physiognomically similar. Such vegetation could have favored diversification of wind-pollinated shrubs and trees, although many shrubs and trees of the modern short deciduous forest flower (as well as leaf out) during the wet season.

One aspect of these middle-latitude Late Cretaceous floras deserves emphasis: coniferous bisaccate pollen was common and at times even dominant in the palynofloras. The analogy to Southern Hemisphere megathermal (MAT >20°C) forests is apparent; trees of genera such as *Podocarpus* and *Agathis* can occur as emergents in such present-day wet forests. Even in modern megathermal dry vegetation some angiosperms mimic the physiognomy (especially foliar) of conifers; for example, *Jacquinia* (Theophrastaceae), which occurs in both the short deciduous forest and the Sonoran Desert.

At high latitudes, conifers and other gymnosperms played a major role in the vegetation until near the end of the Cretaceous [15–17], and many of these gymnosperms appear to have been deciduous. Thin leaf texture and taphonomy also suggest that deciduousness was the common habit among the woody angiosperms. The angiosperm deciduous trees and shrubs were, however, not of the Normapolles type but rather represent primitive stocks of Hamamelididae [33], such as the trochodendroids (Trochodendraceae, Tetracentraceae, Cercidiphyllaceae), hamamelids (Hamamelidaceae), and platanoids (Platanaceae). These primitive hamamelidids were diversifying in parallel to the "higher" hamamelidid (Normapolles) diversification at lower latitudes. The selection for deciduousness and other dormancy mechanisms could be related to winter cold, low winter light, or both [17].

Based on CLAMP analyses of Coniacian and Turonian leaf assemblages from the Arctic, Herman and Spicer [12] suggested that the major limiting environmental factor was light, because winter temperatures were generally sufficiently high to allow survival of broad-leaved evergreens. A CMMT as high as about 6°C is inferred at paleolatitude 75°N in areas influenced by the seaway that extended from low latitudes, through the western interior of North America, and into the Arctic. While some groups (e.g., platanoids) were shared between low and high latitudes, emphasized here is that high-latitude areas characterized by low winter light generally had different dominant groups than low-latitude areas characterized by seasonal rainfall. Present evidence indicates that the "higher hamamelidids" were not a significant element in high-latitude floras until the Tertiary [34].

13.2.2 Cretaceous–Tertiary Boundary

The most significant environmental and probably worldwide series of events in the history of the angiosperms was the disruption at the Cretaceous–Tertiary (K-T) boundary 65 myr ago. Unfortunately, in most regions of the world thus far investigated, time gaps of varying amounts are present in the depositional record, and thus what occurred in these regions is very speculative. The only region where deposition in nonmarine environments continued across the K-T boundary and where the fossil plant record has been studied is in the Western Interior of North America, from New Mexico north into Saskatchewan and Alberta. Even in the Western Interior, only some of the broader outlines of environmental change have been inferred at most localities, especially from palynological analyses.

At one K-T boundary site, Teapot Dome in Wyoming, the fossil plant record is particularly excellent and is being studied in detail by Margaret Collinson, Katherine Roucoux, and myself [35]. The record clearly indicates mass kill at the boundary among the aquatic plants. This evidence of mass kill occurs in beds that also contain spherules (altered microtektites), shock-metamorphosed quartz, and high levels of iridium. All strongly indicate that some layers in which the plants are preserved represent fallout from one or more bolide (e.g., meteoritic, cometary) impacts, one of which was surely the impact that caused the huge Chicxulub Crater in the Yucatan Peninsula of Mexico [36].

The major impact would have been followed by an "impact winter," a period of perhaps a few to several weeks of subfreezing temperatures by attenuation of solar radiation [37,38]. Our work thus far indicates that the "impact winter" occurred during the Northern Hemisphere growing season, so the greatest vegetational effects should have been in the Northern Hemisphere, with lesser effects in the Southern Hemisphere. Following the "impact winter," a greenhouse effect should have occurred, and paleobotanical data indicate that a major temperature increase occurred in the Western Interior [39]. Added to this, at least some areas that had supported subhumid vegetation during the latest Cretaceous now had very humid climate.

The profound environmental changes would have favored selection of some kinds of plants over other kinds during periods of a few to many hundreds or thousands of years. Indeed, aquatic plants and terrestrial plants with well-developed dormancy mechanisms (e.g., a deciduous habit) appear to have survived the K-T boundary well, as noted by Upchurch [40]. Although leaf physiognomy strongly suggests that some areas of the Western Interior had temperatures following the boundary events that would have selected for broad-leaved evergreens, these areas had an abundance of many kinds of deciduous dicotyledons. Apparently many broad-leaved evergreens in these areas became extinct, and the woody deciduous dicotyledons dominated by default.

Actual mass kill and extinction of some, perhaps many, kinds of plants should have occurred. Evaluation of the palynological record indicates that 20%–30% of the pollen "species" were victims [41,42], but, because pollen "species" can represent entire genera, the 20%–30% figure is probably generic-level extinction; actual species-level extinction may have been as high as 75% in some areas of the Western Interior [40,43].

Some plant extinctions, either at the K-T boundary or during the early Paleocene, represent Magnoliidae (e.g., Laurales were especially victimized). Other extinctions, however, occurred among nonmagnoliid dicotyledon groups of uncertain affinities. For example, most kinds of *Aquilapollenites*, a diverse palynomorph group that characterized northwestern North America and Siberia, failed to survive the K-T boundary events, but the botanical affinities of this pollen type within the nonmagnoliid dicotyledons are conjectural; Frederiksen [44] suggested that *Aquilapollenites* and some other associated pollen types represented herbs.

13.3 Tertiary

13.3.1 Paleocene Recovery and Diversification

Most of the "classic" Paleocene leaf floras, such as the Fort Union [45], Atanekerdluk [46], and Mull [47], represent vegetation at paleolatitudes of 45° to 60°N. These floras are generally of low diversity, as are floras known from higher

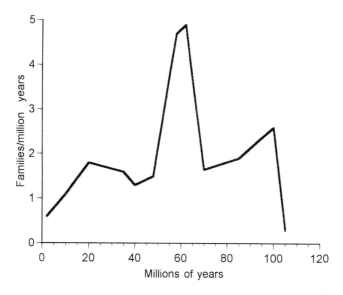

FIG. 13.2. Rates of origination based on first appearances of angiosperm families from the mid-Cretaceous through the Tertiary. Rates decline during the Late Cretaceous then markedly increase across the Cretaceous–Tertiary boundary (65 million years [myr]). Rates again decline during the warm and mild Eocene and then again increase during the Neogene (1.8–23 myr), largely from diversification of terrestrial herbs

paleolatitudes in Siberia, Alaska, and Canada (e.g., the Ravenscrag flora de-scribed by McIver and Basinger [48]). At paleolatitudes of 30° to 45°N, represen-tative leaf floras are those of the type Paleocene in Belgium and France [49], the Midway and lower Wilcox in the southeastern United States [50], and the upper Raton in New Mexico and Colorado [51]. The Ravenscrag, Atanekerdluk, lower Fort Union (Tullock), Midway, and upper Raton floras are all of early Paleocene (Danian) age, but the others mentioned either have imprecise dating or are of late Paleocene age. Note that the Paleocene European plant megafossil record, which comprises the type Paleocene and Isle of Mull leaf floras and the primarily diaspore flora of the Reading Beds [52–54], is entirely of late Paleocene age.

The early Paleocene floras—relative to either microfossils or megafossils—appear to be of low diversity, especially at the level of species. During the later Paleocene, many taxa were added to the terrestrial flora, and, by the early Eocene, floral diversity appears to be well beyond the Maastrichtian level. Orig-ination of new families (Fig. 13.2) during the Paleocene proceeded at a very high rate. This phenomenon probably resulted from (1) diversification in response to filling in ecological niches vacated by extinction of Late Cretaceous taxa, (2) diversification in response to niche filling in the greatly expanded rain forest environments, (3) diversification into seasonally dry environments that expanded near the end of the Paleocene, and (4) an increase of about 25% in land area from the Maastrichtian to the Paleocene [25,55]. Further, as the thermal optimum of the early Eocene was approached, megathermal rain forests would also have expanded. Studies of Paleocene floras should prove significant in understanding the evolutionary relationships of especially megathermal clades. Caesalpinoid Leguminosae, for example, appear in the pollen record of now-tropical areas in the late Paleocene [56], as well as in the megafossil record in southeastern North America (paleolatitude ~30°N) at about the same time [57].

13.3.2 Eocene Warmth

During the early Eocene (~50–54 myr ago), temperatures reached high levels [58], and megathermal rain forests reached their greatest poleward extent [59]; in some coastal areas, megathermal (or marginally so) vegetation occurred at paleolatitudes as high as 70°N. During this warm period Australia was still less than 10° of latitude from Antarctica as was South America from Antarctica; some floristic interchange between austral megathermal areas might have been possible, but most of Antarctica was too far south to have provided megathermal environments. In the Northern Hemisphere, however, narrow gaps separated Europe from Greenland and, in turn, from North America equatorward of 60°N, and these land masses were joined by continuous land at about 70°N; Tiffney [60,61] emphasized the North Atlantic route as the most probable migratory route for megathermal clades during the early Eocene, although the Beringian region was also marginally megathermal, as demonstrated by Budantsev [62]. Rather than the present-day floristic distinction between the paleotropics and neotropics, the Eocene distinction was between the Northern and Southern hemispheres [63]. The widespread boreotropical flora has been erroneously called a "paleotropical" flora; although many Eocene boreotropical taxa still survive in southeastern Asia, some also survive in Central and South America [64].

The early Eocene megathermal rain forest flora is particularly well known from the classic work of Reid and Chandler [65] and Chandler [53] on the diaspores of the London Clay of southern Britain. Noteworthy is that in some groups, such as Menispermaceae, not only are many extant genera represented but so also are extinct genera; i.e., the Menispermaceae may have been more diverse than at present. This does not necessarily imply that early Eocene megathermal rain forest was collectively more diverse than today, although the great geographic extent of megathermal rain forest during the early Eocene might favor high diversity. The higher diversity of some families, such as Menispermaceae, in part reflects the fact that Magnoliidae and Ranunculidae played more important roles in Eocene megathermal rain forests than, for example, woody Asteridae, which were just starting to diversify.

The early Eocene thermal maximum had a profound effect on microthermal (MAT <13°C) vegetation, because this vegetation became geographically highly restricted. As noted earlier, the Late Cretaceous high-latitude microthermal flora was largely composed of groups such as lower Hamamelididae, and this dominance continued into the late Paleocene and probably into the early Eocene. By the late early and early middle Eocene (~48–50 myr), however, a significant cooling occurred concomitant with the appearance of diversity in now-microthermal clades, e.g., Betulaceae, Ulmaceae, Rosaceae, and Aceraceae. These clades were not, however, appearing primarily at high latitudes at this time but at middle latitudes in what were upland areas. Two major environmental events, the cooling at the end of the early Eocene thermal optimum and the uplift of a large part of western North America during the early Eocene, combined to create an area of diverse microthermal habitats, an area of rugged topography, including many different soil types (but especially volcanic), and bounded by lower altitude mesothermal vegetation. Although other Northern Hemisphere, middle-latitude areas had some uplands, no evidence suggests that these were as great in geographic extent or as high (>2–3 km) as those in western North America. Thus, in western North America the Eocene upland floras contain an ever-increasing sectional and specific diversity in genera such as *Acer* [66] and an ever-increasing generic and specific diversity in families such as Rosaceae.

After the cooling near the early to middle Eocene boundary, the middle and late Eocene saw two renewed warm intervals (~43–46 myr and 34–37 myr) separated by a cool interval (~38–42 myr). The middle Eocene warm interval was not as warm as the early Eocene. In areas such as southeastern Alaska, vegetation was mesothermal with no exclusively megathermal floristic elements, and, by analogy, the North Atlantic land connection was at too high a paleolatitude for easy interchange of megathermal floristic elements. Possible long-distance dispersal cannot, of course, be eliminated, but the similarities of the middle Eocene Clarno diaspore flora in Oregon [67] to the early Eocene London Clay diaspore flora most probably reflect early Eocene dispersals.

A major change in precipitation distribution and intensity occurred in low-middle latitude areas by the middle Eocene. Although precipitation appears to have been abundant during the Paleocene (and probably early Eocene) in areas such as southeastern North America and southern China, by the middle Eocene these areas had markedly seasonally dry climate [6,68,69]. Because of the present-day proclivity of woody Leguminosae in dry megathermal climates it is not surprising that this family, which had appeared by the late Paleocene in southeastern North America, showed an increasing diversity in this region as climates dried during the first half of the Eocene. This largely megathermal family (at least relative to the woody members) also had expanded into microthermal climates by the end of the Eocene [70].

The Northern Hemisphere high-latitude flora of the Paleocene and Eocene gives no support to the concept of an "Arcto-Tertiary Geoflora," that is, a group of early Tertiary communities that evolved in the Arctic and gradually migrated to middle latitudes as climate cooled during the mid-Tertiary. Note that Engler [71], who first used the term "arcto-tertiär," was referring only to the groups of trees and shrubs that now inhabited middle latitudes of the Northern Hemisphere; only later did paleobotanists propose that these groups had been long associated in the Arctic and subsequently migrated equatorward. The arcto-tertiary element in Arctic floras does not appear significant until the middle Eocene at about 45 myr, some 5–7 myr after the high-altitude, middle-latitude appearance discussed previously. Generally, members of the arcto-tertiary element in low-altitude, middle-latitude floras are more likely to be related to the upland arcto-tertiary plants than to the high-latitude arcto-tertiary plants.

13.3.3 "Terminal Eocene" Events

Epochal terms such as Eocene or Oligocene are largely based on local gaps in the stratigraphic record in regions such as the Paris Basin, and these gaps may have little or no relevance to major climatic events. This is certainly valid relative to the Eocene–Oligocene boundary, because a major temperature deterioration occurred at about 33 myr, 1 myr after the end of the Eocene as defined on the basis of the rock stratigraphic record and thus in the earliest Oligocene. Other significant events during the Oligocene were a drop of about 70 m in the world-wide sea level [55], which may be correlated with a significant accumulation of ice on Antarctica, and the collision of the Indian tectonic plate with the Asian continent. Further, the north-south seaway just east of the Urals became dry land by 30 myr [25].

The temperature deterioration at 33 myr was major in North America, especially at middle to high latitudes. Mean annual temperature dropped approximately 6°–8°C [72,73]. The main temperature component to decline was winter temperature, and this markedly affected the flora and vegetation. Megathermal vegetation was affected especially in North America. The seasonally dry megathermal vegetation of southeastern North America [68] would have been eliminated from that region, and the palynological work by Ager (Owens et al.) [74] indicates disappearance of the Eocene megathermal taxa and vegetation and their replacement by a forest dominated by oaks, presumably a vegetation type not radically different from the present-day vegetation. In western North America, wet, megathermal broad-leaved evergreen forest that extended to about 50°N disappeared; in the lowlands of the Pacific Northwest, some broad-leaved evergreens survived, but the vegetation was dominantly deciduous, especially with even a moderate increase in elevation. Mesothermal broad-leaved evergreen forest during the late Eocene and earliest Oligocene had extended to 60°N and some broad-leaved evergreens had occurred in vegetation as far north as almost 70°, but after the deterioration, broad-leaved evergreens (except for very small leaved plants) were generally restricted to latitudes equatorward of 50°N.

Extinction of lineages was very high at high latitudes as a result of the 33-myr deterioration, but, although significant, extinction was lower at middle latitudes (Fig. 13.3). Postdeterioration Oligocene floras are unknown in areas such as central and southern California, but the fact that some predeterioration, broad-leaved evergreen lineages reappear in late Oligocene floras along the Oregon coast [75] and in Neogene floras such as the Weaverville in northern California [76] indicates that the wet broad-leaved evergreen vegetation probably survived to the south. However, most of the wet megathermal taxa (e.g., *Paleophytocrene*, mastixioids) that had occurred in Eocene and earliest Oligocene floras in western North America did not recur when climate again warmed during the late Oli-

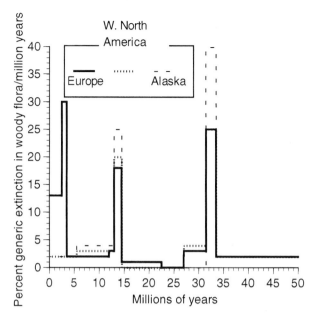

FIG. 13.3. Rate of generic regional extinction in the woody angiosperm flora of selected middle- to high-latitude regions of the Northern Hemisphere. Graph is based largely on published sources, but, especially for Alaska, includes unpublished data. The extinction rates at the 33-myr and 13-myr events are based on the assumption that the extinctions occurred during 2-myr periods

gocene and the middle Miocene. Although Mai [77] has emphasized the reappearance of the Paleogene "paleotropical" element in the European Miocene warm intervals, some of the most characteristic taxa (e.g., members of Icacinaceae) did not reappear and were presumably eliminated from Europe by the 33-myr deterioration, just as analogous taxa were eliminated from western North America.

At least 50% of the woody dicotyledon genera were eliminated at middle latitudes in western North America and Europe; note that Fig. 13.3 is based on an assumption that the 33-myr deterioration occurred during a period of 2 myr and portrays the rate of extinction for that period. Some typically Eocene genera (e.g., *Paleophytocrene*, *Florissantea*) survived the 33-myr deterioration but then became extinct within a few more million years.

The major decrease in winter temperatures at middle latitudes in the Northern Hemisphere at 33 myr created large land areas for which broad-leaved deciduous forest would be the best-adapted physiognomy. The flora for these broad-leaved deciduous forests was largely derived from the lineages that had been understory trees and shrubs in the dominantly coniferous forests of the Eocene uplands. However, as emphasized elsewhere [70], many of the broad-leaved deciduous lineages of the upland Eocene flora also became extinct. For example, sections of *Acer* (e.g., Macrantha, Platanoidea) that had originated in the western American

Eocene uplands disappeared from western North America and survived only in Eurasia, although some reentered North America during the Miocene.

13.4 Post-Eocene

13.4.1 Diversification of Terrestrial Herbs

As might be expected, angiospermous aquatic herbs are well represented as megafossils during the Tertiary, and several families (e.g., Nymphaeaceae, Nelumbonaceae) extend back into the Late Cretaceous. In contrast, one of the major lacks in the megafossil record of the angiosperms is the almost total lack of terrestrial herbs. Taphonomic settings that preserve megafossils of terrestrial herbs are notable for their rarity [78], and our knowledge of extinct herbaceous groups of Late Cretaceous and Paleogene age is almost nonexistent. The palynological record of the Late Cretaceous and Paleogene contains many angiosperm taxa of unknown affinities other than that these taxa were nonmagnoliid, and many of these could represent terrestrial herbs. Ferns may have occupied many of the terrestrial herbaceous habitats during the Late Cretaceous and Paleogene; ferns and angiosperms, however, today share these habitats, and they also did in at least one Late Cretaceous setting [78].

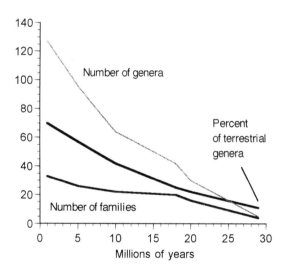

FIG. 13.4. Increase in diversity among extant families and genera of terrestrial angiosperm herbs following the major increase in winter temperatures at 33–34 myr. This compilation is based only on the Eurasian published diaspore record. Note the rate increase following the cooling at 14–14 myr and the drying at 5–6 myr. The herbs in microthermal climates became more diverse in genera than the woody members during the late Miocene (5–10 myr)

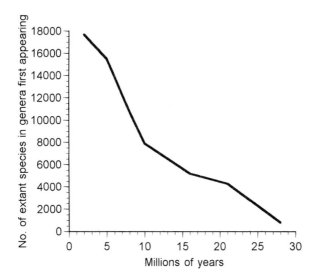

Fɪɢ. 13.5. Number of extant species in genera of terrestrial angiosperm herbs first appearing during the post-Eocene based only on the Eurasian published diaspore record. This does not represent the actual number of species known in the diaspore record. The genera that now have the largest number of species appeared largely during the late Miocene and Pliocene (1.8–10 myr)

Not until the Oligocene—after the 33-myr deterioration—do angiosperm genera and families that are now exclusively or predominantly composed of terrestrial herbs appear. Some of these records are palynological, for example, the well-documented appearance of pollen of Compositae near the Oligocene–Miocene boundary (~23 myr ago). The palynological record [21] does record the general expansion in both diversity and numbers of terrestrial herbaceous angiosperms during the Oligocene and Neogene, but probably the most striking documentation of this expansion is in the fruit and seed record.

Fossil fruits and seeds constitute much of the fossil record of the European Tertiary, and diaspores are also abundant in the post-Eocene sediments of Asia, especially of Russia. Unlike most pollen, fruits and seeds can typically be determined to the specific and generic levels and thus provide an excellent index of diversification. What is notable in the Eurasian diaspore record (Figs. 13.4, 13.5) is the low diversity of extant terrestrial herbaceous taxa in the Oligocene but the ever-increasing diversity through the Neogene. This diversification occurred in many families, subfamilies, and genera, so that an obvious conclusion is that the diversification was environmentally driven. The diversification continued unabated during periods of renewed warming.

Winter cold became widespread at middle and high latitudes of the Northern Hemisphere after the 33-myr deterioration, and this factor was probably the most significant in the early diversification of the extant terrestrial herb groups. What better way to survive cold winters than to become herbaceous? The extant perennial herbaceous groups were present in the Oligocene, while the extant annual herbaceous groups began to appear by the early Miocene.

Additional factors that promoted the herbaceous habit were added during the Neogene. Some drying in continental interiors was apparently associated with the 33-myr deterioration [79,80], and this would also select for the herbaceous habit. However, the collision of the Indian tectonic plate with Asia during the Oligocene and consequent elevation of the Himalayas during the Neogene would create large areas that were both dry and cold to the north of the Himalayas, and both factors would have accelerated the development and diversification of terrestrial herbs.

Climates favoring tundra vegetation would, of course, have been increasing in area during the Neogene, especially during the later Neogene as summer temperatures cooled at high latitudes [81]. High altitudes at low-middle- to middle-latitude areas may also have played a role in the evolution of tundra herbs. For example, the subalpine vegetation in southern Colorado during the late Oligocene contains at least one taxon—a *Potentilla* (Rosaceae)—that has affinities to taxa in alpine and Arctic vegetation of Canada, and also present are leaves of a presumably woody plant that appears ancestral to the extant subalpine and alpine perennial herb *Luetkea* (Rosaceae).

13.4.2 Neogene Climates

Notable climatic events of the Neogene are (1) the climatic optimum (warming) that started by the end of the early Miocene and continued through the early part of the middle Miocene (~14–18 myr), (2) cooling at about 13 myr, probably concomitant with marked growth in the Antarctic ice sheet, (3) the drying associated with temporary closure of the Mediterranean (Messinian event, ~5–6 myr), and (4) the first major Northern Hemisphere glaciation at 2.5 myr during the Pliocene.

The mid-Miocene climatic optimum allowed vegetation that had a large, if not dominant, broad-leaved evergreen element to spread into middle latitudes in the Northern Hemisphere. At the same time, some broad-leaved deciduous arcto-tertiary taxa were able to spread across Beringia. As noted earlier, some sections of *Acer* were able to attain a bicontinental distribution, as did some species of *Fagus*. Although many arcto-tertiary disjunctions are associated with the 33-myr deterioration, some are associated with the 13- to 14-myr deterioration in the middle Miocene.

The middle Miocene saw the end of widespread broad-leaved deciduous forests in high-middle and low-high latitudes of the Northern Hemisphere. As well, broad-leaved evergreen forests became restricted to low-middle and low latitudes. Regional generic-level extinctions were high; about 40% of the woody broad-leaved genera disappeared from Europe and western North America, extinctions that occurred over about 2 myr. At high latitudes, the extinction levels were even higher, and the woody flora began to take on a much more modern aspect. At high latitudes, where dominantly coniferous forest, especially boreal forest (taiga), became widespread and tundra also expanded, woody genera such as *Salix* diversified to fill many new niches.

In western North America, extinction continued at a moderate rate during the later Miocene, in part because of decreasing growing season precipitation. Many of the woody genera eliminated from western North America during the later Miocene and the Pliocene remained in Europe. However, the first widespread Northern Hemisphere glaciation (~2.5 myr) eliminated many of these genera from Europe (see Fig. 13.3). The glaciations of the Quaternary (0–1.8 myr) continued to eliminate woody genera, but, because so many had been eliminated by the first glaciation during the Pliocene, the extinction rate was lower.

13.5 Conclusions

The groups of angiosperms at the level of subclass and superorder diversified during the Late Cretaceous, a period characterized by climates that were generally mild. After the medial Cretaceous, appearance of new orders occurred at an almost uniform rate. After an initially rapid diversification of families, during the Late Cretaceous the rate of familial diversification tapered off, but the rate again soared following the Cretaceous–Tertiary boundary (65 myr). The rate again declined during the warm and mild Eocene but increased following the Eocene, largely from the origination of herbaceous families.

Many events, both climatic and geographic, have occurred during the history of the angiosperms. The geographic events especially represent trends over millions of years, but some climatic events, such as those at the Cretaceous–Tertiary boundary, represent short-term changes of weeks or months. The well-documented climatic oscillations of the Pliocene and Quaternary occurred over periods of thousands of years. Each event or series of events requires careful evaluation and documentation as to its intensity in different regions of the earth, as well as inferences relative to the part of the biota most affected.

The angiosperms have responded in different ways to these different events. In some instances (e.g., the Cretaceous–Tertiary boundary events), many extinctions resulted followed by a prolonged recovery characterized by major diversification at the level of genus or even family to fill niches left vacant by extinction. In yet other instances (e.g., the 33-myr deterioration), the angiosperms responded by a gradual but major diversification within a given life form, that of terrestrial herbs. These events resulted in both major extinctions and major diversifications.

The first Northern Hemisphere glaciation of the Pliocene eliminated many genera in regions such as Europe, but most genera survived in a limited fashion in regions such as eastern Asia. Whether the woody angiosperms can respond to these more recent climatic events by renewed diversification is, of course, unknown. Overall, as the woody angiosperms decreased in diversity during the Oligocene and Neogene, the herbaceous angiosperms increased in diversity, possibly at an even higher rate than the decrease in woody diversity.

Simplistic concepts of mass movements of communities during long periods of time, with little floristic change, must be viewed as rejecting evolution. Recall that the geofloral concept [82] envisions groups of communities that originated in particular regions and, over tens of millions of years, the lineages constituting these communities have maintained their associations while migrating over thousands of kilometers. Thus, the "Arcto-Tertiary Geoflora" supposedly originated in the early Tertiary (or even Late Cretaceous) of the Arctic region and gradually migrated southward, where it is found during the Neogene at middle latitudes; similarly, the "Paleotropical Tertiary Geoflora" supposedly originated in the paleotropics (especially southeastern Asia), migrated northward into Europe by the Eocene, and, with some repeated reintroductions into middle latitudes during Miocene warm intervals, gradually became restricted to southeastern Asia. This concept ignores paleogeographic knowledge: southeastern Asia and Africa were separated during the Late Cretaceous and most of the Tertiary by Tethys, and southeastern Asia was also separated from Europe during the Late Cretaceous, as well as during the early Eocene thermal optimum until the end of the Eocene. More significantly, when viewed in detail in light of the known fossil record of Asia [83], Europe [84], Central America [85], or North America [86], the development of the flora in each of these areas proves to be historically complex and related to both geographic and climatic changes.

References

1. Cronquist A (1981) An integrated system of classification of flowering plants. Columbia University Press, New York
2. Bailey IW, Sinnott EW (1915) A botanical index of Cretaceous and Tertiary climates. Science 41:831–834
3. Richards PW (1952) The tropical rain forest. Cambridge University Press, Cambridge
4. Spicer RA (1989) Physiological characteristics of land plants in relation to climate through time. Trans R Soc Edinb 80:321–329
5. Dolph GE, Dilcher DL (1979) Foliar physiognomy as an aid in determining paleoclimate. Palaeontogr Abt B Palaeophytol 170:151–172
6. Wolfe JA (1993) A method for obtaining climatic parameters from leaf assemblages. U.S. Bull 2040, Geological Survey, Washington, DC
7. Dilcher DL (1973) A paleoclimatic interpretation of the Eocene floras of southeastern North America. In: Graham A (ed) Vegetation and vegetational history of northern Latin America. Elsevier, Amsterdam, pp 39–59
8. Wheeler EF, Baas P (1991) A survey of the fossil record for dicotyledonous wood and its significance for evolutionary and ecological wood anatomy. Int Assoc Wood Anat Bull 12:275–332
9. Wolfe JA (1995) Paleoclimatic estimates for Tertiary leaf assemblages. Annu Rev Earth Planet Sci 24:119–142
10. ter Braak CJF (1992) CANOCO—a FORTRAN program for canonical correspondence ordination. Microcomputer Power, Ithaca, NY
11. ter Braak CJF, Prentice IC (1988) A theory of gradient analysis. Adv Ecol Res 18:271–371

12. Herman AB, Spicer RA (1996) Palaeobotanical evidence for a warm Cretaceous Arctic Ocean. Nature 330:330–333

13. Doyle JA, Donoghue MJ (1986) Seed plant phylogeny and the origin of angiosperms: an experimental cladistic approach. Bot Rev 52:321–431

14. Axelrod DI (1952) A theory of angiosperm evolution. Evolution 6:29–30

15. Crane PR (1987) Vegetational consequence of the angiosperm diversification. In: Friis EM, Chaloner WG, Crane PR (eds) The origin of angiosperms and their biological consequences. Cambridge University Press, Cambridge, pp 107–144

16. Crane PR (1989) Paleobotanical evidence on the early radiation of nonmagnoliid dicotyledons. Plant Syst Evol 162:165–191

17. Upchurch GR, Wolfe JA (1993) Cretaceous vegetation of the Western Interior and adjacent regions of North America. Geol Assoc Canada Spec Pap 39:243–281

18. Hickey LJ, Wolfe JA (1975) The bases of angiosperm phylogeny: vegetative morphology. Ann MO Bot Gard 62:538–589

19. Wolfe JA, Doyle JA, Page VM (1975) The bases of angiosperm phylogeny: paleobotany. Ann MO Bot Gard 62:801–824

20. Doyle JA, Hickey LJ (1976) Pollen and leaves from the mid-Cretaceous Potomac group and their bearing on early angiosperm evolution In: Beck CB (ed) Origin and early evolution of angiosperms. Columbia University Press, New York, pp 139–206

21. Muller J (1981) Fossil pollen records of extant angiosperms. Bot Rev 47:1–142

22. Stockey RA, Hoffman GL, Rothwell GW (1997) The fossil monocot *Limnobiophyllum scutatum* (Dawson) Krassilov: resolving phylogeny of the Lemnaceae. Am J Bot 84:355–368

23. MacNeal DL (1958) The flora of the Upper Cretaceous Woodbine Sand in Denton County, Texas. Monogr Acad Nat Sci Phila 10:1–152

24. Newberry JS (1895) The flora of the Amboy clays. Monograph 26. U.S. Geological Survey, Washington, DC

25. Smith AG, Smith DG, Funnell BM (1994) Atlas of Mesozoic and Cenozoic coastlines. Cambridge University Press, Cambridge

26. Wolfe JA, Upchurch GR (1987) North American nonmarine climates during the Late Cretaceous. Palaeogeogr Palaeoclimatol Palaeoecol 61:33–77

27. Crepet WL, Friis EM, Nixon KC (1991) Fossil evidence for the evolution of biotic pollination. Philos Trans R Soc Lond B Biol Sci 333:187–195

28. Upchurch GR, Dilcher DL (1990) Cenomanian angiosperm leaf megafossils, Dakota Formation, Rose Creek locality, Jefferson County, southeastern Nebraska. U.S. Bull 1915, Geological Survey, Washington, DC

29. Doyle JA (1969) Cretaceous angiosperm pollen of the Atlantic Coastal Plain and its evolutionary significance. J Arnold Arbor Harv Univ 50:1–35

30. Friis EM (1983) Upper Cretaceous (Senonian) floral structure of juglandalean affinity containing *Normapolles* pollen. Rev Palaeobot Palynol 39:161–188

31. Wolfe JA (1974) Fossil forms of Amentiferae. Brittonia 25:334–355

32. Wolfe JA (1976) Stratigraphic distribution of some pollen types from the Campanian and lower Maestrichtian rocks (Upper Cretaceous) of the Middle Atlantic States. U.S. Prof pap 977, Geological Survey, Washington, DC

33. Herman AB (1994) Late Cretaceous Arctic platanoids and high latitude climate. In: Boulter MC, Fisher HC (eds) Cenozoic plants and climates of the Arctic. Springer, Berlin, Heidelberg New York, pp 151–159

34. Spicer RA, Wolfe JA, Nichols DJ (1987) Alaskan Cretaceous-Tertiary floras and Arctic origins. Paleobiology 13:73–83

35. Wolfe JA (1991) Palaeobotanical evidence for a June "impact winter" at the Cretaceous-Tertiary boundary. Nature 352:420–423

36. Alvarez W, Claeys P, Kieffer SW (1995) Emplacement of Cretaceous-Tertiary boundary shocked quartz from Chicxulub Crater. Science 269:930–935

37. Sharpton VL, Ward PD (eds) (1990) Global catastrophes in Earth history. Geol Soc Am Spec Pap 247

38. Silver LT, Schultz PH (eds) (1982) Geological implications of large asteroids and comets on the Earth. Geol Soc Am Spec Pap 190

39. Wolfe JA (1990) Palaeobotanical evidence for a marked temperature increase following the Cretaceous-Tertiary boundary. Nature 343:153–156

40. Upchurch GR (1989) Terrestrial environmental change and extinction patterns at the Cretaceous-Tertiary boundary, North America. In: Donovan SK (ed) Mass extinctions: processes and evidence. Columbia University Press, New York, pp 195–216

41. Lerbekmo JF, Sweet AR, St Louis RM (1987) The relationship between the itidium anomaly and palynological floral events at three Cretaceous-Tertiary boundary localities in western Canada. Geol Soc Am Bull 99:325–330

42. Nichols DJ, Fleming RF (1990) Plant microfossil record of the terminal Cretaceous event in the western United States and Canada. Geol Soc Am Spec Pap 247: 445–455

43. Johnson KR, Hickey LJ (1990) Megafloral change across the Cretaceous/Tertiary boundary in the northern Great Plains and Rocky Mountains, USA. Geol Soc Am Spec Pap 247:433–444

44. Frederiksen NO (1989) Changes in floral diversities, floral turnover rates, and climates in Campanian and Maastrichtian time, North Slope of Alaska. Cretaceous Res 10:249–266

45. Brown RW (1962) Paleocene flora of the Rocky Mountains and Great Plains. Prof pap 375, U.S. Geological Survey, Washington, DC

46. Koch BE (1963) Fossil plants from the Lower Paleocene of the Agatdalen (Angmârtussut area, central Nûgssuaq Peninsula, northwest Greenland. Meded Grønl 172(5):1–120

47. Boulter MC, Kvacek Z (1989) The Palaeocene flora of the Isle of Mull. Palaeontol Assoc Spec Pap Palaeontol 42

48. McIver EE, Basinger JF (1993) Flora of the Ravenscarg Formation (Paleocene), southwestern Saskatchewan, Canada. Palaeontogr Can 10

49. Saporta G, Marion AF (1878) Révision de la floré heersiennne de Gelinden. Acad R Belg Mém Cour Sav 41

50. Berry EW (1916) The lower Eocene floras of southeastern North America. Prof pap 91, U.S. Geological Survey, Washington, DC

51. Knowlton FH (1917) Fossil floras of the Vermejo and Raton formations of Colorado and New Mexico. US Geol Surv Prof Pap 101:223–455

52. Chandler MEJ (1961) The Lower Tertiary floras of southern England. I. Br Mus (Nat Hist)

53. Chandler MEJ (1964) The Lower Tertiary floras of southern England. IV. Br Mus (Nat Hist)

54. Crane PR (1981) Betulaceous leaves and fruits from the British Upper Palaeocene. Bot J Linn Soc 83:103–136

55. Haq BU, Hardenbol UJ, Vail PR (1988) Mesozoic and Cenozoic chronostratigraphy and cycles of sea-level change. In: Wilgus CK, Hastings BS, St C Kendall CG, Posa-

mentier HW, Ross CA, Wagoner JC (eds) Sea-level changes: an integrated approach. Soc Econ Paleontol Mineral Spec Publ 42:71–108

56. Adegoke OS, Jan du Cheêne RE, Agumanu AE (1978) Palynology and age of the Kerri-Kerri Formation, Nigeria. Rev Esp Micropaleontol 10:267–283

57. Herendeen PS, Crepet WL, Dilcher DL (1992) The fossil history of the Legumi-nosae: phylogenetic and biogeographic implications In: Herendeen PS, Dilcher DL (eds) Advances in legume systematics. Part 4. The fossil record. Royal Botanical Gardens, Kew, pp 303–316

58. Rea DK, Zachos JC, Owen RM, Gingerich PD (1990) Global change at the Paleocene-Eocene boundary: climatic and evolutionary consequences of tectonic events. Palaeogeogr Palaeoclimatol Palaeoecol 79:117–128

59. Wolfe JA (1985) Distribution of major vegetational types during the Tertiary. In: Sundquist ET, Broecker WS (eds) The carbon cycle and atmospheric CO_2: natural variations Archean to present. Monogr 32. American Geophysical Union, Washing-ton, DC, pp 357–376

60. Tiffney BH (1985) Perspectives on the origin of the floristic similarity between eastern Asia and eastern North America. J Arnold Arbor Harv Univ 66:73–94

61. Tiffney BH (1985) The Eocene North Atlantic land bridge: its importance in Tertiary and modern phytogeography of the Northern Hemisphere. J Arnold Arbor Harv Univ 66:243–273

62. Budantsev LY (1994) The fossil flora of the Paleogene climatic optimum in northeast-ern Asia. In: Boulter MC, Fisher HC (eds) Cenozoic plants and climates of the Arctic. Springer, Berlin Heidelberg New York, pp 297–313

63. Wolfe JA (1975) Some aspects of plant geography of the Northern Hemisphere during the Late Cretaceous and Tertiary. Ann MO Bot Gard 62:264–279

64. Lavin M, Luckow M (1993) Origins and relationships of tropical North America in the context of the boreotropical hypothesis. Am J Bot 80:1–14

65. Reid EM, Chandler MEJ (1933) The flora of the London Clay. Br Mus (Nat Hist)

66. Wolfe JA, Tanai T (1987) Systematics, phylogeny, and distribution of *Acer* (maples) in the Cenozoic of western North America. J Fac Sci Hokkaido Univ Ser V Bot 22:1–246

67. Manchester SR (1994) Fruits and seeds of the middle Eocene Nut Beds flora, Clarno Formation, Oregon. Palaeontogr Am 58:205

68. Graham A, Dilcher DL (1995) The Cenozoic record of tropical dry forest in northern Latin America and the southern United States. In: Bullock SH, Mooney HA, Medina E (eds) Seasonally dry tropical forests. Cambridge University Press, Cambridge, pp 124–145

69. Guo S (1980) Late Cretaceous and Eocene floral provinces. Academia Sinica (Nan-jing) Inst Geol Paleontol Rep, p 9

70. Wolfe JA (1988) An overview of the origins of the modern vegetation and flora of the northern Rocky Mountains. Ann MO Bot Gard 74:785–803

71. Engler A (1882) Versuch einer Entwicklungsgeschichte der extratropischen florenge-biete der südlichen Hemisphäre und der tropischen gebiete. Engelmann, Leipzig

72. Wolfe JA (1992) Climatic, floristic, and vegetational changes near the Eocene/Oli-gocene boundary in North America. In: Prothero DR, Berggren WA (eds) Eocene-Oligocene climatic and biotic evolution. Princeton University Press, Princeton, pp 421–436

73. Wolfe JA (1994) Tertiary climatic changes at middle latitudes of western North America. Palaeogeogr Palaeoclimatol Palaeoecol 108:95–105

74. Owens JP, Bybel LM, Paulachok G, Ager TA, Gonzalez VM, Sugarman PJ (1988) Stratigraphy of the Tertiary sediments in a 945-foot-deep corehole near Mays Landing in the southeastern New Jersey Coastal Plain. Prof pap 1484. U.S. Geological Survey, Washington, DC

75. McClammer JU (1978) Paleobotany and stratigraphy of the Yaquina Flora (latest Oligocene–earliest Miocene) of western Oregon. Master's thesis, University of Maryland, College Park

76. MacGinitie HD (1937) The flora of the Weaverville beds of Trinity County, California. Carnegie Inst Washington Publ 465:84–156

77. Mai DH (1981) Entwicklung und klimatische Differenzierung der Laubwaldflora Mitteleuropas im Tertiär. Flora (Jena) 171:525–582

78. Wing SL, Hickey LJ, Swisher CC (1993) Implications of an exceptional fossil flora for Late Cretaceous vegetation. Nature 363:342–344

79. Leopold EB, Liu G, Clay-Poole S (1992) Low-biomass vegetation in the Oligocene? In: Prothero DR, Berggren WA (eds) Eocene-Oligocene climatic and biotic evolution. Princeton University Press, Princeton, pp 382–398

80. Retallack GJ (1992) Paleosols and changes in climate and vegetation acreoss the Eocene/Oligocene boundary. In: Prothero DR, Berggren WA (eds) Eocene-Oligocene climatic and biotic evolution. Princeton University Press, Princeton, pp 382–398

81. Wolfe JA (1994) A preliminary analysis of Neogene climates in Beringia. Palaeogeogr Palaeoclimatol Palaeoecol 108:107–115

82. Chaney RW (1959) Miocene floras of the Columbia Plateau, composition and interpretation. Carnegie Inst Washington Publ 617:1–134

83. Tanai T (1992) Tertiary vegetational history of East Asia. Mizunami Fossil Mus Bull 19:125–163

84. Kvacek Z (1994) Connecting links between the Arctic Palaeogene and European Tertiary floras. In: Boulter MC, Fisher HC (eds) Cenozoic plants and climates of the Arctic. Springer, Berlin Heidelberg New York, pp 251–166

85. Graham A (1995) Development and affinities between Mexican/Central American and northern South American lowland and lower montane vegetation during the Tertiary. In: Churchill SP, et al (eds) Biodiversity and conservation of neotropical montane forests. New York Botanical Garden, New York, pp 11–22

86. Wolfe JA (1994) Alaskan Palaeogene climates as inferred from the CLAMP database. In: Boulter MC, Fisher HC (eds) Cenozoic plants and climates of the Arctic. Springer, Berlin Heidelberg New York, pp 223–237

87. Takhtajan AL (1980) Outline of the classification of flowering plants (Magnoliophyta). Bot Rev 46:225–359

CHAPTER 14

Modes and Mechanisms
of Speciation in Pteridophytes

CHRISTOPHER H. HAUFLER

14.1 Introduction

Curiosity about the origin and maintenance of organic diversity is one of the forces that has propelled the science of biology. From direct observation of the remarkable variety that surrounds and supports us as one of the millions of species on Earth to experiments designed to reproduce the circumstances that generate this variation, natural historians, population biologists, and systematists are driven to discover how and why there are so many species.

With the knowledge explosion that has accompanied this great curiosity, individual scientists can no longer grasp the totality of information that relates to the many facets of speciation. Thus, contemporary researchers concerned with the diversity of life tend to fall into different categories, including ecologists who study and describe the remarkable ways that organisms respond to their environments, population biologists who provide quantitative parameters for analyzing evolutionary change, and systematists who ponder the genealogical relationships of species. Biologists belonging to each of these categories depend on information from their colleagues to help formulate and test their hypotheses. But just as systematists seldom appreciate the nuances of ecological modeling, ecologists frequently ignore the importance of history.

It is with some trepidation, therefore, that I attempt to summarize the current understanding of pteridophyte speciation. Instead of developing a comprehensive treatise, I shall (1) introduce some of the challenges facing students of pteridophyte evolution, (2) discuss the *modes* that appear to pertain to pteridophyte speciation, and (3) focus on what we know about the *mechanisms* that have operated to generate the evolutionary patterns that we detect among extant pteridophytes.

Department of Botany, Haworth Hall, University of Kansas, Lawrence, KS 66045, USA

14.2 The Challenges and Constraints

Some of the difficulties facing scientists who study speciation are common to all organisms, not just to pteridophytes. For example, when discussing "modes and mechanisms" of speciation, one is attempting to characterize evolutionary events by considering the ultimate results of those events (the modes or patterns) and the proximate causes of the events (the mechanisms or processes). In a majority of cases, however, because of the considerable complexities involved, individual studies can focus on one component or the other but not both. To date, most studies generate hypotheses or predictions stating that one or another mechanism *could* have produced the pattern that is elucidated, or that a given mechanism *should* result in a particular mode of evolutionary change. Usually it is difficult for single investigations to forge direct links between the ways that genetic changes *can* take place and the actual observed results of historical events. Without a time machine, we can only infer what process yielded the patterns that we observe or whether new lineages will actually emerge following a given genetic change.

This is not to imply that there is anything inherently "wrong" or "improper" about the way that students of evolutionary biology do their science. Posing and testing hypotheses *is* the way that science progresses. And, certainly, most other biological disciplines depend on working hypotheses about the history of life on earth to interpret their results. I simply believe it is important to recognize the significant challenges that we face as we attempt to build and test hypotheses about evolutionary history and the inferred results of ancient events. Other impediments are imposed by the particular features of the organisms under investigation, and pteridophytes have erected some of the more daunting obstacles.

14.2.1 Pteridophyte Lineages are the Longest Among the Vascular Land Plants

As reviewed recently by Rothwell [1], pteridophytes trace their phylogenetic "roots" back to the Silurian, more than 400 million years ago. Such ancient origins pose at least two major obstacles to researchers seeking to reconstruct the events of evolutionary history. First, the time available to accumulate great diversity in form and genetic variability is vast. Since their origin, pteridophytes have had ample opportunity to undertake a huge array of evolutionary "experiments," in such aspects of their biology as leaf form, reproductive biology, habit, and habitat. Remnants of those experiments are still extant and continue to provide riddles because the links to their origins are extinct. For example, how many times has heterospory arisen? Among pteridophytes, not only are there modern representatives from diverse lineages (e.g., *Selaginella* and *Marsilea*) but fossil evidence shows that heterospory was present among extinct groups. And yet it is generally accepted that at least one of the heterosporous "experiments" was essential to the origin of the seed plants. Considering another feature, how many different kinds of arborescent habit have evolved? Again, both extant and extinct groups provide good evidence of multiple origins. These observations call into question hypotheses of the homology of characters that are more poorly preserved, such as leaf form and venation. In attempting to use the patterns of diversity to develop hypotheses about the origins of species, having a clear understanding of character homologies is crucial for correctly interpreting evolutionary change. Even though most evidence indicates that the leptosporangiate ferns originated relatively recently [2], determining character homologies is difficult even in this "young" clade. Because of the numerous "evolutionary experiments" that preceded the origin of the leptosporangiate ferns, determining which "experiments" are homologous remains challenging. Evidence that analysis of pteridophyte evolution may be fraught with homoplasy greatly complicates the application of modes to predict mechanisms.

Second, at the same time that the pteridophytes have been diversifying, extinction has been occurring. And, of course, the longer the lineage, the larger the number of extinctions. This significant loss of information about pteridophyte ancestry has a profound impact on correctly interpreting evolutionary relationships. Because of the insufficiency of the fossil record [2,3], evidence regarding the connections between extant groups is lacking, and in most cases it is difficult to test hypotheses regarding origins of lineages and the evolutionary relationships between groups. Especially given the possibility that allopolyploidy, diploid extinction, and gene silencing have been prominent components of the evolutionary history of pteridophytes (see following), the loss of information through extinction is particularly problematic in tracing pteridophyte phylogenies [4].

14.2.2 "Obvious" Morphological Features Provide Confusing or Conflicting Phylogenetic Clues

As just mentioned, one of the primary problems in trying to develop hypotheses about pteridophyte evolutionary history is identifying features that are not homoplastic. The best hypotheses relating to species origins are those that can be linked directly to the speciation mechanism itself. In some of the flowering plants, visually apparent floral modifications represent qualitative changes that modify the pollination system and may be linked to speciation events [5,6]. However, in pteridophytes it may be difficult to separate changes that may have phylogenetic significance from those that are part of an adaptational response subsequent to a speciation event. We know very little about the evolution of features that may be linked directly to speciational processes, such as the differentiation of chemical attractants for sperm. However, given the changing views that have been developed for characters that had been considered phylogenetically significant, we should be very careful in drawing conclusions about how modes or mechanisms of speciation may relate to morphological features. For example, even such seemingly significant differences such as the presence or absence of an indusium may represent either an ancestral absence (plesiomorphy) or a derived loss (apomorphy). In the Polypodiaceae, absence of indusia was viewed as an ancestral trait, tying this family to the Gleicheniaceae [7,8]. Revised analyses of morphological and molecular features, however, have demonstrated that lack of indusia in the Polypodiaceae represents a derived loss [9,10]. This kind of discovery indicates that great care must be exercised when employing morphology as clues to changes that accumulate in conjunction with speciation events.

14.2.3 High Chromosome Numbers

Ever since Manton [11] provided the first large compendium of chromosome counts in pteridophytes and demonstrated that pteridophytes had extraordinarily high numbers, scientists have speculated about the significance of such genetic features among homosporous species. Klekowski ([12], and references therein) used this observation to develop new hypotheses about fundamental aspects of the genetics of pteridophytes. He predicted that high chromosome numbers indicated that homosporous pteridophytes had many duplicated genes and would therefore have different patterns of expression of traits and different rates of response to change than plants with lower numbers. If this prediction was accurate, standard population genetic models would require modification in developing hypotheses of evolutionary change in pteridophytes, and those attempting to develop ideas about speciation mechanisms would need to invent new models based on assumptions of wholesale gene duplication. However, electrophoretic investigations [13–15] demonstrated that some of Klekowski's predictions were not accurate. Even though pteridophytes have high chromosome numbers, they follow standard diploid models of gene expression. Thus, the dynamics of extant populations can be studied without worrying about complications relating to genetic duplication, but characterizing more ancient events, such as the origin of species, cannot ignore the evolutionary consequences of having genetically diploid individuals with high chromosome numbers.

As I have discussed previously [16], if polyploidization followed by diploidization is a regular feature of the evolutionary history of homosporous pteridophytes, any summary of common modes and mechanisms of speciation must address how such processes will influence the origins and fates of species. It may be that homospory has a different combination of genetic and evolutionary consequences than heterospory. Although there are numerous questions about the significance of genome size in organisms [17] and we have much to learn about the fluidity of nuclear contents, data from recent molecular studies support the hypothesis that polyploidy and gene duplication played a prominent role in the history of pteridophytes [18,19]. It is hoped that during the consideration of speciation pathways we can also generate new and improved perspectives on the apparent paradox of genetic diploidy despite high chromosome numbers.

14.2.4 Alternation of Independent Generations

Understanding the population dynamics and reproductive biology of species is critical to formulating accurate hypotheses about patterns and processes of speciation. Distinguishing pteridophytes from other vascular land plants is that they have independent gametophyte and sporophyte generations. In considering the biology of species and speciation, the lightweight spores used for dispersal and the delicate, ephemeral gametophytes required for establishment of new populations places parameters on the processes of colonization and biogeographic patterns that differ from most seed plants. Further complicating the situation is that two environmentally different habitats are commonly required for initial colonization by gametophytes and ultimate establishment of successful sporophyte populations. Recent reviews [20,21] have emphasized the importance of disturbance in opening habitats adequate for establishing gametophyte populations, while sporophytes typically are found in mature communities. Combined with the availability of spore banks [22] and the probable chemical interaction between aboveground and subterranean gametophytes [23], recent studies have provided hypotheses for promoting crossing between compatible gametophytes and the maintenance of genetic variation within species. These kinds of studies are helping us to appreciate the population biology within pteridophyte species.

Hybridization between species is also a significant component of pteridophyte biology. Because gametophytes of diverse species colonize disturbed habitats (there is little or no evidence that gametophytes of different species have specialized growth requirements), events promoting disturbance (e.g., glaciation, erosion) provide opportunities for interspecific hybridization. In addition to such ecological considerations, pteridophyte species do not appear to enjoy the complex, prezygotic isolating mechanisms that are available to seed plant lineages, and the simplicity of pteridophyte reproductive biology may promote higher rates of interspecific hybridization than is common in higher groups.

If these summaries of the basic parameters of the pteridophyte biology are accurate, they may help to explain why reticulate (allopolyploid) speciation is such a prominent feature of pteridophyte evolution, especially in regions that regularly experience disturbance. Open habitats that provide the opportunity for the gametophytes of diverse species to occur sympatrically may lead to high levels of interspecific hybridization. Enhancing possibilities for gene exchange between incipient new lineages would reduce the rate of pteridophyte speciation. If lineages are only isolated by establishing postzygotic barriers, divergence to the species level may require long periods of time to accumulate the necessary genetic differentiation.

14.3 The Modes

Modes are common or typical ways of accomplishing a particular result, in this case, the formation of species. Grant [24] suggested two major modes of speciation: *primary*, the divergence of diploid populations to the level of species, and *secondary*, the origin of new species involving hybridization or polyploidy within or between established species; and I have added a third category, *tertiary* speciation, the divergence among populations of secondary species to initiate new lineages [16]. Because the fundamental mechanisms associated with these three major modes are qualitatively different, I think it is useful to separate and discuss them independently. Within each of these major modes are subcategories that describe different ways of achieving speciation. Figure 14.1 indicates how the different modes may be integrated in developing a composite picture of pterido-phyte speciation.

14.3.1 Primary Speciation

Primary speciation has been used to encompass the most basic and commonly recognized ways of initiating new species, the divergence of diploid populations to the level of species. Having a clear and well-supported phylogeny for the group of taxa being studied is particularly important in developing hypotheses about primary speciation. Unless sister taxa are compared, erroneous conclusions about the processes involved will be obtained. Within this major mode are more specific categories including *allopatric speciation*, the divergence of populations to the species level through isolation by geographic separation, *parapatric speciation*, divergence of populations to the species level even though populations maintain

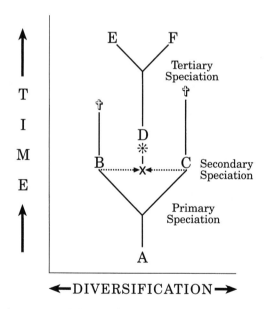

Fɪɢ. 14.1. Coordinating some of the possible modes and mechanisms of speciation to obtain a composite hypothesis that may help to depict the evolutionary history of pterido-phytes. The three different modes of speciation (primary, secondary, and tertiary) are characterized by qualitatively different mechanisms. (See text for more detail.) *Solid lines* are separate lineages; *dotted lines* indicate interactions between distinct species; *letters* designate separate species. X, hybridization; ✳, polyploidization; ✞, lineage extinction

contiguous but nonoverlapping geographic distributions, and *sympatric specia-tion*, divergence of populations to the species level even though the populations occupy the same geographic region. Given the complex biotic and behavioral interactions that have been associated with sympatric speciation [25] and the high probability that simple isolating mechanisms characterize pteridophytes, it seems unlikely that they speciate sympatrically at the diploid level.

Discussions of primary speciation in pteridophytes commonly allude to allo-patric models [20,26–29]. Because homosporous pteridophytes disperse light, wind-borne spores that can travel long distances via wind currents, these species were considered to be ideally suited to long-distance migration. The clearest model of allopatric speciation involves migration to islands [26], and there are fern species that are among the earliest colonists on new volcanic islands. Factors cited as contributing to successful migration included whether the source was wide ranging and had a broad ecological amplitude [27]. It was also proposed that single spores could initiate new populations because individual homosporous pteridophyte gametophytes are potentially hermaphroditic [26]. Thus, a single spore wafted great distances into a new habitat could initiate a new population. However, with the discovery that outcrossing rather than inbreeding predomi-nates in most fern populations [15], at least two spores are probably necessary to found new populations of a majority of homosporous pteridophyte species. It has been suggested that self-fertilization by individual gametophytes (intragameto-phytic selfing sensu Klekowski [12]), rather than representing a common or ancestral trait of homosporous species, may be a specialized adaptation for rapid colonization which has arisen only in certain derived taxa [14]. Given new per-ceptions of the population biology of homosporous pteridophytes, there is no compelling reason to consider that the majority of species is particularly well suited to allopatric speciation or island hopping.

Turning to primary speciation involving continental populations, the differen-tiation between allopatric and parapatric modes becomes less clear cut. Especial-ly because we are forced to consider only a single slice through time, we do not know precisely what has resulted in the geographic patterns that are extant. Certainly there appear to be instances where sister taxa have overlapping ranges, but there are other situations in which they do not. But in most cases we can only guess at how these different distributional patterns originated. If sister species are sympatric for a portion of their ranges, does such a region represent the point of origin of the species or did they originate allopatrically and subsequently expand their ranges into partial sympatry? We cannot necessarily infer different modes of primary speciation from different biogeographic patterns, and this lack of knowledge clearly indicates that additional studies of primary divergence are desperately needed.

Based on the observation that pteridophyte taxa have wider intercontinental distributions than do flowering plants, Smith [26] suggested that they also contain less endemism. However, ongoing studies are demonstrating that some of these so-called intercontinental species are actually complexes of two or more separate species [30–33]. As may be true in considering long-distance dispersal, the distinction between pteridophytes and flowering plants may relate to our misperception of their patterns of morphological diversity rather than to actual evolutionary differences. In recognizing species and other taxa, it may be important to begin placing greater weight on cryptic features [4] so that more realistic evolutionary units can be circumscribed and compared accurately.

14.3.2 Secondary Speciation

When it can be demonstrated that the speciation under investigation involved genomic-level changes, such as hybridization or polyploidy, a separate mode is proposed. The magnitude of genetic modification in secondary speciation often can be characterized, and it appears to be qualitatively different from that caused by the more incremental changes that are typical of primary speciation. Further, secondary speciation usually involves interactions *between* distinct and separate lineages that remain intact (autopolyploidy is the exception). These interactions result in the production of a new lineage that is reproductively isolated from its progenitors, shares significant portions of its genome with them, and is usually intermediate in morphology between them. Thus, instead of a single lineage evolving into two new lineages (as in primary speciation), two lineages interact to yield a third lineage, and all three lineages persist.

Characterization of a variety of patterns provides circumstantial evidence of different kinds of secondary speciation. When different ploidy levels are detected among individuals that are morphologically uniform, *autopolyploidy* is suspected. Some summaries of speciation have used autopolyploidy as an example of "sympatric" speciation. However, autopolyploidy involves genome duplication, a mechanism that is quite different from those leading to the origin of diploid lineages. As reviewed by Gastony [34], speciation by chromosome doubling within pteridophyte species has been largely overlooked as a significant mechanism. In some groups, however, especially when accompanied by apomixis, autopolyploidy may occur frequently.

When three distinct species are observed and one of the three appears intermediate between the other two, it is possible that *allohomoploidy* (hybrid speciation that does not involve chromosome doubling) has generated this pattern. Obviously a similar pattern could evolve through primary speciation, but allohomoploidy is especially likely if all three species overlap in range. The most extensively documented example of allohomoploidy involves tree ferns in the Caribbean [35]. Given the difficulty of detecting patterns that have resulted through allohomoploidy, it may be that this mode of speciation is more common than currently appreciated.

Allopolyploidy is a mode that has been associated frequently with pteridophytes. As discussed earlier for autopolyploidy, because of the distinctly different mechanisms involved, I have not included allopolyploidy as an example of sympatric speciation among diploids. Detecting the patterns produced through allopolyploidy begins with a consideration of morphology. Allopolyploid species usually have features that are intermediate between those of its presumed progenitors. If the putative allopolyploid is fertile and contains twice as many chromosomes as its progenitors, the allopolyploid hypothesis is supported. When more than three species (two diploid progenitors and the allopolyploid derivative) are involved, the term *reticulate evolution* is used to describe the complex patterns of interactions and evolutionary outcomes that are obtained. Beginning with Manton [11] and popularized by Wagner [36], pteridophytes provide some of the best models of allopolyploid speciation and reticulate evolution (see the diagrams in *Flora North America*, Vol. 2, 1993, for some current examples).

Identifying allopolyploid patterns becomes more difficult when the putative progenitors are very similar morphologically. In recent investigations, isozyme variants have been powerful tools in reconstructing intricate webs of reticulate evolution [37], even in genera whose species are strikingly similar to each other. In some groups, long-standing open questions of allopolyploid origins have been answered (e.g., *Polypodium* [38], *Polystichum* [39]), in others new species have been recognized (e.g., *Adiantum* [32], *Cryptogramma* [30], *Gymnocarpium* [33]), and in others extinct taxa have been proposed and characterized genetically to account for the origins of extant polyploids (e.g., *Cystopteris* [40,41], *Dryopteris* [42]). These recent studies demonstrate clearly that we still have much to learn about modes of allopolyploidy in pteridophytes.

14.3.3 Tertiary Speciation

Polyploid species, whether arising through autopolyploidy or allopolyploidy, have long been considered temporary entities with little evolutionary significance [43,44]. After isozymic surveys demonstrated that pteridophytes having high chromosome numbers were genetically diploidized [13], hypotheses were developed whereby polyploids could diversify through reciprocal gene silencing [45]. Haufler [14] suggested that these discoveries provided the foundation for a third speciational mode, which he called tertiary speciation. Thus, instead of being evolutionarily viscous because gene duplications would buffer against the expression of new mutations, polyploids were proposed as pools of latent genetic diversity that would speciate as they expanded their geographic range. Citing examples from *Cystopteris*, *Dryopteris*, and *Polypodium*, Haufler noted that in some groups the polyploids were the most geographically successful species, far outdistancing the range of the diploids. He proposed that as the diploids became extinct, the polyploids of some groups represented the wave of future success. Although there is little empirical evidence for this proposed mode of speciation, groups that are dominated by polyploids may provide testing grounds for further exploration of these hypotheses.

14.4 The Mechanisms

14.4.1 Primary Speciation

Although primary speciation is commonly considered to represent the pathway that has yielded the bulk of the diversity of pteridophytes on the planet, we have little or no empirical evidence to document the actual mechanisms by which this diversification has occurred. Hypotheses involving allopatry and isolation by distance resulting in the accumulation of new morphological features have been provided [46], evidence suggests that isolated populations do, indeed, accumulate mutational differences from each other [47,48], and hypotheses have been developed concerning the activity of mechanisms for generating morphological and physiological variants [28]. However, no studies of how ecological pressures or selective regimes could influence morphological change or physiological responses have been published. Clearly, this is an area that demands more investigation and hypothesis testing.

As I have suggested elsewhere [4], there may be two distinctly different modes of primary speciation among pteridophytes. In regions having a reduced ecological amplitude, such as temperate areas and perhaps also the lowland tropics, there may be also a reduced rate of speciation. I think this may owe to the simple genetic system that pteridophytes have—it may be that they lack the great variety of chromosomal gimmicks, incompatibility factors, and genetic mechanisms that have been demonstrated for angiosperms. In the absence of ecological barriers to gene exchange, pteridophytes would speciate by accumulating high levels of genetic divergence through mutations. Wide dispersal of spores coupled with common gametophyte habitats and a lack of barriers to fertilization (sperms may be able to penetrate the eggs of distantly related species) should result in high rates of gene flow among genetically compatible individuals [49,50]. Thus, regular mechanisms for homogenization of genomes may prevent the fixation of polymorphisms. If true, this could be one factor that would limit the rate at which speciation could occur among pteridophytes.

But a different model may hold for some tropical habitats. Although the number of studies is small and the number of reliable hypotheses about sister taxon relationships is even smaller, there seem to be contrasting trends in the level of genetic divergence between species in temperate regions and those in tropical habitats. Data suggest that there is either more active speciation taking place in the tropics than in temperate regions or that the process of speciation in tropical zones differs from that in temperate regions. The average genetic identity between "temperate" sister species is 0.542 whereas that between tropical congeners is 0.941 (see Table 1 in Haufler [4]). This discrepancy may be caused by differences in speciation opportunities between temperate and tropical regions.

Perhaps there are more pronounced shifts in ecological zones in tropical regions than in temperate parts of the world [51,52]. Especially in tropical upland regions, the changes in species composition that are evident as you ascend mountains is striking [53,54]. Such rapid and distinct ecological shifts could establish more opportunities for speciation than in temperate regions and could lead to ecological isolation of species that simply does not occur in temperate regions. Just as there appear to be different ecological relationships among sporophytes, gametophytes as well may be more selective in their habitat preferences. If ecological barriers to gene exchange are not as effective in temperate climates, more genetic divergence must accumulate before incipient new taxa are isolated sufficiently to become separate lineages. Assuming this model holds, there would be more opportunity for rapid fixation of morphological polymorphisms in some tropical systems than in temperate ones and our ability to recognize tropical species would be enhanced. We certainly need more information about the physiological ecology of sporophytes and gametophytes, and ultimately it may be necessary to adopt a more ecologically based species concept for tropical pteridophytes [55].

Clearly, the topic of primary speciation in pteridophytes requires much further study and the accumulation of many more data, and it is a critical area that should continue to be actively explored. For example, perhaps the reason for high levels of genetic divergence among temperate pteridophyte species is that there has been more extinction in these regions than in the tropics. I pose these possibilities not as dicta but as hypotheses that may be tested to bring us closer to understanding the patterns and processes of primary speciation.

14.4.2 Secondary Speciation

Although some information is available from analyses of crop plants [56], mechanisms that result in the duplication of genomes among natural populations are poorly understood [57]. Whether polyploidy arises primarily in somatic tissues [58] or whether meiotic mistakes predominate [59] is unknown. It is likely that barriers to intragametophytic selfing break down in polyploids [60], and they may be better colonists than diploids. Studies of how the "mistakes" that result in polyploid cells are initiated and what environmental and genetic pressures may promote polyploid origins will enhance our understanding of secondary speciation.

Polyploidization Within a Lineage (Autopolyploidy)

The most clearly and completely documented examples of autopolyploid mechanisms are triploid apomicts derived from diploid progenitors [61], but some studies have also showed that mechanisms are operating that could generate fertile tetraploids through autopolyploid processes [34,59,62].

Hybridization (Allohomoploidy)

The first demonstration of apparent allohomoploid speciation was in *Pteris* [63,64]. By analyzing chromosome behavior and morphology, Walker showed that diploid species may be ecologically isolated, but were not always reproductively isolated. Under some circumstances, hybrid swarms have developed and apomictic species are frequent. Unfortunately, Walker's perceptive and seminal hypotheses have not been tested by the application of molecular methods such as isozymes. Nonetheless, Walker's work demonstrates the great evolutionary complexity that can be revealed through biosystematic studies of ferns and current generations of pteridologists should consider testing his hypotheses.

The most extensively documented case of allohomoploidy involves species of *Alsophila* in the Caribbean [35]. This model begins with primary speciation that isolates species rapidly based on different ecological preferences (see foregoing). Isozyme studies have demonstrated that although these tree fern species are morphologically distinct, they have very high congeneric genetic identities (averaging 0.912 [65]), and they are interfertile. Without disturbance, they might remain ecologically isolated from each other and eventually accumulate more genetic differentiation. Conant and Cooper-Driver [35] used a combination of morphological and biochemical data to demonstrate, however, that the barriers to species isolation had broken down and fertile hybrids had formed. These hybrids were sufficiently morphologically distinct from their progenitors that they had been accorded species status and, at least in some circumstances, were ecologically isolated from their progenitors. Given that this is an island system, the most provocative component of this analysis is the possibility that the hybrid-derived species could migrate to a neighboring island which could contain only one or neither of the parental species.

Hickok and Klekowski [66] provide additional evidence of an actual mechanism by which fertile interspecific hybrids could become isolated from their progenitors. An artificial hybridization experiment yielded clues to chromosomal changes that could occur during hybridization and generate progeny having partial reproductive isolation from their parental species.

Allohomoploidy, therefore, emphasizes the role of ecology in promoting and maintaining diversity despite only minor genetic modification. If this mechanism is prominent in tropical habitats, accurate characterization of species and speciation in such regions will be especially challenging. At the same time, demonstration that such a mechanism *is* operating provides important clues to the special complexity of tropical systems. Cytological and isozymic investigations may yield evidence of cryptic divergence in the parental species that results in hybrid progeny having distinctive features yet remaining at least partially interfertile.

Hybridization and Polyploidy (Allopolyploidy)

It has been suggested [14] that the basic qualities of cryptogams contribute to the prominence of allopolyploid complexes in pteridophytes. As discussed, there are probably few barriers to the initiation of interspecific zygotes and field studies have demonstrated a high frequency of vigorous but sterile hybrids in some complexes [67]. The persistence of these perennial herbs and the huge number of aborted sporangia that they contain provide the opportunity for initiating polyploid offspring through meiotic "mistakes." Because of redundant copies of genes, polyploid gametophytes may be more tolerant of intragametophytic selfing than their diploid progenitors [60] and thus allopolyploid sporophytes may be initiated relatively frequently in pteridophytes. Isozyme [68] and chloroplast DNA analyses [69] have provided evidence of recurring allopolyploid origins, indicating that the variability from progenitor diploids can be pumped into derivative polyploid lineages.

Although it is an accepted and frequent mode of speciation, there are certainly many open questions about so-called species complexes that involve allopolyploidy. Results from recent investigations of such well-studied groups as *Cystopteris* [70], *Dryopteris* [42], *Gymnocarpium* [33], *Polypodium* [38], and *Polystichum* [39] have illustrated how much we can still learn through application of new technologies to long-standing questions. Perhaps most surprisingly, in several of these cases, testing hypotheses of allopolyploid origins has led to significant revisions of species boundaries at the diploid (primary speciation) level. Surveys of chromosome numbers in tropical habitats [71] have demonstrated that polyploidy is frequent, and we should be alert to the possibility of allopolyploid complexes as we expand biosystematic and molecular studies into the tropics.

14.4.3 Tertiary Speciation

We have only questions and hypotheses about the possibility that secondary species could diversify after they originate. Are polyploids subject to the same kinds of pressures as primary species, or can they accumulate mutations more rapidly because they have duplicate genes for each critical function? Do allopolyploids diverge at the polyploid level to become new and (probably) cryptic species? Are polyploids returning to the diploid level through gene silencing? Are they speciating through reciprocal gene silencing [45] to form what have been called tertiary species [16]? Is understanding polyploids one of the keys to synthesizing information about the evolutionary history of fern species? These are open questions that new generations of pteridologists need to address if we are to develop improved perspectives on pteridophyte species.

14.5 Summary and Synthesis

Speciation is at the same time one of the most important and one of the most poorly understood events in organic evolution. We can clearly see that a great diversity of different species exists on earth and we have identified numerous mechanisms that may account for this diversity. Yet, especially among primary species, we do not have a clear understanding of the mechanisms by which species actually originate. Pteridophytes present both unique constraints and unique opportunities for studying speciation. Having a limited range of morphological complexity reduces the number of characteristics that are available for recognizing and charting change over time. Having to consider the origin of high chromosome numbers as well as the origin of species gives pteridologists an extra burden and an extra concern in developing models of diversification. Having separate gametophyte and sporophyte generations with different roles and ecological conditions provides an added level of complexity.

However, these apparent limitations can also be viewed as opportunities. Working with speciation systems that are defined primarily by physical factors of geology and geography constrains the range possibilities for making predictions about the origin of evolutionary changes and simplifies the development of realistic models. Working with species that have a distinct separation between gametophyte and sporophyte provides opportunities for distinguishing clearly the contribution of each to the processes governing evolutionary modification. Working with species in which allopolyploidy is playing (and probably has played) a major role in generating genetic novelty gives pteridologists a distinct advantage in focusing on this important element in the history of life on earth.

Thus, in terms of both the modes and patterns that can be observed and the mechanisms which can be inferred and tested, understanding the evolution of pteridophytes will benefit from additional investigations. Especially at the level of primary speciation, we need to have more and better models of how speciation occurs, and we need to develop ways of testing these models in the field and laboratory. At the same time that such investigations enhance our knowledge of pteridophytes, testing hypotheses about speciation among ferns and other free-sporing vascular plants will result in opportunities to expand our appreciation of the possibilities and probabilities for evolutionary change. It seems likely that pteridophytes are uniquely suited to study how secondary speciation in general and polyploidy in particular contribute to the origin and manipulation of genetic variability. We should, building on the solid foundation provided by past investigators, generate innovative ways to interpret and appreciate the evolutionary history of pteridophytes and use pteridophyte systems to illuminate general aspects of the modes and mechanisms of speciation.

References

1. Rothwell GW (1996) Pteridophytic evolution: an often underappreciated phytological success story. Rev Palaeobot Palynol 90:209–222
2. Harris TM (1973) What use are fossil ferns? In: Jermy AC, Crabbe JA, Thomas BA (eds) The phylogeny and classification of ferns. Academic, London, pp 41–44
3. Lovis JD (1977) Evolutionary patterns and processes in ferns. Adv Bot Res 4:229–415
4. Haufler CH (1996) Species concepts and speciation in Pteridophytes. In: Camus JM, Gibby M, Johns RJ (eds) Pteridology in perspective. Roy Bot Gard, Kew, pp 291–305
5. Adams VD (1983) Temporal patterning of blooming phenology in *Pedicularis* on Mount Rainier. Can J Bot 61:786–791
6. Grant V (1994) Modes and origins of mechanical and ethological isolation in angiosperms. Proc Natl Acad Sci USA 91:3–10
7. Holttum RE (1949) The classification of ferns. Biol Rev Camb Philos Soc 24:267–296
8. Pichi Sermolli REG (1977) Tentamen Pteridophytorum genera in taxonomicum ordinem redigendi. Webbia 31:313–512
9. Jarrett FM (1980) Studies in the classification of the leptosporangiate ferns. I. The affinities of the Polypodiaceae and Grammitidaceae. Kew Bull 34:825–838
10. Haufler CH, Ranker TA (1995) *rbcL* sequences provide phylogenetic insights among sister species of the fern genus *Polypodium*. Am Fern J 85:359–372
11. Manton I (1950) Problems of cytology and evolution in the Pteridophyta. Cambridge University Press, Cambridge
12. Klekowski EJ (1979) The genetics and reproductive biology of ferns. In: Dyer AF (ed) The experimental biology of ferns. Academic, London, pp 133–170

13. Haufler CH, Soltis DE (1986) Genetic evidence indicates that homosporous ferns with high chromosome numbers may be diploid. Proc Natl Acad Sci USA 83:4389–4393

14. Haufler CH (1987) Electrophoresis is modifying our concepts of evolution in homosporous pteridophytes. Am J Bot 74:953–966

15. Soltis DE, Soltis PS (1989) Polyploidy, breeding systems, and genetic differentiation in homosporous pteridophytes. In: Soltis DE, Soltis PS (eds) Isozymes in plant biology. Dioscorides Press, Portland, pp 241–258

16. Haufler CH (1989) Towards a synthesis of evolutionary modes and mechanisms in homosporous pteridophytes. Biochem Syst Ecol 17:109–115

17. Rees H (1984) Nuclear DNA variation and the homology of chromosomes. In: Grant WF (ed) Plant biosystematics. Academic, Toronto, pp 87–96

18. Pichersky E, Soltis DE, Soltis PS (1990) Defective chlorophyll *a/b*-binding protein genes in the genome of a homosporous fern. Proc Natl Acad Sci USA 87:195–199

19. McGrath JM, Hickok LG, Pichersky E (1994) Assessment of gene copy number in the homosporous ferns *Ceratopteris thalictroides* and *C. richardii* (Parkeriaceae) by restriction fragment length polymorphisms. Plant Syst Evol 189:203–210

20. Barrington DS (1993) Ecological and historical factors in fern biogeography. J Biogeogr 20:275–280

21. Schneller JJ (1995) Aspects of spore release of *Asplenium ruta-muraria* with reference to some other woodland ferns: *Athyrium filix-femina, Dryopteris filix-mas* and *Polystichum aculeatum*. Bot Helv 105:187–197

22. Dyer AF, Lindsay (1992) Spore banks of temperate ferns. Am Fern J 82:89–123

23. Haufler CH, Welling CB (1994) Antheridiogen, dark spore germination, and outcrossing mechanisms in *Bommeria* (Adiantaceae). Am J Bot 81:616–621

24. Grant V (1981) Plant speciation. Columbia University Press, New York

25. Bush GL (1994) Sympatric speciation in animals: new wine in old bottles. Trends Ecol Evol 9:285–288

26. Smith AR (1972) Comparison of fern and flowering plant distributions with some evolutionary interpretations for ferns. Biotropica 4:4–9

27. Tryon RM (1970) Development and evolution of fern floras of oceanic islands. Biotropica 2:76–84

28. Tryon RM (1972) Endemic areas and geographic speciation in tropical American ferns. Biotropica 4:121–131

29. Tryon RM (1986) The biogeography of species, with special reference to ferns. Bot Rev 52:117–156

30. Alverson ER (1989) *Cryptogramma cascadensis*, a new parsley fern from western North America. Am Fern J 79:95–102

31. Haufler CH, Windham MD (1991) New species of North American *Cystopteris* and *Polypodium*, with comments on their reticulate relationships. Am Fern J 81:6–22

32. Paris CA ,Windham MD (1988) A biosystematic investigation of the *Adiantum pedatum* complex in eastern North America. Syst Bot 13:240–255

33. Pryer KM, Haufler CH (1993) Electrophoretic and chromosomal evidence for the allotetraploid origin of the common oak fern *Gymnocarpium dryopteris* (Dryopteridaceae). Syst Bot 18:150–172

34. Gastony GJ (1986) Electrophoretic evidence for the origin of fern species by unreduced spores. Am J Bot 73:1563–1569

35. Conant DS, Cooper-Driver G (1980) Autogamous allohomoploidy in *Alsophila* and *Nephelea* (Cyatheaceae): a new hypothesis for speciation in homoploid homosporous ferns. Am J Bot 67:1269–1288

36. Wagner WH (1954) Reticulate evolution in the Appalachian aspleniums. Evolution 8:103–118
37. Werth CR (1989) The use of isozyme data for inferring ancestry of polyploid pteridophytes. Biochem Syst Ecol 17:117–130
38. Haufler CH, Windham MD, Rabe EW (1995) Reticulate evolution in the *Polypodium vulgare* complex. Syst Bot 20:89–109
39. Soltis PS, Soltis DE, Wolf PG, Riley JM (1989) Electrophoretic evidence for hybridization in *Polystichum*. Am Fern J 79:7–13
40. Haufler CH (1985) Pteridophyte evolutionary biology: the electrophoretic approach. Proc R Soc Edinb 86:315–323
41. Paler MH, Barrington DS (1995) The hybrid *Cystopteris fragilis* × *C. tenuis* (Dryopteridaceae) and the relationship between its tetraploid progenitors. Syst Bot 20:528–545
42. Werth CR (1991) Isozyme studies on the *Dryopteris* "spinulosa" complex. I. The origin of the log fern *Dryopteris celsa*. Syst Bot 16:446–461
43. Stebbins GL (1950) Variation and evolution in plants. Columbia University Press, New York
44. Wagner WH (1969) The role and taxonomic treatment of hybrids. Bioscience 19:785–789
45. Werth CR, Windham MD (1991) A model for divergent, allopatric speciation of polyploid pteridophytes resulting from silencing of duplicate gene expression. Am Nat 137:515–526
46. Tryon RM (1971) The process of evolutionary migration in species of *Selaginella*. Brittonia 23:89–100
47. Masuyama S, Watano Y (1994) Hybrid sterility between two isozymic types of the fern *Ceratopteris thalictroides* in Japan. J Plant Res 107:269–274
48. Ranker TA, Floyd S, Windham MD, Trapp P (1994) Historical biogeography of *Asplenium adiantum-nigrum* (Aspleniaceae) in North America and implications for speciation theory in homosporous pteridophytes. Am J Bot 81:776–781
49. Soltis PS, Soltis DE (1987) Population structure and estimates of gene flow in the homosporous fern *Polystichum munitum*. Evolution 41:620–629
50. Wolf PG, Sheffield E, Haufler CH (1991) Estimates of gene flow, genetic substructure and population heterogeneity in bracken (*Pteridium aquilinum*). Biol J Linn Soc 42:407–423
51. Tuomisto H, Ruokolainen K (1994) Distribution of Pteridophyta and Melastomataceae along an edaphic gradient in an Amazonian rain forest. J Veg Sci 5:25–34
52. Poulsen AD, Nielson IH (1995) How many ferns are there in one hectare of tropical rain forest? Am Fern J 85:29–35
53. Janzen DH, Ataroff M, Farinas M, Reyes S, Rincón N, Soler A, Soriano P, Vera M (1976) Changes in the arthropod community along an elevational transect in the Venezuelan Andes. Biotropica 8:193–203
54. Wolda H (1987) Altitude, habitat and tropical insect diversity. Biol J Linn Soc 30:313–323
55. Van Valen L (1976) Ecological species, multispecies, and oaks. Taxon 25:233–239
56. Harlan JR, deWet JMJ (1975) On Ö. Winge and a prayer: the origins of polyploidy. Bot Rev 41:361–390
57. deWet JMJ (1980) Origins of polyploids. In: Lewis WH (ed) Polyploidy: biological relevance. Plenum, New York, pp 3–16

58. Butters FK, Tryon RM Jr (1948) A fertile mutant of a *Woodsia* hybrid. Am J Bot 35:132
59. Haufler CH, Windham MD, Britton DM, Robinson SJ (1985) Triploidy and its evolutionary significance in *Cystopteris protrusa*. Can J Bot 63:1855–1863
60. Masuyama S (1979) Reproductive biology of the fern *Phegopteris decursive-pinnata*. I. The dissimilar mating systems of diploids and tetraploids. Bot Mag Tokyo 92:275–289
61. Gastony GJ (1983) The *Pellaea glabella* complex: electrophoretic evidence for the derivations of the agamosporous taxa and a revised taxonomy. Am Fern J 78:44–67
62. Rabe EW, Haufler CH (1992) Incipient polyploid speciation in the maidenhair fern (*Adiantum pedatum*; Adiantaceae). Am J Bot 79:701–707
63. Walker TG (1958) Hybridization in some species of *Pteris* L. Evolution 12:82–92
64. Walker TG (1962) Cytology and evolution in the fern genus *Pteris* L. Evolution 16:27–43
65. Barrington DS, Conant DS (1989) Breeding system, genetic distance, and hybridization in *Alsophila* (abstract). Am J Bot 76(suppl):201
66. Hickok LG, Klekowski EJ (1974) Inchoate speciation in *Ceratopteris*: an analysis of the synthesized hybrid *C. richardii* × *C. pteridoides*. Evolution 28:439–446
67. Carlson TJ (1979) The comparative ecology and frequencies of interspecific hybridization of Michigan woodferns. Mich Bot 18:47–56
68. Werth CR, Guttman SI, Eshbaugh WH (1985) Recurring origins of allopolyploid species in *Asplenium*. Science 228:731–733
69. Haufler CH, Soltis DE, Soltis PS (1995) Phylogeny of the *Polypodium vulgare* complex: insights from chloroplast DNA restriction site data. Syst Bot 20:110–119
70. Haufler CH, Windham MD, Ranker TA (1990) Biosystematic analysis of the *Cystopteris tennesseensis* (Dryopter idaceae) complex. Ann Missouri Bot Gard 77:314–329
71. Jermy AC, Walker TG (1985) Cytotaxonomic studies of the ferns of Trinidad. Bull Br Mus (Nat Hist) Bot 13:133–276

CHAPTER 15

Speciation and Morphological Evolution in Rheophytes

RYOKO IMAICHI[1] and MASAHIRO KATO[2]

15.1 Introduction

It is widely accepted that species are defined to be reproductively isolated groups within which natural populations interbreed [1,2]. According to this biological species concept, if reproductive isolation is established among populations, valid species could be produced. The implication is that speciation need not be accompanied by any detectable morphological changes [3]. In plants with limited movement, however, speciation is sometimes associated with distinct morphological changes, which enable plants to invade new ecological niches. The mode of speciation may hold in the case for the evolution of rheophytes.

Some rivers are regularly subject to flash floods after heavy rains. Rheophytes are plant species confined to the beds of swift-running streams and rivers, and growing there up to flood level, but not beyond the reach of regularly occurring flash floods [4]. They are also characterized by having particular morphological characters such as narrow lanceolate leaves or leaflets; matted root systems; short erect, ascending, or creeping rhizomes tightly attached to streambed substrates; and flexible stems and petioles. Dryland species generally do not occur in the flooded zone where rheophytes occur, while rheophytes do not occur in dryland habitat where dryland plants thrive. Thus there is distinct habitat segregation between rheophyte and dryland species, especially in the humid tropics.

Rheophytes, except for the entirely or almost entirely rheophilous families Hydrostachyaceae and Podostemaceae, are found scattered among various groups of different systematic affinities from bryophytes to angiosperms [4]. This suggests that rheophytes have evolved from dryland species in parallel or by convergence under the strong natural selection pressure of the environment.

The most conspicuous characteristic among several adaptive morphological traits of rheophytes is the narrow leaf or leaflet (the stenophyll; Fig. 15.1), which

[1] Department of Chemical and Biological Sciences, Faculty of Science, Japan Women's University, 2-8-1 Mejirodai, Bunkyo-ku, Tokyo 112, Japan
[2] Department of Biological Sciences, Graduate School of Science, University of Tokyo 7-3-1 Hongo, Bunkyo-ku, Tokyo 113, Japan

Fig. 15.1a,b. Silhouettes of pinnules of *Osmunda lancea* (**a**, rheophyte) and *O. japonica* (**b**, dryland plant). *Scale bar* = 1 cm

seems to decrease resistance to swift-running water when submerged. Questions relevant to rheophyte speciation include how stenophyllization occurred during the evolutionary process, how adaptive the narrow leaf is anatomically and physiologically, and how distinct habitat segregation is maintained between rheophytes and dryland plants. This chapter reviews recent information on the role of stenophyllization in morphological evolution and speciation of rheophytes.

15.2 Anatomical Adaptation of Fern Rheophytes

15.2.1 Anatomy of Narrow Leaves

A comparative anatomical study showed, in 12 pairs of fern rheophytes and closely related dryland species, that there is a strong correlation between gross morphology and anatomy of leaves [5]. The narrow leaves of the rheophytes contain smaller mesophyll cells than do the broad leaves of the dryland plants, and the intercellular spaces surrounded by lobes of mesophyll cells are smaller in the rheophytes than in the dryland plants (Fig. 15.2). Epidermal cells, except guard cells, were also reduced in the narrow leaves of many fern rheophytes examined. The lobes of amoeboid epidermal cells are shorter in the rheophytes than in the related dryland plants.

Rheophytic stenophylls have a thicker cuticular layer and denser epicuticular wax deposits on the epidermis than do the broad leaves of related dryland plants. The thicker cuticular layer as well as the fine mesophyll cell network (tightly packed mesophyll cells) may make rheophytic leaves more tolerant to crumpling and tearing when violently flexed than the leaves of dryland plants [5]. It is not surprising that the thallus-like plant body of the torrenticolous rheophytic angiosperm family Podostemaceae, which inhabits rock surfaces in riverbeds and is almost always subject to running water of high velocity, has very few intercellular spaces [6] and a thick cuticular layer (Ichiba et al., unpublished data).

It is argued that there is a correlation between anatomical and physiological features, one such example being the amount of intercellular space in the mesophyll tissue, which may affect photosynthetic rate [7]. Because small intercellular spaces and stenophylls with a small total leaf area are generally correlated with low photosynthetic rate [8], Kato and Imaichi [5] suggested that the small intercellular space of rheophytes may be a disadvantage in photosynthesis. Rheophytes, with a presumably low photosynthetic rate, may barely survive in dryland habitats where dryland species, which are photosynthetically much more produc-

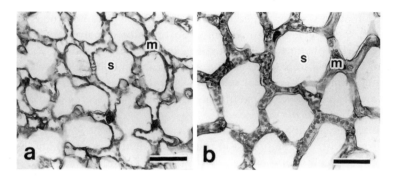

FIG. 15.2a,b. Paradermal sections of mature pinnules of *Osmunda lancea* (**a**) and *O. japonica* (**b**) show mesophyll cells (*m*) and intercellular space (*s*). *Scale bars* = 50 μm

tive, can grow vigorously [5]. This may lead to restriction of rheophyte distribution to the streambeds. Comparative physiological research between rheophytes and related dryland plants, as well as in situ ecological examination, is needed to test this hypothesis.

Rheophytes are apt to occur in more or less sunny riverbeds in forest gaps or margins. A somewhat thicker cuticular layer and denser epicuticular wax deposits on the epidermis may play a role in protecting leaves from desiccation by inhibiting transpiration when the leaves are exposed to air. Epicuticular crystalline wax also repels water and protects plants from biotic attack and mechanical injury [9].

15.2.2 Adaptation in Gametophytes and Young Sporophytes of Fern Rheophytes

Pteridophytes have free-living sporophytes and gametophytes. Based on an expectation that the gametophytic generation of fern rheophytes also shows its own adaptive strategy, as does the sporophytic generation, Hiyama et al. [10] compared the gametophyte development of the rheophilous *Osmunda lancea* Thunb. ex A. Murray with the related dryland *O. japonica* Thunb. ex A. Murray. The species are closely related phylogenetically, and it is believed that *O. lancea* was derived from an ancestor resembling *O. japonica* [11]. In agar culture, *O. lancea* gametophytes grew more rapidly and reached reproductive maturity earlier than those of *O. japonica*, although there were no prominent morphological differences in developing and mature gametophytes between the two species [10]. The *O. lancea* gametophytes might be adapted to the periodically flooded habitat by shortening its life span, a different strategy from that of the sporophyte, which has evolved particular morphological adaptations.

The fact that *O. japonica* gametophytes exhibit almost the same morphology as *O. lancea* suggests that not only gametophytes but also very young sporophytes of *O. japonica* may survive in rheophytic zones together with those of *O. lancea*.

This presumption was confirmed by field observation [12]. The young, probably 1-year-old, sporophytes of *O. japonica* do occasionally grow in the "wrong" habitat, the rheophytic zone, although adults never do. Similarly young sporophytes of *O. lancea* sometimes grow in the "wrong" habitat, the dryland region. The implication is that natural selection operates mainly on the young sporophyte stage of the life cycle, resulting in a distinct habitat segregation between *O. lancea* and *O. japonica*.

In cultivation, the first to the eighth youngest leaves of *O. lancea* differ considerably from those of *O. japonica* in the cuneate leaf base and pinna base (Fig. 15.3), with relatively short petioles with thin-walled epidermal cells and obliquely ascending (not horizontal) lamina [12]. After about the fifth leaf stage, roughly equivalent to 1-year-old leaves in the field, *O. lancea* seems to be increasingly adapted to the streambed. This increasing flood tolerance is seen in that the leaf of *O. lancea* shows earlier lamina partitioning than that of *O. japonica* (Fig. 15.3). Young sporophytes of *O. lancea* show anatomically more adaptive features than those of *O. japonica*, such as denser epicuticular wax deposits on the epidermis after the first leaf stage and a finer mesophyll network after the fourth leaf stage.

15.3 Morphogenesis and Evolution of the Stenophyll

Kato and Imaichi [5] hypothesized that stenophyllization associated with weakly expanded mesophyll cells is produced by modification of leaf ontogeny. Imaichi and Kato [12] compared pinnule development between the rheophilous *Osmunda lancea* and the related dryland *O. japonica*. Growth analysis of pinnules of the plants cultivated in the same phytotron showed that, although growth pattern of pinnules is fundamentally the same in the two species, pinnule growth period is shorter in *O. lancea* than in *O. japonica* (Fig. 15.4). The maximum growth rate in pinnule length is similar between both species (Fig. 15.4a), while growth rate in pinnule width is much less in *O. lancea* than in *O. japonica* (Fig. 15.4b). Therefore, the pinnules of *O. lancea* are less elongated and much less broadened (Fig. 15.4c).

The period of leaf development consists of two successive phases: (1) increment of cell number by cell divisions and (2) cell enlargement. Shortening of the growth period in *O. lancea* takes place mainly in the cell enlargement phase rather than in the cell increment phase and produces weakly expanded mesophyll

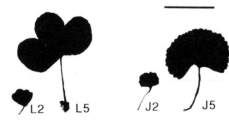

Fɪɢ. 15.3. Silhouettes of the second and the fifth youngest leaves of *Osmunda lancea* (L2, L5) and of *O. japonica* (J2, J5). *Scale bar* = 1 cm. (From [12], with permission)

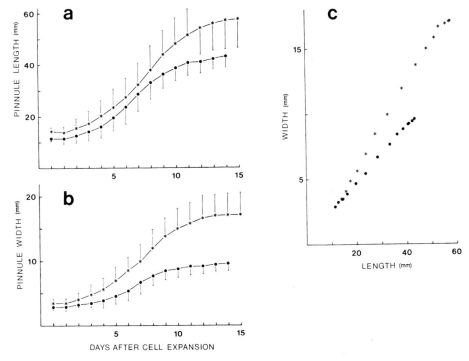

FIG. 15.4a–c. Growth curves for pinnule length (**a**), pinnule width (**b**), and relative growth in length and width of pinnules (**c**) of *Osmunda lancea* (*circles*) and *O. japonica* (*asterisks*) during cell expansion. (From [11], with permission)

cells, resulting in the smaller and narrower mature leaves (Fig. 15.5). The mature narrow leaves of *O. lancea* are anatomically similar to the immature leaves of *O. japonica*.

The stenophyllization to yield the rheophilous *O. lancea* seems to be associated with, or mainly the result of, the shortened growth period. This mode of leaf evolution is thought to be a kind of progenesis, one of six major heterochronic modes. Progenesis is an evolutionary process in which, as growth period decreases, the descendant has morphology at maturity that resembles that at juvenile stages in the ancestor [13,14].

15.4 A Broad-Leaved Variant of Rheophyte: Toward Understanding the Genetic Basis of Heterochrony

There is a presumption that morphological differences between species reflect a large number of genetic changes, and that such differences evolve through accumulation of small mutations over a period of time. Recently, however, it has been pointed out that morphological changes may be brought about by a relatively

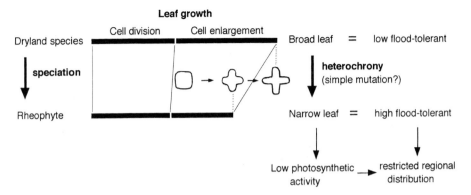

Fɪɢ. 15.5. Hypothesis of morphological evolution and adaptation of the narrow leaf in fern rheophytes. Stenophyllization of dryland species to yield rheophytes presumably occurred mainly through shortening of the cell enlargement phase rather than the cell increment phase of leaf development. Ontogenetic changes of mesophyll cell shapes of rheophytes and dryland plants are shown in the *middle row*

small number of genes [15,16]. Simple genetic changes may be the cause for heterochronic changes, whose underlying genetic mechanism still remains unknown. Many heterochronic mutants have been described for the nematode *Caenorhabditis elegans* [17], and it is worthy of note that the heterochronic gene of *C. elegans* is postulated to have contributed to speciation of related species [18].

There is, however, very little information about the genetic mechanism underlying stenophyllization. A broad-leaved variant of the rheophilous fern *Tectaria lobbii* (Hooker) Copel., which was collected in Borneo, may provide a good material for answering this question. This variant shows characteristics not typical of rheophytes but those typical of dryland species, such as large mesophyll cells (Fig. 15.6), a thin cuticular layer, and few epicuticular wax deposits [19]. It is possible that the genetic and developmental changes involved in the presumed origin of the broad-leaved variant from the typical narrow-leaved *T. lobbii* are similar to those which took place in the course of evolution of rheophytes from dryland species, although the direction of the change was opposite in the two cases.

15.5 A Facultative Angiosperm Rheophyte, *Farfugium japonicum* var. *luchuense* (Compositae)

Van Steenis [4] recognized obligate and facultative rheophytes. Facultative rheophytes contain conspecifically both dryland and rheophytic ecotypes or populations. The facultative rheophyte *Farfugium japonicum* var. *luchuense*

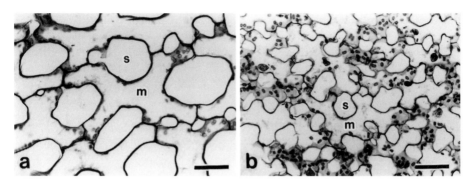

FIG. 15.6a,b. Paradermal sections of mesophylls of the broad-leaved variant (**a**) and typical form (**b**) of *Tectaria lobbii*. *m*, mesophyll cell; *s*, intercellular space. *Scale bars* = 50 μm

(Masamune) Kitamura (Compositae), distributed in the Ryukyu Islands, occurs in both streambeds and inland habitats, and shows various leaf forms ranging from cuneate- to truncate- and to cordate based, i.e., from narrow to broad leaves (Fig. 15.7).

In this rheophyte, the divergence angle of lamina base is a function of the number of primary veins and the intervein distance [20]. An anatomical study showed that the divergence angle is strongly correlated with the number of cells intervening between veins [20]. This makes a marked contrast to obligate fern rheophytes already reported in which the leaf shape is strongly correlated with cell size rather than cell number [5,12]. It suggests that in *F. japonicum* var. *luchuense* the cell increment phase changed greatly between the narrow- and broad-leaved plants, while changes at the cell enlargement phase were quite small (Fig. 15.8). The narrow leaves of the rheophyte might result from a precocious maturation involving reduction in the duration of the cell increment phase. Thus the narrowing of leaves has a different basis in *F. japonicum* var. *luchuense* compared to many fern rheophytes.

There is a rough tendency for plants that grow closer to streams to have smaller blade base divergence angles than those more distant from streams. This is perhaps a result of clinal selection pressures of habitats along a gradual decrease of flooding frequency from streambed to inland. The progeny from seeds produced on single wild plants show some variations in leaf shape under similar conditions of cultivation (Imaichi and Kato, unpublished data). Usukura et al. [20] suggested this variation may result partly from fertilization by genetically different sperm cells. It seems likely that the occurrence of the variable phenotypes results from a variety of genotypes, that is, a kind of polymorphism, which might be a preadaptation for speciation. This facultative rheophytic *F. japonicum* var. *luchuense* might be at an initial stage of rheophyte evolution, making it useful for better understanding of morphological evolution and speciation.

FIG. 15.7. Leaf silhouettes of the facultative rheophyte *Farfugium japonicum* var. *luchuense* show a wide range of leaf shape. *Scale bar* = 5 cm. (From [20], with permission)

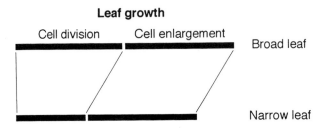

FIG. 15.8. Difference in leaf development between broad and narrow leaves of *Farfugium japonicum* var. *luchuense*. Narrowing of leaves presumably resulted from reduction in the duration of the cell increment phase rather than the cell enlargement phase of leaf development

15.6 Concluding Remarks

The speciation of rheophytes from dryland species is associated with obvious morphological evolution, e.g., stenophyllization. Such a speciation pattern recalls the debate between punctuated equilibrium and gradualism, which has attracted most evolutionary biologists and has been discussed by many authors [21]. For the punctualist, major morphological changes mostly occur during speciation [22], while for gradualists, morphological change is not necessarily confined to speciation events [21] and is the result of accumulation of small mutations or changes over a long period [15]. As discussed by Stebbins and Ayala [23] and by Ayala [24], paleontologists recognize species on the basis of different structures preserved in fossil records, and speciation events yielding few or no morphologically different products go totally unrecognized. Furthermore, in the eyes of paleontologists, many thousands of years are instantaneous [23]. The distinction between punctuated and gradual changes is in some instances not well defined [21]. If facultative rheophytes like *Farfugium japonicum* var. *luchuense* are considered the first step in rheophytic speciation, speciation of certain rheophytes can be interpreted in terms of gradualism.

Narrow-leaf or leaflet mutants of dwarf cultivated plants, e.g., *Kalmia latifolia* f. *angustata* Rehd. and *Pisum sativum* L., whose leaves might be comparable with rheophytic stenophylls, have been reported to be controlled by single genes

[25,26]. This suggests that stenophyllization might have possibly been produced by simple mutation (see Fig. 15.5). *Osmunda × intermedia* (Honda) Sugimoto, which is a natural hybrid between *O. japonica* and *O. lancea* [27,28], should provide useful material to analyze the genetic basis of stenophyllization. The narrow-leaf mutant *angustifolia* of the model plant, *Arabidopsis thaliana* (L.) Heynh., has been examined by Tsuge et al. [29]. In mutant *angustifolia* the number of cells per leaf blade is not changed, but cell shape changes in consequence of restricted cell elongation in the leaf-width direction and enhanced elongation in the leaf-thickness direction. Anatomically, mutant *angustifolia* does not seem to be the same as fern rheophytes; nonetheless, *A. thaliana* will be useful for clarifying the genetic basis of evolutionary change in leaf shape.

Acknowledgments. We are grateful to Dr. G. Bharathan, University of California, Davis, for her critical reading and correcting the English of the manuscript. Part of this study was supported by funding from the Fujiwara Natural History Foundation and a Grant-in-Aid for Scientific Research from the Ministry of Education, Science and Culture, Japan.

References

1. Mayr E (1970) Populations, species, and evolution. Harvard University Press, Cambridge
2. Grant V (1981) Plant speciation, 2nd edn. Columbia University Press, New York
3. Stanley SM (1982) Speciation and the fossil record. In: Barigozzi C (ed) Mechanisms of speciation. Liss, New York, pp 41–49
4. Van Steenis CGGJ (1981) Rheophytes of the world. Sijthoff & Noordhoff, Alphen aan den Rijin
5. Kato M, Imaichi R (1992) Leaf anatomy of tropical fern rheophytes, with its evolutionary and ecological implications. Can J Bot 70:165–174
6. Metcalfe CR, Chalk L (1950) Anatomy of the dicotyledons, vol 2. Clarendon, Oxford
7. Nobel SP, Zaragoza LJ, Smith WK (1975) Relation between mesophyll surface area, photosynthetic rate, and illumination level during development for leaves of *Plectranthus parviflorus* Henckel. Plant Physiol (Bethesda) 55:1067–1070
8. Koike T (1988) Leaf structure and photosynthetic performance as related to the forest succession of deciduous broad-leaved trees. Plant Species Biol 3:77–87
9. Juniper BE, Jeffree CE (1983) Plant surfaces. Arnold, London
10. Hiyama T, Imaichi R, Kato M (1992) Comparative development of gametophytes of *Osmunda lancea* and *O. japonica* (*Osmunda*ceae): adaptation of rheophilous fern gametophyte. Bot Mag Tokyo 105:215–225
11. Imaichi R, Kato M (1992) Comparative leaf development of *Osmunda lancea* and *O. japonica* (Osmundaceae): heterochronic origin of rheophytic stenophylly. Bot Mag Tokyo 105:199–213
12. Imaichi R, Kato M (1993) Comparative leaf morphology of young sporophytes of rheophytic *Osmunda lancea* and dryland *O. japonica*. J Plant Res 106:37–45
13. Gould SJ (1977) Ontogeny and phylogeny. Harvard University Press, Cambridge

14. Alberch P, Gould SJ, Oster GF, Wake DB (1979) Size and shape in ontogeny and phylogeny. Paleobiology 5:296–317
15. Hilu KW (1983) The role of single-gene mutations in the evolution of flowering plants. Evol Biol 16:97–128
16. Gottlieb LD (1984) Genetics and morphological evolution in plants. Am Nat 123:681–709
17. Ambros V, Horvitz HR (1984) Heterochronic mutants of the nematode *Caenorhabditis elegans*. Science 226:409–416
18. Ambros V, Fixsen W (1987) Cell lineage variation among nematodes. In: Raff RA, Raff EC (eds) Development as an evolutionary process. Liss, New York, pp 139–159
19. Kato M, Imaichi R (1992) A broad-leaved variant of the fern rheophyte, *Tectaria lobbii*. Int J Plant Sci 153:212–216
20. Usukura M, Imaichi R, Kato M (1994) Leaf morphology of a facultative rheophyte, *Farfugium japonicum* var. *luchuense* (Compositae). J Plant Res 107:263–267
21. Futuyma DJ (1986) Evolutionary biology, 2nd edn. Sinauer, Sunderland
22. Eldredge N, Gould SJ (1972) Punctuated equilibria: an alternative to phyletic gradualism. In: Schopf TJM (ed) Models in paleobiology. Freeman, San Francisco, pp 82–115
23. Stebbins GL, Ayala FJ (1981) Is a new evolutionary synthesis necessary? Science 213:967–971
24. Ayala FJ (1982) Gradualism versus punctualism in speciation: reproductive isolation, morphology, genetics. In: Barigozzi C (ed) Mechanisms of speciation. Liss, New York, pp 51–66
25. Jaynes RA (1981) Inheritance of ornamental traits in mountain laurel, *Kalmia latifolia*. J Hered 72:245–248
26. Bassett MJ (1981) Inheritance of a lanceolate leaf mutation in the common bean. J Hered 72:431–432
27. Shimura Y (1972) Study of reproduction of *Osmunda* × *intermedia* Sugimoto (in Japanese with English summary). J Geobot 20:38–42
28. Shimura Y, Matsumoto S (1977) On the chromosome association in meiosis of *Osmunda* × *intermedia*. J Jpn Bot 52:377–378
29. Tsuge T, Tsukaya H, Uchimiya H (1996) Two independent and polarized processes of cell elongation regulate leaf blade expansion in *Arabidopsis thaliana* (L.) Heynh. Development (Camb) 122:1589–1600

Subject Index

ABC model 102, 184
Abietoideae 71, 81
Acer 284, 279
Aceraceae 279
Aconitum 214
  A. pterocaule 214
  A. sanyoense 214
Acorus 162
Adansonia 100
Adoketophyton 5, 8, 9
adventitious embryony 225
adventitious root 32, 33
Afropollis 141
AG 180, 190, 192, 193
agamospermous polyploid 210
agamospermy 225
agamosporous polyploid 210
AGAMOUS see *AG*
AGL1 191, 192
AGL11 190
AGL12 192
AGL15 192
AGL17 192
AGL2 183, 189
AGL3 183
AGL4 183
AGL5 191, 192
AGL6 187–189
AGL8 188
Aglaophyton 3, 4, 5
allohomoploidy 301, 302
allopatric speciation 296, 297
allopolyploid 255, 293, 299, 302, 304
  hybrid 214
allozyme diversity 256

Alphonsea 139
Alsinidendron 258
Alsophila 301
Ambitisporites sp. 14
Amborella 162
  A. trichopoda 129
Amborellaceae 125, 129
Amentiferae 274
amphidiploidization 210
anagenesis 100
Androglandula tennessensis 127
Anemopsis 144
aneuploid 210
aneurophyte 9
Aneurospora sp. 15
Angiopteris 35
angiosperm(s) 38, 285
  divergence 269
  diversification 144, 202
  origin and diversification 121
  relationships 157
Angiospermopsida 34
angiospermy 102
angustifolia 317
Annona 139
Annonaceae 135, 136, 139
Anonaspermum 139
Anthocephale chuchlensis 125
Anthoceros formosae 53
anthophyte hypothesis 202
anthophytes 102
Antirrhinum 102, 182, 188–191, 231
AP1 181, 188, 189, 193
AP2 191
AP3 181, 189, 190, 192

APETALA1   see *AP1*
apical cell   28, 30, 33–37
  theory   35
apical meristem   31, 35
  evolution   34
apical plug   29
apomorphy   293
Appomattoxia   143–146
  A. ancistropohora   143
aquatic herbs   282
Aquilapollenites   273, 276
Arabidopsis   102, 180, 182, 183, 189, 190
arabinogalactan-protein   107
Archaeanthus   136–138, 145, 146
  A. linnenbergii   137
Archaeopteris   9
Archaeozonotriletes   see *Laevolanchis*
Archaepetala
  A. beekeri   137
  A. obscura   137
Archicupressus   65, 68
Arcto-Tertiary Geoflora   279, 286
Argyranthemum   251, 254, 256, 258
Aristolochia   123
  A. cochica   123
  A. africanii   123
Aristolochiaceae   122, 123
Aristolochiales   122, 124
Aristolochioxylon prakashii   123
Aristrochiaceae   158
Armeria   227
Ascarina   129, 130
Asimina   139
Asplenium   255
Asteridae   158, 278
Asteropollis   130, 131
Asteroxylon   35
Austrobaileya   162
  A. scandens   135
Austrobaileyaceae   124, 125, 135, 136
autogamy   225
automatic selection advantage   228
autopolyploidy   254, 298, 299
autosomal chromosome   212
axial organs   37

B chromosome   211
barley   47

bell pepper   48
Baker's law   227
banding pattern   212
Baragwanathia   7
  flora   10
Barclaya   141
Barclayaceae   141
Barclayopsis urceolata   142
barinophytes   5, 8
base modification   51
base substitution   51
BELL1   191
Belliolum   141
Bennettitales   135, 136
Betulaceae   279
Bidens   251, 254, 256, 257
biological species concept   309
biparental inbreeding   225, 230
bivalent   215
boreotropical flora   278
Botrychium   35
Botryopteris antiqua   32
branching
  dichotomous,   34
  endogenous,   34
  exogenous,   34
Brasenia   141, 142
Braseniella   142
Brassica   240
Bromus mollis   236
bryophyte   15, 18, 37
Bubbia   141

C to U   51
Cabomba   142
Cabombaceae   141, 142
Caesalpinoid   278
CAL   188, 189
Calamites   36
Caltha palustris   227
Calycanthaceae   125, 128, 129, 145
Calycanthus   128, 129
calyptra   29
cambium   30
Canary Islands   256
Canella   140
Canellaceae   135, 140
Carex   210

carpel 38
carpellary flank 106
Caryophyllaceae 129
Catenalis digitata 9
CAULIFLOWER see *CAL*
CCA (Canonical Correspondence
    Analysis) 272
CENP 210, 212
Centaurodendron 254
central dogma 45
centrarch 9, 10
centromere 210
Ceratophyllaceae 123, 141
Ceratophyllales 123
Ceratophyllum 123, 162, 201
Ceratopteris 193
Cercidiphyllaceae 161, 275
Charopsida 6
Chimonanthus 128, 129
Chlamidomonas 193
chlL 53
Chloranthaceae 121, 122, 125, 129, 130,
    132, 144, 145, 158, 203, 204, 216
Chloranthistemon 133–135, 146
    C. alatus 134
    C. endressii 133, 134
    C. crossmanensis 135
chloranthoid
    fruits 146
    pollen 131
Chloranthus 129–135, 145
chloroplast transcript 53
choripetaly 110
chromatin 209
chromoplast 52
chromosome
    behavior in meiosis 209
    complement 209
    number 209
    paring in meiosis 209
    shape 209, 212
Cinnamodendron 140
Cinnamomum
    C. felixii 127
    C. prototypum 127
Cinnamosma 140
Circaeaster 144
Circaeasteraceae 144
cladogenesis 100

CLAMP (Climate-Leaf Analysis
    Multivariate Program) 271, 273,
    275
Clarkia 225, 250, 253
Clavatipollenites 130, 140, 203, 204
cleistogamous flower 107
cleistogamy 225
climatic changes 270
climatic events 285
CMA 212
CMB2 190
CMMT (Cold Month Mean
    Temperature) 272
Coleochaete
    C. mauldinensis 130
    C. orbicularis 6
    C. pertoni ssp. *pertoni* 14
    C. pertoni 5, 19
    C. sp. 130
    C. synorispora 14
Coleoptera 101
Collinsia heterophylla 238
collumelae 201
combined data set 172
comparative study 100
Compositae 283
concomitant pollen tube selection 102
conducting tissue 9, 19
congruence of multiple data sets 165
conifer 61
coniferous forest 284
constraint tree 78, 84
Cooksonia 4, 7, 18
Coreopsis grandiflora 235
corm 30
Couperites 130, 203
Crassidenticulum 125, 132
crassinucellar ovule 103
Cretaceaisporites 129
Cretaceous 69, 81, 86, 101, 269, 272, 282,
    285
Cretaceous-Tertiary boundary see K-T
    boundary
Crinopolles 202
cryptospore 11, 13–15
Cunninghamiostrobus 64, 68
Cunnunghamia 68
Cupressaceae 62, 67, 84, 86
cupular wall 106

cuticle   17, 19
Cymbohilates   15
Cystopteris   299, 303
cytological character   209

Dactylis   255
DAL1   187
DAL2   190, 192
Daphniphyllum   125
DAPI   212
deamination   51
deciduousness   275
Deeringothamnus   139
DEF   180
DEFICIENS   see *DEF*
Degeneria   200
   D. roseiflora   139
   D. vitiensis   139
Degeneriaceae   135, 139
Demergotheca   8
Dendroseris   251, 254, 256, 258
Densinervum   125
   D. kaulii   132
   D. patagonica   141
   D. qujingensis   7
   D. sp.   11
   D. spinaeformis   7
diaspores   283
Dictyotriletes emsiensis   7
dioecy   226
diploid   214
diploidization   294
Diptera   101
Discalis   8
Drimys   141
Dryopteris   299, 303
Dubautia   251, 252
Dusembaya   142
dyad(s)   11, 13
Dyadospora   15

early sympetaly   109
editing reaction   51
effective population size   234
Eichhornia paniculata   227, 228
ektexine   201
Elmerrillia   137

embryophytes   6, 11, 13
embryotega   141, 142
enations   35, 38
endemic species   251
endexine   201
endohydric conducting system   19
entomophilous flower   123
entomophily   134
environment   269
environmental
   change   272
   factor   271
   pressures   19
   thermal optimum   286
Eocene   270, 278–280
Eophyllophyton bellum   9
epicontinental seaway   270
epistatic interaction   236
Equisetopsida   27, 34, 36
Equisetum   33, 36
erroneous sequence   171
ethereal oils   158
etioplast   52
Eucalyptus geinitzii   25, 127
eudicots   158
Eupomatia   200
   E. bennettii   140
   E. laurina   140
Eupomatiaceae   129, 135, 140
Eupteleaceae   161
Euryale   142, 143
eutracheophytes   5
exarch   8
exine
   atectate-amorphous   200
   macromolecular structure   202
   non-collumelar   200
   collumelar   200
   granular   200
Exospermum   141
exospore   11, 13
extinctions   284
extra-sporal material   14

Fagus   284
Farfugium japonicum var. *luchuense*
   314
FBP1   190

FBP3   190
ferns   33, 38
Filicopsida   27
first-branching angiosperms   162
floral bud protection   101
floral color change   101
floral diversity   100
floral meristem identity gene   183
floral organ identity gene   183
floral tube   109
floral ultraviolet reflection   101
Florissantea   281
floristic change   286
Foveomorphomonocolpites
    humbertoides   139
frequency-dependent advantage   238
functional protein   48

Galapagos Islands   258
Galbulimima   140
gamete fertility   209, 214
gametophyte   37
gametophytic apomixis   225
geitonogamy   225
gene duplication   299
gene silensing   293
genome analysis   209
genomic sequence   48
Gilia   225
glaciation   284, 285
glucosinolate clade   161, 163
Gnetales   106
Gnetum   190
   G. parvifolium   186
Goethe   111
Gomortega nitida   128
Gomortegaceae   125, 128
Gondwana   7
Gosslingia   9
grade   158
Grevilleophyllum   125
Gumuia   8
Gunnera   256
Gymnocarpium   303
Gymnospermopsida   34
gymnosperms   33
Gymnotheca   144
Gyrocarpus   128

Haborosequoia   64, 68
Hamamelididae   275, 278
Harrisipollenites   141
Hawaii   251, 252, 254, 258
Hazomalania   128
Hedyosmum   121, 129–131, 145, 146
Helianthus   259
hepatics   11
herbaceous   284
   magnoliids   122
Hernandia   128
Hernandiaceae   125, 128
heterochromatic segment   212
heterochrony   313
heterospory   292, 294
heterostyly   226
heterozygosity   235
Hibiscus   201
Himantandeaceae   135, 140
histone   217
holocentric chromosome   210
homeotic gene   179
homoiohydry   4, 19
homospory   294
homozygosity   234
Horneophyton   3
Hornwort   47
horsetails   38
Hostinella   4, 14
Houttuynia   144
Huia recurvata   9
Huyvena   5

Icacinaceae   281
Ideogram   213
Idiospermaceae   125
Idiospermum   128
Illiciaceae   122, 124, 135, 158
Illiciales   124, 125
Illicium   124, 162
Illigera   128
impact winter   276
inbreeding depression   226, 229, 230, 231,
    233
INCENP   210, 212
incompatibility system   239
initiation condon   47
insect pollination   102

integument   103, 106
intercalary growth   109
intragametophylic selfing   297
introgression   255
Ipomoea purpurea   228
Ipomopsis   201
Irthyshenia   143
Isoetales   33
Isoetes   30, 35–37, 52, 53
isomorphic alteration of generations   3
ITS   107

Jacquinia   274
Juan Fernandez Islands   251, 254, 256, 258
Juglandales   274

K box   180, 181, 183
K-T (Cretaceous-Tertiary) boundary   275, 276
Kadsura   124
Kalmia latifolia f. *angustata*   316
Kalymmanthus walkeri   137
karyomorphology   209, 216
Kazakhstan palaeocontinent   10
Kmeria   137

Lactoridaceae   123–125, 135, 158
Lactoridales   122, 124
Lactoripollenites africanus   125
Lactoris fernandeziana   124
Ladonia   125
Laevolanchis (Archaeozonotriletes)   14
   L. cf. *divellomedium*   15
   L. divellomedia   14
land bridge   270
late sympetaly   109
Lauraceae   125, 126, 145
Laurales   124, 125, 129, 143
Laureae   126
Laurelia   128
Lauricoideae   71, 82
Laurophyllum   125
Laurussia   7, 8, 10
leaf physiognomy   271, 272
leaf   27, 38

LEAFY   see *LFY*
Leguminosae   273, 279
lepidodendrids   29
Lepidoptera   101
leptopsorangiate fern   293
Lesqueria   136
   L. alata   139
   L. elocata   135
   L. fossulatus   136
   L. germanica   139
   L. protogea   139
Lethomasites   202
LFY   187
Liliacidites   203
Lilium   172
Limonium   227
Linanthus parryae   234
*Lindera rottensi*s   127
Liriodendroidea   138
Liriodendroideae   137
Liriodendron   138, 139, 145
   L. chinense   137
   L. tulipifera   137
Liriodendroxylon   137
Liriophyllum   136
   L. kansense   137, 138
   L. populoides   138
Litocarpon   145
   L. beardii   138
   Litseopsis rottensis   127
liverworts   17
Lolium perenne   238–240
long-distance dispersal   297
Longstrethia varidentata   124
Loranthacites   273
Luetkea   284
Lycophytes   8, 32
Lycophytina   34
Lycopodiales   4, 33
Lycopodium   35–37
lycopods   37, 38
Lycopsida   27, 28, 34–38

Macaronesian Islands   251, 254, 255, 258
MADS box   180, 186
MADS gene   103, 182, 189–193
MADS gene family   179, 260
Magnolia   121, 125, 137, 138, 145, 146

Magnoliaceae   121, 135, 136, 138, 145,
    202
  Cretaceous record   137
  woody,   121
Magnoliaceoxylon   137
Magnoliaephyllum   125
  M. sp.   127
Magnolialean hypothesis   203
Magnoliales   123, 124, 135, 203
Magnoliidae   121, 136, 274, 278
magnoliids   123, 273
Magnolioideae   137
Magnolioxylon   137
maize   47
male gamete   209
male gametephyte selection   109
Malmea   39
Malus   233
Manglietia   137, 139
MAP   210, 212
Marattiaceae   35
Marchantia polymorpha   53
Marsilea   292
mass kill   275
MAT (Mean Annual Temperature)   272
mating system   224–356, 240
Maulidinia   127, 145, 146
  M. mirabilis   126
MCM1   180, 183,
mechanical tissue   146
MEF2   180, 181
Megaceros enigmaticus   53
megaphyllous leaves   38
megaphylls   9
megathermal
  rain forest   278
  vegetation   280
meiosis   14, 15
meiotic abnormality   14
Menispermaceae   278
meristem   30
  zoned,   37
mesothermal broad-leaved evergreen
    forest   280
messenger RNA   45
  mature,   49
methylation   217
Michelia   137
microphyllous leaves   38

microphylls   8
microspore   215
microthermal flora   278
microtubule   210
migratory route   278
Mimulus   250, 259
  M. guttatus   228, 235
  M. micranthus   235
Miocene   270, 286
mixed mating   226, 229
model phylogeny   173
molecular clock   251
Monimiaceae   125, 127
monocots   123
monoecy   226
monosulcate pollen   200
monosulcates   158
multivalent   215
mutagen   210
Myricanthium amentaceum   126
Myristicaceae   135, 139
Myristicarpum   140
Myristicoxylon   139
Myrtophyllum   125

nad1   51
nad2   51
Narthecium ossifragum   227
ndhA   51
ndhB   51
Nelumbonaceae   158, 282
Nematophytales   4
Nematothallus   4
Neogene   283, 284
neutral mutation   211
Nicotiana tabacum   240
Nikitinella   143
nitrogen-fixing clade   161, 163
nontenuinucellate ovule   158
nonterminate growth   8
NOR   212
Normapolles   274, 275
Nothia   3
nucleolar organizing region   212
nucleotide substitution   51
Nuphar   142, 143, 201
Nymphaea   142, 143
  N. zanzibarensis   142

Nymphaeaceae   141, 142, 282
  Cretaceous record   142
Nymphaeales   122, 123, 141

Obirastrobus   72, 83
oceanic islands   249, 256
Oenothera organensis   238, 239
Old Red Sandstone   7
Oligocene   280, 283
Ondinea   142
operculum   141
Ophioglossaceae   35
Ophrys apifera   227
Osmunda
  O. japonica   311
  O. lancea   311
  O. x *indtermedia*   317
Outcrossing   229
ovule formation   102

Pabiania   125
Pachylanax   137
Pademophyllum   125
Palaeanthus problematicus   136
Palaeoeuryale   143
Palaeonitella   6
Palaeonymphaea   142
Paleocene   276
paleoenvironment   270
paleoherb   101, 122, 123, 204
  hypothesis   203
paleopalynology   199
Paleophytocrene   280, 281
Paleotropical Tertiary Geoflora   286
Panamanian isthmus   270
Pandaniidites   273
Papaver rhoeas   238, 239
parapatric speciation   296
Paraphyllanthoxylon   126
Parataiwania   64, 68
parsimony jackknife approach   161
parthenogenesis   209
pathogens   18
Paurodendron   30, 33
PE zone   see polygonalis-emsiensis
    zone
Peperomia sibirica   144

Periporopollenites
  P. demarcatus   129
  P. fragilis   129
perispore   14
permineralized cone   61
Perseanthus crossmanensis   127
Perseeae   126
Pertica   9
Pertonella langii   14
Petunia   190
Phaeophyceae   9
Phlox   238
  P. cuspidata   238
  P. drummondii   238
phosphorylation   217
phyllotaxis   37
phylogenetic analysis   63, 72
Physcomitrella   193
physiognomy   271
PI   181, 192
Picea abies   186, 190
Piceoideae   72, 83
Pinaceae   69, 72, 84
Pinnatiramosus   11
Pinoideae   71, 81, 83
Pinus   71, 81, 83
  P. coulteri   234
  P. radiata   186
Piper   144
Piperaceae   143, 144
Piperales   122–125, 129, 143, 144
Piptocalyx   129
PISTILATA   see *PI*
Pisum sativum   316
Pityostrobus   72, 83, 84
plant mitochondria   46
plasmodesmata   106, 107
Platanaceae   158
PLE   191
PLENA   see *PLE*
Pleodendron   140
plesiomorphy   293
ploidy level   210
PMADS1   190
PMADS2   190
PMC   215
Poa   210
pollen
  analysis   199

development   201
  record   273
  limitation   226
  mother cell   214
  tube competition   108, 109
  tube guidance   106, 107
  tube transmitting tissue   106
  tube transmitting tract   107
pollination   145, 146, 223
  system   293
pollinator attraction   101
Polygenic variation   236
polygonalis-emsiensis (PE) zone   7
polypetalous corolla   158
polyploidization   210, 294
polyploidy   210, 254, 258, 295, 298, 301
Polypodium   299, 303
polysporangiophytes   5
Polystichum   303
population biology   214
postgenital fusion   104, 106
postmating isolation   214
Potamogeton   172
potential ancestral   8
Potentilla   284
PR400   186
precipitation   279
premature RNA   51
primexine   201
Prisca reynoldsii   126, 127
progymnosperms   27
proplastid   52
protective plug   30
Proteophyllum   125
Protobarclaya   142
protogynous   101
Protolepidodendron scharyanum   8
Protomonimia kasai-nakajhongii   136
Prototaxites   4
protracheophytes   5
psbF   52
psbL   52
Pseudoaraucaria   72, 82, 83
pseudodyad   11
Pseudodyadospora   11
Pseudoeuryale   143
pseudomonopodial branching   9
Pseudowintera   141
  P. pollis   141

Psilophytales   5
Psilophyton   5, 9, 32
  P. crenuatum   32
  P. dawsonii   32
Psilopsida   27, 32, 35, 38
psilotalean rhizome   35–38
Psilotales   30
Psilotum   30, 32
  P. nudum   34, 30, 37
pteridophytes   7, 33–38
  axial organs   34
  classification system   38
Pteris   301
Pterophytina   34
Pteropsida   27, 34–37

QTL's   252
quantitative trait loci   252
Quaternary   285

Ranunculidae   278
ranunculids   158, 162
Ranunculus
  R. bulbosus   227
  R. cantoniensis   214
  R. chinensis   14
  R. flammula   227
  R. sirelifolous   214
Raphanus sativus   240
rbcL   157, 183, 188
rDNA
  18S,   157
  repetitive,   212
recombinational speciation   259
Renalia   5
reproductive isolation   214
Resilitheca   15, 19
reticulate evolution   299
reticulate speciation   295
Retimonocolpites   203
Retusotriletes cf. *coronadus*   14
rheophyte   309
rhizome   30
  psilotalean,   30, 31
rhizomorph   29
rhizophore   28, 29, 34, 37, 38
Rhynia gwynne-vaughanii   5, 37

Rhyniales 5
Rhynie chert 3, 4, 6, 14
Rhyniophytina 5
rhyniophytoids 11, 17, 18
RNA editing 50
RNA-binding proteins 51
Robinsonia 254, 258
root 27, 28, 36–38
  branching patterns 33
Rosaceae 279
Rosidae 158
rRNA 188

S allele 233, 240
Sabiaceae 158
Sabrenia 142
Salix 285
Salopella marcensis 15
saprobe 18
Sarcandra 129, 132
Sarchorhachis 144
Sargentodoxa 171, 172
Saururaceae 143, 144
Saururopsis nipponensis 144
Saururus 144
  S. biloba 144
Sawdonia 8, 9
Scalesia 258
Schiedea 258
Schisandra 124, 162
Schisandraceae 122, 124, 135, 158
Schrankipollis 141
seclusion 106
secondary constriction 212
secretory 107
seed-plant(s) 9, 33, 36, 38
Segestrespora membranifera 11
Selaginella 28, 30, 34, 35, 53, 292
  S. plana 29, 36
  S. caudata 36
  S. delicatula 29, 35, 36
  S. glabra 124
  S. kraussiana 35
  S. martensii 35
  S. speciosa 35
  S. uncinata 28, 29, 33, 35, 36
Selaginellales 33
self-compatibility 109

self-fertilization 224–226, 228, 235, 297
self-incompatibility 108, 226, 229, 230,
  231, 233, 238–240
Sennicaulis 5
sexual reproduction 224
Silene latifolia 188
silversword alliance 251, 253, 256, 257
Sindora 273
Sinocalycanthus 128
SLM4 188
SLM5 189
snapdragon 47
Solanum
  S. carolinense 238, 239
  S. chacoense 240
somatic cell division 209
somatic metaphase chromosome 209
Sonchus 254, 258
South China plate 9, 10
Sparattanthelium 128
Sphaerocarpales 11
Sphenophyllum 36
spinach 47
sporangium(a)
  bivalved, 15
  Salopella-like, 15
sporophytes 18, 37
sporopollenin 11, 14, 202
Sprengel 111
SQUA 188
SQUAMOSA see *SQUA*
SRF 180, 181
Stachyophyton 8, 9
stage-specific modulation 48
stem 27, 34, 37, 38
stenophyllization 310, 314
Stephanocolpites fredericksburgensis 133
Stephanomeria 250, 259
sterome 18
stigma 107, 108
stigmatic secretion 101
Stockmansella 5
stomata 19
stop codon 48, 50
SUPERMAN 191
supratectal elements 201
sympatric speciation 297, 298
sympetalous corolla 158
sympetaly 101, 105, 109

syncarpy   104, 105
Synorisporites verrucatus   14

taiga   285
Taiwania   68
Tambourissima ficus   127
Tavdenia   143
Taxodiaceae   62, 84
Tectaria lobbii   314
tectum   201
telomere   212
telomes   34
temperature deterioration   270, 280
tenuinucellar ovule   104
tenuinucellate ovule   158
terrestorial herbs   282–285
Tertiary   269, 276
Tetracentraceae   161, 275
tetrads   13
    earliest   6
Tetrahedraletes medinensis   13, 15
Tetramolopium   251, 252, 254, 256, 258
tetraploid   214
Theophrastaceae   274
Thysanoptera   101
tissue-specific modulation   48
TM5   183
Tmesipteris   30, 32
Tobacco   47
Tolpis   255
Tomskiella   143
Tortilicaulis   18, 19
tracheid(s)   18
    G-type,   5, 8, 10
    lateral pits   19
    S-type,   5, 19
Tracheophyta   9, 34
Tragopogon   255
transmitting tissue-specific proteins   107
transcription factor   179
Transitoripollis   143
    T. similis   143
translocation   210
tree-thinking   100
Trianthera eusideroxyloides   127
triaperturate pollen   104, 158
tricolpate   158
Trimenia   129

Trimeniaceae   124, 125, 129
Trimerophytales   9, 10
Trimerophytina   5
Trimerophytopsida   34
Triticum   209
Trochodendraceae   161, 275
trochodendroids   275
Tucanopollis   143
tundra   285

U to C   51
Ulmaceae   279
uniaperturate pollen   158
unitegmic ovule   104
Uskiella spargens   5
Uvaria   139

valvate dehiscence   101
vascular plants   18
    homosporous,   4
    early,   11
    primitive,   27
vegetative reproduction   225
Verrucosisporites polygonalis   7
Victoria   142
Virginianthus   145, 146
    V. calycanthoides   128
Virola   140

Walkeripollis gabonensis   141
Warburgia   140
water-conducting cells   5
Wikstroemia   256
Williamsonia
    W. elocata   see *Lesqueria elocata*
    W. recentior   136
wind pollination   274
Winteraceae   122, 124, 135, 140, 145
wood evolution   274

xylem anatomy   5

Yezosequoia   64, 68
Yubaristrobus   65, 68

Yunia dichotoma   10

Z*AG3*   186
ZAG5   186
ZAP1   188, 189
Zippelia   144
Zonosulcites
    Z. parvus   143

Z. scollardensis   142
Zosterophyllophytina   5
Zosterophyllopsida   4, 35
zosterophylls   9, 18
Zosterophyllum   8
    Z. australianum   7, 10
    Z. myretonianum   10
    Z. yunnanicum   7
Zygogynum   141